全国高等医药院校药学类实验教材

微生物学实验

（第二版）

主　编　徐　威
副主编　马晓楠　苏　昕
编　者　（以姓氏笔画为序）
　　　　马晓楠　刘晓辉　苏　昕
　　　　周丽娜　徐　威　徐慰倬
　　　　蔡苏兰

中国医药科技出版社

内 容 提 要

本书为全国高等医药院校药学类实验教材之一。全书分为四章，分别为微生物学实验的基础知识、微生物学基础性实验、研究性与设计性实验、综合性实验。为适应教育国际化的要求，增加了英文对照内容，以便于学生在阅读英文文献、撰写英文论文时参考。

本书可作为药学及相关专业教材使用，也可供实验科研人员参考。

图书在版编目（CIP）数据

微生物学实验/徐威主编．—2版．—北京：中国医药科技出版社，2014.8
全国高等医药院校药学类实验教材
ISBN 978-7-5067-6929-7

Ⅰ．①微… Ⅱ．①徐… Ⅲ．①微生物学-实验-医学院校-教材 Ⅳ．①Q93-33

中国版本图书馆 CIP 数据核字（2014）第 170678 号

美术编辑	陈君杞
版式设计	郭小平
出版	中国医药科技出版社
地址	北京市海淀区文慧园北路甲 22 号
邮编	100082
电话	发行：010-62227427　邮购：010-62236938
网址	www.cmstp.com
规格	787×1092mm 1/16
印张	19
字数	348 千字
初版	2006 年 3 月第 1 版
版次	2014 年 8 月第 2 版
印次	2022 年 5 月第 5 次印刷
印刷	三河市百盛印装有限公司
经销	全国各地新华书店
书号	ISBN 978-7-5067-6929-7
定价	39.00 元

本社图书如存在印装质量问题请与本社联系调换

全国高等医药院校药学类规划教材常务编委会

名誉主任委员　邵明立　林蕙青
主　任　委　员　吴晓明（中国药科大学）
副主任委员　（按姓氏笔画排序）
　　　　　　　　刘俊义（北京大学药学院）
　　　　　　　　匡海学（黑龙江中医药大学）
　　　　　　　　朱依谆（复旦大学药学院）
　　　　　　　　朱家勇（广东药学院）
　　　　　　　　毕开顺（沈阳药科大学）
　　　　　　　　吴少祯（中国医药科技出版社）
　　　　　　　　吴春福（沈阳药科大学）
　　　　　　　　张志荣（四川大学华西药学院）
　　　　　　　　姚文兵（中国药科大学）
　　　　　　　　高思华（北京中医药大学）
　　　　　　　　彭　成（成都中医药大学）
委　　　员　（按姓氏笔画排序）
　　　　　　　　王应泉（中国医药科技出版社）
　　　　　　　　田景振（山东中医药大学）
　　　　　　　　李　高（华中科技大学同济药学院）
　　　　　　　　李元建（中南大学药学院）
　　　　　　　　李青山（山西医科大学药学院）
　　　　　　　　杨　波（浙江大学药学院）
　　　　　　　　杨世民（西安交通大学药学院）
　　　　　　　　陈思东（广东药学院）
　　　　　　　　侯爱君（复旦大学药学院）
　　　　　　　　娄红祥（山东大学）
　　　　　　　　宫　平（沈阳药科大学）
　　　　　　　　祝晨蔯（广州中医药大学）
　　　　　　　　柴逸峰（第二军医大学药学院）
　　　　　　　　黄　园（四川大学华西药学院）
　　　　　　　　朱卫丰（江西中医药大学）
秘　　　书　夏焕章（沈阳药科大学）
　　　　　　　　徐晓媛（中国药科大学）
　　　　　　　　沈志滨（广东药学院）
　　　　　　　　浩云涛（中国医药科技出版社）
　　　　　　　　赵燕宜（中国医药科技出版社）

第二版前言

《微生物学实验》自 2004 年 10 月出版以来，受到了药学类各专业师生的欢迎。它不仅在学生掌握微生物学实验基本操作技术方面提供了有益帮助，更重要的是提高了学生的实验动手能力，掌握了实事求是的科学作风。

本书自第一版以来，10 个年头过去了。随着微生物学的发展，实验内容不断更新、实验方法不断改进、实验教学条件大幅度改善等，实验教学必须与时俱进；同时，在我们的自身教学过程中，也发现了首版教材中存在的不足。这些都促使我们着手《微生物学实验》的再版工作。

近年来，教师坚持以研究促教改，通过承担各类级别的高等教育教学改革项目等方式，投身教学改革，并将教学改革的成果固化在教材中，不断提高教学效果。此次推出的药学专业双语实验教材《微生物学实验》，是我校教师长期钻研实验教学课程体系，改革实验教学内容，实现教育创新的重要成果。本书适用于药学类院校各专业的本科生教学的要求，同时可作为药学类研究生和相关研究人员的参考书。这套教材具有以下特点：

1. 精选教学内容，体现药学特色。在编写工程中，为配合《微生物学与免疫学》理论课教学内容，在原有实验内容上有所增加和删减。本书实验共分 4 部分，52 个实验。为了体现药学专业的基本特点，我们安排了 39 个典型的微生物学基本操作技能实验，目的是强化学生对基本实验技能的掌握。

2. 以能力培养为核心，通过综合性、设计性实验的开设，引导学生创新思维。随着我国经济体制和教育体制的改变，我国的基础教育正在从应试教育向素质教育转变。为了满足创新人才的培养，我们安排了 12 个研究型与设计性实验及综合性实验，内容比第一版丰富得多。这些实验的设立有利于学生全面了解和综合掌握本门课程的教学内容，激发同学创新的愿望和培养他们的创新能力。

3. 教材采用双语体系编写，为实验课程改革构建外语教学平台，有利于提高学生的科技英语水平。

4. 强调实用性和可操作性。实验内容在编写单位已经经过了多次试用。

本书是一份集体创作，经过近一年的努力，《微生物学实验》（第二版）现已完稿。在教材编写过程中，中国医药科技出版社的领导给予极大的支持，何红梅编辑在教材的审编方面给予了悉心指导，在这里一并表示衷心感谢。

由于采用双语体系编写药学教学实验教材尚属首次，缺乏足够经验，在选材、实验方法和实验内容的编排上错误在所难免，敬请读者提出宝贵意见，以臻逐步完善与提高。

编者

2014 年 5 月 26 日

目 录

第一章 微生物学实验的基础知识 (1)
第一节 微生物学实验目的和要求 (1)
第二节 微生物实验室设置及主要仪器设备 (2)
第三节 微生物学实验常用器皿的洗涤和包扎技术 (11)
第四节 微生物实验室常用物品的消毒灭菌和无菌操作技术 (16)
第五节 实验报告的书写要求 (18)
第六节 验证性、综合性、设计性实验概述 (18)

第二章 微生物学基础性实验 (20)
Chapter 2　Basic Experiments in Microbiology (20)
第一节 显微镜和显微技术 (20)
Section 1　Microscope and Microscopic Techniques (20)
实验一　普通光学显微镜的结构、使用与维护 (20)
Experiment 1　Bright-Field Light Microscope (24)
实验二　相差显微镜的结构与使用 (28)
Experiment 2　Phase-Contrast Light Microscope (30)
实验三　荧光显微镜的结构与使用 (31)
Experiment 3　Fluorescence Microscope (33)
实验四　细菌运动性观察 (35)
Experiment 4　Bacterial Motility Observation (36)
实验五　微生物测微技术 (38)
Experiment 5　Microscopic Measurement of Microbes (40)
第二节 微生物染色、形态结构与菌落特征观察 (42)
Section 2　Microorganisms Stain, Morphology and Colony Characteristics (43)
实验六　细菌的单染色法与细菌菌体形态观察 (43)
Experiment 6　Simple Staining (45)
实验七　细菌的革兰染色 (46)
Experiment 7　Gram Staining (49)

实验八　细菌的芽孢染色法 ……………………………………………………… (51)
　Experiment 8　Spore Staining ……………………………………………… (53)
实验九　细菌荚膜、鞭毛的染色观察 …………………………………………… (54)
　Experiment 9　Capsule and Flagella Stain ……………………………… (55)
实验十　放线菌形态观察 ………………………………………………………… (56)
　Experiment 10　Actinomyces Morphology ……………………………… (59)
实验十一　酵母菌形态观察 ……………………………………………………… (60)
　Experiment 11　Morphology of Yeast …………………………………… (62)
实验十二　霉菌形态及观察 ……………………………………………………… (64)
　Experiment 12　Morphology of Molds …………………………………… (65)
实验十三　细菌、放线菌、酵母菌、霉菌菌落的特征观察 …………………… (66)
　Experiment 13　Colonies Characteristics of Bacteria, Actinomyces, Yeasts
　　　　　　　　　and Molds ……………………………………………… (67)

第三节　培养基的制备、灭菌和除菌技术 …………………………………………… (68)
Section 3　Media Preparation, Sterilization and Disinfection Techniques … (68)
　实验十四　微生物常用培养基的配制 ………………………………………… (68)
　　Experiment 14　Preparation of Commonly Used Microbial Media ……… (71)
　实验十五　培养基的灭菌和灭菌验证 ………………………………………… (73)
　　Experiment 15　Culture Media Sterilization and Sterility Test ………… (78)

第四节　微生物的接种技术 …………………………………………………………… (82)
Section 4　Microbial Inoculation Technology ………………………………… (82)
　实验十六　微生物的接种技术 ………………………………………………… (82)
　　Experiment 16　Microbial Inoculation Technology ……………………… (89)

第五节　微生物的纯种分离与培养技术 ……………………………………………… (94)
Section 5　Isolation and Cultivation of Pure Cultures ……………………… (94)
　实验十七　微生物的平板划线分离法 ………………………………………… (95)
　　Experiment 17　Isolation of Pure Culture-Streak Plate Technique ……… (99)
　实验十八　涂布平板和倾注平板分离法 ……………………………………… (101)
　　Experiment 18　Spread Plate Technique and Pour Plate Technique …… (105)
　实验十九　厌氧微生物的培养技术 …………………………………………… (107)
　　Experiment 19　Cultivation for Anaerobic Bacteria ……………………… (111)
　实验二十　噬菌体的分离与纯化 ……………………………………………… (114)
　　Experiment 20　Isolation and Purification of Bacteriophages …………… (117)

第六节　微生物菌种保藏技术 …………………………………………（120）
Section 6　Technology of Preservation of Pure Cultures ……………（121）
　　实验二十一　菌种保藏 ……………………………………………（121）
　　Experiment 21　Preservation of Pure Cultures ………………………（126）
第七节　微生物的生长繁殖技术 ………………………………………（131）
Section 7　Microbial Growth Techniques ………………………………（131）
　　实验二十二　微生物显微镜直接计数法——血球计数板计数法 …（131）
　　Experiment 22　Direct Microscopic Count——Blood Counting Chamber
　　　　　　　　　Method …………………………………………（134）
　　实验二十三　微生物间接计数法——平板菌落计数法 ……………（136）
　　Experiment 23　Indirect Microscopic Count——Plate Count ………（138）
　　实验二十四　细菌生长曲线的测定——比浊法 ……………………（140）
　　Experiment 24　Bacterial Growth Curve ……………………………（142）
　　实验二十五　霉菌生长的测定——重量法 …………………………（143）
　　Experiment 25　Mold Growth Measurement – Weighing Method ……（145）
第八节　环境微生物的分布规律 ………………………………………（146）
Section 8　Distribution of Enviromental Microorganisms ……………（146）
　　实验二十六　空气中微生物检查 ……………………………………（146）
　　Experiment 26　Distribution of Microbes in the Air ………………（147）
　　实验二十七　人体表面及口腔中的微生物检查 ……………………（149）
　　Experiment 27　Distribution of Microbes on Human Body …………（150）
第九节　细菌的生理生化反应 …………………………………………（151）
Section 9　Biochemical Tests of Bacteria ………………………………（151）
　　实验二十八　糖类发酵试验 …………………………………………（151）
　　Experiment 28　Utilization of Carbohydrates ………………………（153）
　　实验二十九　吲哚实验 ………………………………………………（155）
　　Experiment 29　Indole Test …………………………………………（157）
　　实验三十　甲基红实验 ………………………………………………（158）
　　Experiment 30　MR Test ……………………………………………（159）
　　实验三十一　乙酰甲基甲醇实验 ……………………………………（161）
　　Experiment 31　V. P Test ……………………………………………（162）
　　实验三十二　枸橼酸盐利用试验 ……………………………………（164）
　　Experiment 32　Citrate Utilization Test ……………………………（165）

· 3 ·

  实验三十三 淀粉水解试验 ⋯⋯⋯⋯⋯⋯⋯⋯⋯⋯⋯⋯⋯⋯⋯⋯⋯ (167)

  Experiment 33 Starch Hydrolysis Test ⋯⋯⋯⋯⋯⋯⋯⋯⋯⋯⋯⋯⋯ (168)

  实验三十四 产硫化氢试验 ⋯⋯⋯⋯⋯⋯⋯⋯⋯⋯⋯⋯⋯⋯⋯⋯⋯ (169)

  Experiment 34 Hydrogen Sulfide Production Test ⋯⋯⋯⋯⋯⋯⋯⋯ (171)

  实验三十五 明胶液化试验 ⋯⋯⋯⋯⋯⋯⋯⋯⋯⋯⋯⋯⋯⋯⋯⋯⋯ (172)

  Experiment 35 Gelatin Hydrolysis Test ⋯⋯⋯⋯⋯⋯⋯⋯⋯⋯⋯⋯ (174)

 第十节 免疫学实验技术 ⋯⋯⋯⋯⋯⋯⋯⋯⋯⋯⋯⋯⋯⋯⋯⋯⋯⋯⋯ (175)

 Section 10 Immunological Experimental Techniques ⋯⋯⋯⋯⋯⋯⋯⋯ (175)

  实验三十六 E 玫瑰花环试验 ⋯⋯⋯⋯⋯⋯⋯⋯⋯⋯⋯⋯⋯⋯⋯⋯ (175)

  Experiment 36 Erythrocyte Rosette Test ⋯⋯⋯⋯⋯⋯⋯⋯⋯⋯⋯ (176)

  实验三十七 淋巴细胞增殖试验——体内法 ⋯⋯⋯⋯⋯⋯⋯⋯⋯⋯ (178)

  Experiment 37 Lymphocyte Proliferation Assay ⋯⋯⋯⋯⋯⋯⋯⋯ (179)

  实验三十八 凝集反应 ⋯⋯⋯⋯⋯⋯⋯⋯⋯⋯⋯⋯⋯⋯⋯⋯⋯⋯⋯ (181)

  Experiment 38 Coagulation Test ⋯⋯⋯⋯⋯⋯⋯⋯⋯⋯⋯⋯⋯⋯⋯ (183)

  实验三十九 沉淀反应——琼脂双向扩散实验 ⋯⋯⋯⋯⋯⋯⋯⋯⋯ (186)

  Experiment 39 Precipitation Reaction Test ⋯⋯⋯⋯⋯⋯⋯⋯⋯⋯ (187)

第三章 研究性与设计性实验 ⋯⋯⋯⋯⋯⋯⋯⋯⋯⋯⋯⋯⋯⋯⋯⋯⋯ (189)

 Chapter 3 Research and Designing Experiment ⋯⋯⋯⋯⋯⋯⋯⋯⋯⋯ (189)

  实验四十 土壤中抗生素产生菌的分离 ⋯⋯⋯⋯⋯⋯⋯⋯⋯⋯⋯⋯ (189)

  Experiment 40 Isolation of Antibiotic–Producing Microorganisms from Soil

    ⋯⋯⋯⋯⋯⋯⋯⋯⋯⋯⋯⋯⋯⋯⋯⋯⋯⋯⋯⋯⋯⋯⋯⋯⋯⋯⋯ (192)

  实验四十一 紫外线对枯草芽孢杆菌产淀粉酶的诱变效应研究 ⋯⋯⋯ (195)

  Experiment 41 Screening and Ultraviolet Mutation Breeding of *Bacillus*

    subtilis Strains Producing Amylase ⋯⋯⋯⋯⋯⋯⋯⋯⋯⋯⋯⋯ (197)

  实验四十二 细菌氨基酸营养缺陷型菌株的筛选及鉴定 ⋯⋯⋯⋯⋯⋯ (200)

  Experiment 42 Isolation and Identification of Amino Acid Auxotrophic

    Strain by UV ⋯⋯⋯⋯⋯⋯⋯⋯⋯⋯⋯⋯⋯⋯⋯⋯⋯⋯⋯⋯⋯ (205)

  实验四十三 微生物培养条件的优化 ⋯⋯⋯⋯⋯⋯⋯⋯⋯⋯⋯⋯⋯⋯ (210)

  Experiment 43 Optimization of Microbial Culture Conditions ⋯⋯⋯⋯ (211)

  实验四十四 影响微生物生长的物理因素 ⋯⋯⋯⋯⋯⋯⋯⋯⋯⋯⋯⋯ (213)

  Experiment 44 Influence of Physical Factors on Microbial Growth ⋯⋯ (216)

  实验四十五 药物的体外抗菌试验 ⋯⋯⋯⋯⋯⋯⋯⋯⋯⋯⋯⋯⋯⋯⋯ (219)

  Experiment 45 Determination of Antimicrobial Activity ⋯⋯⋯⋯⋯⋯ (223)

实验四十六　抗生素效价的测定 …………………………………………… (227)
Experiment 46　Biological Assay of the Potency of Antibiotics ……………… (229)

第四章　综合性实验 ………………………………………………………………… (233)
Chapter 4　Comprehensive Experiments ………………………………………… (233)

实验四十七　大肠埃希菌噬菌体的分离及效价测定 …………………………… (233)
Experiment 47　Isolation and Titration of *Escherichia coli* Phage …………… (235)
实验四十八　口服药细菌数的测定及大肠菌群的测定 ………………………… (237)
Experiment 48　Detection of Bacteria Counts and Coliforms in Oral
　　　　　　　Medicines …………………………………………………… (239)
实验四十九　利用 Biolog 系统进行的分类鉴定 ………………………………… (242)
Experoiment 49　Microbial Identification and Classification with Biolog
　　　　　　　Analysis System ……………………………………………… (244)
实验五十　药物无菌检查法 ……………………………………………………… (247)
Experiment 50　Sterility Test of Drugs ………………………………………… (251)
实验五十一　药物的微生物限度检查法 ………………………………………… (256)
Experiment 51　Microbial Limit Tests of Drugs ………………………………… (260)

附录 ……………………………………………………………………………………… (265)

一、微生物实验常用培养基配方 ………………………………………………… (265)
二、微生物实验常用菌种 ………………………………………………………… (282)
三、常用染色液的配制 …………………………………………………………… (284)
四、常用试剂和溶液的配制 ……………………………………………………… (286)
五、微生物限度标准 ……………………………………………………………… (288)
六、微生物实验室废品处理 ……………………………………………………… (289)
七、实验室常用的化学消毒剂分类及其应用 …………………………………… (290)

第一章 微生物学实验的基础知识

第一节 微生物学实验目的和要求

微生物学实验的目的在于通过循序渐进的常规实验，使学生验证和巩固所学的微生物学基本理论，掌握基本操作技能；通过综合实验了解较为先进的科研方法与技能，培养学生联想及综合分析问题的能力；通过设计性实验，培养学生独立思考及创新的能力，建立实事求是、严谨的科学态度，提高解决实际问题的能力，为后续课程的学习和毕业后从事相关方面科研工作奠定良好的基础。为了达到这些目的，要求学生做到如下几点。

【实验前】

1. 实验前务必做好预习，仔细阅读实验教材，了解实验目的、基本原理、实验方法、操作步骤和注意事项。
2. 结合实验内容，复习有关微生物学方面的理论知识，做到充分理解。

【实验时】

1. 实验器材的放置力求稳当、整齐。
2. 严格按照实验指导上的步骤进行操作，称量药品要准确，珍惜实验材料。
3. 在实验中要坚持严肃性、严格性和严密性，培养、训练良好的科学作风，根据实验的特点注意科学地分配和运用时间。
4. 仔细耐心地观察实验过程中出现的各种现象，实事求是地记录实验时间、实验现象，联系课堂讲授内容进行思考，借以培养学生的科学思维方法。
5. 在实验过程中遇到疑难之处，先要自己设法解决，如一时无法解决，应当向指导教师说明情况，请求教师协助解决。对于实验室的贵重仪器，在未熟悉其性能之前，不可轻易调试。
6. 实验过程中保持安静、整洁。使用药品后，需要用原瓶盖（塞）塞好，公用药品和器材使用后请归位。

【实验后】

1. 将实验用的仪器擦拭干净，登记使用记录本，清点整理后放到指定位置，如有损坏、缺少，应及时报告指导教师；将实验中使用的带菌物品放在实验室指定位置，统一进行消毒灭菌处理。
2. 认真整理实验记录，经过分析与思考，撰写实验报告，按时上交实验报告。
3. 做好实验室的清洁卫生工作。

（徐 威）

第二节　微生物实验室设置及主要仪器设备

　　微生物学实验室的常规工作包括器具的洗涤、灭菌、微生物形态观察和计数、培养基的配制和灭菌、微生物生理生化反应的测定等。常用仪器包括显微镜、恒温培养箱、冰箱、超净工作台、高压蒸汽灭菌锅、摇床、电泳仪、凝胶成像仪等。为了保证实验人员人身安全、环境安全、样本安全及不污染环境等目的，从事这些工作必须有专门的仪器设备和操作房间。根据实验室的安全要求和使用要求，微生物实验室除了有学生操作实验室外，还需要配有其他辅助功能用房，包括：准备室、洗刷室、灭菌室、摇床室、无菌室、培养室和操作观察室等。这些房间的共同特点是陈设不宜过多，便于打扫卫生，不积灰尘，以保证实验室的卫生、安全质量，得到令人信服的实验数据。

一、微生物学实验室的设置

　　1. 准备室　准备室用于配制培养基和实验用品等。室内设有试剂柜、药品柜、储物柜、实验台、冰箱、电源、加热器、上下水道及紧急冲洗池等专用设备。

　　2. 灭菌和洗刷室　灭菌室，主要用于培养基的灭菌和各种器具及废弃物的灭菌，室内应备有高压蒸汽灭菌锅、烘箱等。洗刷室是用于洗刷器皿的场所，由于使用过的器皿和培养基需要灭菌后处理，因此，灭菌室内最好设置洗刷室，或者以软间壁将两室隔开。室内应备有加热器、上下水和洗刷台等。

　　3. 无菌室　无菌室是无菌操作的专用实验室。要求室内空气无杂菌，因此，房间不宜太高，天棚和四壁要涂漆或镶瓷砖，不留缝隙，室外要有双重过道，用以缓冲空气的作用。室内除放置超净工作台、凳子、酒精灯外，不应有其他物品，以防染菌。无菌室及其过道均要安装紫外线杀菌灯，每次工作前须用紫外灯照射，必要时可用5%的来苏水喷洒全室。工作结束时应立刻清扫，再用紫外灯照射。

　　4. 培养室　培养室是微生物生长繁殖的场所，配有培养箱、冰箱等，室内应安装紫外线杀菌灯。培养室应该定期杀菌消毒。

　　5. 菌种室　为学生实验提供菌种或做菌种保藏的地方，配有冰箱、超净工作台、培养箱等。室内应安装紫外线杀菌灯。

　　6. 摇床室　摇床室是微生物在液体培养基中生长繁殖的场所，摇床室应配有摇床、电源和上下水。室内应安装紫外线杀菌灯。

　　7. 操作观察室　即普通实验室，微生物的观察、计数和生理生化测定等场所。一般均设有实验台、试剂架和实验柜等，有时配备显微镜等常规仪器。为确保实验室洁净，室内应安装紫外线杀菌灯。

二、微生物学实验常用仪器设备

　　微生物学实验所用仪器主要有：高压蒸汽灭菌锅、超净工作台、生物显微镜、培养箱、控温摇床、离心机、电泳仪、紫外分光光度计、PCR扩增仪、凝胶成像仪等。

（一）立式自动压力蒸汽灭菌器

立式自动压力蒸汽灭菌器（图1-1）可杀灭包括芽孢在内的所有微生物，是灭菌效果最好、应用最广的灭菌设备。适用于普通培养基、玻璃容器等物品的灭菌。

A.立式蒸汽灭菌器

B.灭菌锅锅盖

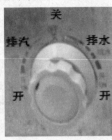
C.排汽、排水阀

D.控制面板

图1-1 立式自动压力蒸汽灭菌器

1. 使用方法

（1）逆时针转动手轮至顶，然后拉起左立柱上的保险销，移开锅盖。接通电源，将电源开关按至"开"处。

（2）开启锅盖，将纯水或蒸馏水直接注入锅内，加水过多时，开启排气阀放去锅中的水。

（3）将包扎的灭菌物品放入灭菌筐内。注意不要堵住安全阀放汽孔。

（4）将横梁全部推入立柱槽内，手动保险销自动下落锁住横梁；顺时针旋转手轮，当连锁灯亮时，显示容器密封到位。

（5）如需调节仪器的运行参数，可按"选择"键，使相应的指示灯点亮，数码管显示该参数值，此时可用"加"和"减"及"左移"键相结合调整该参数至需要的值，按选择键确认贮存（见图中的控制面板）。

（6）当所需温度和时间设定完毕后，开始加热，当温度达到100℃开始排汽，排汽2~3min后，关闭上排汽阀，调小下排汽旋钮，仪器进入自动灭菌循环过程。灭菌结束后，待压力下降至0，或者温度降低到70℃才可打开安全阀。

（7）打开安全阀或将排汽排水总阀旋至"开"，使物品上的残留水蒸气挥发。

（8）向左转动手轮至顶，拉起左立柱上的保险销，移开锅盖。将盖开启，取出已灭菌物品。关闭水源，打开下排水阀，排尽灭菌室的水与水垢，以备下次使用。

2. 注意事项

（1）锅内物品不要堆放过满，灭菌液体时，以不超过容器3/4体积为好。

（2）瓶口选用棉花或纱布塞，切勿使用未打孔的橡胶或软木塞。

（3）灭菌结束后，必须待压力表指针回零位方可排放余气。

（二）超净工作台

超净工作台为微生物接种等无菌操作提供无菌工作环境，通过风机将空气吸入预过滤器，经由静压箱通过过滤介质，将过滤后的空气以垂直或水平气流的状态送出，使操作区域达到百级洁净度，保证环境洁净度的要求。根据气流的方向分为垂直流超净工作台（图1-2）和水平流超净工作台，具体使用方法如下。

图 1-2 双人垂直流超净工作台

1. 使用方法

（1）将无菌室内紫外灯及超净工作台上安装的紫外灯全部打开，紫外线照射 30min。

（2）关闭紫外灯，打开风机开关，调整合适风量，吹风 20min。

（3）用 75% 乙醇浸泡的纱布或毛巾擦工作台面、内壁面。

（4）操作者穿戴好工作服，75% 乙醇消毒手，进行无菌操作。操作过程中，不得有物品阻挡住无菌风的流动。

（5）操作结束后，将带进工作台内的所有物品带出无菌室，用 75% 乙醇浸泡的纱布或毛巾擦工作台面、内壁面。关闭风机和日光灯，打开紫外灯照射 30min。

2. 注意事项

（1）提前 30min 做好超净工作台的灭菌准备工作，之后操作者关闭紫外灯方可入内。

（2）无菌操作中，始终保持一定的净化空气流速（通风量）。

（3）保持洁净度，定期对净化工作台和无菌室进行无菌验证，确保安全、无污染。

（4）定期检测和更换空气过滤介质。

（三）生物显微镜

生物显微镜在细胞学、寄生虫学、肿瘤学、免疫学、遗传工程学、工业微生物学、植物学等领域中应用广泛。主要用于微生物、细胞、组织培养、悬浮体、沉淀物等的观察，可连续观察细胞、细菌等在培养液中繁殖过程等。如图 1-3。

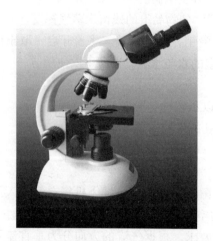

图 1-3 麦克奥迪 SFC-288 生物显微镜

使用方法及注意事项详见本书第二部分第

1 章。

（四）恒温摇床

恒温摇床（图1-4）主要适用于生物、生化、细胞、菌种等各种液态或固态的振荡培养。

1. 使用方法

（1）开机：打开仪器右上侧电源总开关，整机通电。

（2）温度设定：按"修改/确认"键→按"△"键输入密码"3"→按"温度"功能键（按"温度"功能键显示设定温度，再按一下显示实测温度）。

→按"△▽"键改变温度设定值，液晶显示屏显示温度设定值→按"修改/确认"键予以确认。

（3）转速设定：按"修改/确认"键→按"△"键至"3"→按"转速"功能键→设备进入转速参数设定状态，按"△▽"键改变转速设定值，液晶显示屏显示转速设定值→按"修改/确认"键予以确认。

图1-4　QYC-2112型恒温摇床

（4）定时设定：按"修改/确认"键→按"△"键至"3"→按"定时"功能键（按"定时"功能键显示定时设定时间，再按一下显示剩余时间）→设备进入定时参数设定状态，按"△▽"键改变定时设定值，液晶显示屏显示定时设定值→按"修改/确认"键予以确认。

（5）运行和关机：完成上述设定后→按"启动/停止"键，仪器即按照已设定的程序开始运行→在设备运行过程中按"启动/停止"键，可暂停摇板的旋转→按住控制面板的"电源"键2秒，仪器关机，显示屏显示消失→关闭右上侧的电源总开关。

2. 注意事项

（1）开启设备箱门前应确认摇板已处于静止状态，旋紧各摇板螺丝。

（2）设备箱门不宜随意频繁打开，否则会影响恒温效果。

（3）不要在运转过程中拔除插头。否则，会因过热而导致触电或发生火灾。

（4）请勿将异物插入送风口或进风口。若碰触到转动的风扇将导致仪器损坏。

（5）保持箱内外洁净，经常清理杂物、污渍。

（五）高速台式离心机

高速离心机（图1-5）主要用于微量样品快速分离。

1. 使用方法

（1）将样品等量装入离心管中，并将其对称放入转子中。

（2）拧紧转轴螺母，盖好盖门，将仪器通上电源后，打开仪器后面的电源开关，此时数码管显示"0000"，表示仪器已接通电源。

（3）按功能键调节仪器的运行参数（运转时间和运转速度），使相应的指示灯点亮，显示的数码即为该参数值，此时可用"左向"和"▽"及"△"键相结合调整该参数至需要的值，并按"记忆键"确认。

（4）按"开启"键启动仪器，仪器运行过程中数码显示转速。

2. 注意事项

（1）离心盖上不要放置任何物品，每次使用完毕，务必清理内腔和转头。

（2）在离心机未停稳的情况下不得打开盖门。

（3）经常检查转子及实验用的离心管是否有裂纹、老化等现象。

（六）低速台式离心机

低速台式离心机（图1-6）主要用于样品的固液分离。

图1-5　TGL-16G高速台式离心机

A. TDL-40G低速台式离心机　　B. 转子　　C. 操作面板

图1-6　TDL-40G低速台式离心机及操作面板

1. 使用方法

（1）将样品等量倒入在离心管中，天平配平，并将其对称放入转头。

（2）盖好盖门，将仪器接通电源后，打开仪器后面的电源开关，此时数码管显示"0000"，表示仪器已接通电源。

（3）如需调节仪器的运行参数（运转时间和运转速度），可按"选择"键，使相应的指示灯点亮，数码管即显示该参数值，此时可用"左移"和"加"及"减"键相结合调整该参数至需要的值，并按记忆键确认贮存。

（4）按"开启"键启动仪器，仪器运行过程中数码管显示转速，当需要查看其他参数时，可按功能键，使该参数对应的指示灯点亮，数码管即显示该参数值。

2. 注意事项

（1）不得在机器运转过程中或转子未停稳的情况下打开盖门，以免发生事故。

（2）开机前应检查转头安装是否牢固，机腔中有无异物掉入。

（3）样品应先平衡，使用离心筒离心时离心筒与样品应同时平衡。

（4）切勿在样品孔中洒入液体，挥发性或腐蚀性液体离心时，应使用带盖的离心管,并确保液体不外漏以免腐蚀机腔或造成事故。

(5) 除工作温度、运转速度和运转时间外,不要随意更改机器的工作参数,以免影响机器性能,转速设定不得超过最高转速。

(6) 实验结束后,关闭后面的电源开关,拔掉电源插头,做好使用情况记录。

(七) PCR仪

基于聚合酶链式反应原理制造的PCR仪实际就是一个温控设备,能在变性温度、复性温度、延伸温度之间很好地进行循环与控制,应用于对特定基因片段进行体外的大量合成(图1-7)。

- 工作指示灯
- 样品槽选择键(Block)
- 功能选择键(Select)
- 电源指示灯
- 确定键(Proceed)
- 停止键(Stop)
- 取消键(Cancel)
- 小数点/暂停键(Pause)

- RUN------运行程序
- ENTER------输入程序
- LIST------列出程序清单
- EDIT------修改程序
- FILES------文件操作
- SETUP------仪器参数设置

图1-7 PCR仪及操作面板

1. 使用方法

(1) 打开位于机身后面电源开关,机器显示屏会出现自检画面,约10~15s结束,仪器进入初始菜单界面。

(2) 选择样品槽,打开样品槽顶盖,将加好样的PCR管放入样品孔中(放置样品时,应在槽内均匀放置,保证热盖受力均匀),盖好顶盖(向左旋压紧向右旋松开)。

(3) 编写程序:①设置温度、时间(94℃,5min):在Step1中选择TEMP,输入94,再输入时间5:00;②变性温度(94℃,40s):选择TEMP,输入94,再输入时间0:40;③退火温度(55℃,30s):选择TEMP,输入55,再输入时间0:30;④延伸温度(72℃,40s):在Step4中选择TEMP,输入72,再输入时间0:40;⑤循环设置(Goto 2,29 cycles):选择Goto,输入2,再输入循环数29;⑥长延伸温度(72℃,10min):选择TEMP,输入72,再输入时间10:00;⑦End:选择End结束编程;⑧输入完成的程序后,从主菜单中选择Run,按Proceed,开始运行。

(4) PCR反应结束后,取走样品,关闭电源。

(5) 分析实验结果。

2. 注意事项

（1）仪器应放置在水平坚固的平台上，外界电源系统电压要匹配。

（2）仪器内不应进液体，清洗时应关断电源。

（3）仪器应定期清洁维护。

（4）为了延长半导体加热制冷块的寿命，请尽量避免长时间使用，4℃保存。

（八）凝胶成像系统

凝胶成像系统包括凝胶成像仪、电脑和打印机设备（图1-8），应用于DNA/RNA电泳凝胶（EB染色）、蛋白电泳胶、斑点杂交等方面，可以快速获得电泳照片和分析结果。

图1-8　GIS-2020型凝胶成像系统

1. 使用方法

（1）开启总电源，打开凝胶成像仪电源开关。

（2）开启电脑至WINDOWS处于正常工作状态，双击桌面上的"GIS"图标，进入软件系统。

（3）将待观察凝胶放在透射台正中，并关严暗仓门，打开相应灯源的电源开关（如紫外灯源）。

（4）打开拍摄界面，在软件界面中点击摄像按钮，可适当调节光圈，使图像到达合适亮度，调整图像大小及清晰度，待图像最优化时，点击红色按钮使之变绿色，成像保留在屏幕上，经电脑控制进行拍摄、保存图像。

（5）工作完毕，关闭软件窗口；关闭主机上层箱内电源开关；关闭电脑；关闭总电源。

（6）打开暗仓门，取出凝胶，用PE手套包住后处理，并擦洗透射台。

2. 注意事项

（1）开关仓门时防止将染色后凝胶沾在仓门盖上，如不慎沾上，擦干后用水冲洗。

（2）透射板紫外灯寿命有限，调整图像后及时成像，拍摄完毕，请立即关闭灯源电源!

(3) 凝胶应及时清理，防止凝胶固化后贴附在透射板上，造成成像不清晰。

(4) 如需拷贝图片，请将移动盘格式化后再插入。

(5) 图片数据及时处理，不要在电脑中存留，请勿用该电脑处理文档等。

(九) 微量移液器

微量移液器（俗称移液枪或简称"枪"）是进行微生物学实验和分子生物学实验的必备工具，依靠装置内活塞的上下移动，活塞的移动距离是由调节轮控制螺杆结构实现的，推动按钮带动推动杆使活塞向下移动，排除活塞腔内的气体（图1-9）。松手后，活塞在复位弹簧的作用下恢复原位，从而完成一次吸液过程。主要规格有：P2.5（0.1~2.5）；P20（2~20）；P200（20~200）；P1000（200~1000）等等。

A. 微量移液器　　　　B. 微量移液器手持方法　　　　C. 安装枪头方法

图1-9　微量移液器使用方法

1. 使用方法

(1) 选择移液器：根据取用溶液体积选择适当量程的微量移液器。

(2) 设定体积：由低值旋转至高值时，须先超越所欲设定值至少三分之一转后，再反转至设定值；由高值旋转至低值，则直接转至设定值即可。

(3) 安装枪头：移液器顶端插入移液头，轻轻用力下压，装上移液器头。

(4) 吸取溶液：将按钮压至第一段，尽可能保持微量移液器垂直，将微量移液器头尖端浸入溶液2mm左右，再缓慢释放按钮。释放按钮不可太快，以免溶液冲入吸管柱内而腐蚀活塞。

(5) 释放溶液：将微量移液器头与容器壁接触，慢慢压下按钮至第一段，停一两秒再压至第二段，把溶液完全压出。

(6) 打掉枪头，移液器释放按钮回原状。

2. 注意事项

(1) 勿将微量移液器本体浸入溶液中。

(2) 勿将按钮旋转出量程，否则会造成移液器损坏。

(3) 微量移液器不可吸取温度高于70℃的溶液，避免蒸汽侵入腐蚀活塞。

(4) 套有微量移液器头的微量移液器，无论微量移液器头中是否有溶液，均不可平放，需直立架上。

（十）电子分析天平

电子分析天平（图1-10），是实验室进行定量分析工作中最重要、最常用的精密称量仪器。实验室常用的电子分析天平感量有0.1mg、0.01mg和0.001mg三种。

图1-10 BP210S型电子分析天平

1. 使用方法

（1）检查并调整天平至水平位置。

（2）接通电源，预热足够时间后打开天平开关，天平则自动进行灵敏度及零点调节。待稳定标志显示后，可进行正式称量。

（3）称量时将洁净称量瓶或称量纸置于秤盘上，关上侧门，轻下按钮 I/O 去皮键，天平将自动自检，待显示器显示0.0000g后将称量纸放置于秤盘上，然后逐渐加入待称物质，直到所需重量为止。

（4）称量结束应及时除去称量瓶（纸），关上侧门，再按下按钮 I/O，切断电源，并做好使用情况登记。

2. 注意事项

（1）天平应放置在牢固平稳水泥台或木台上，室内要求清洁、干燥及较恒定的温度，同时应避免光线直接照射到天平上。

（2）称量时应从侧门取放物质，读数时应关闭箱门以免空气流动引起天平摆动。前门仅在检修或清除残留物质时使用。

（3）天平箱内应放置吸潮剂（如硅胶），当吸潮剂吸水变色，应立即高温烘烤更换，以确保吸湿性能。

（4）电子天平在初次接通电源或者在长时间断电之后，至少需要30min的预热时间，只有这样，天平才能达到所需要的工作温度。

（十一）可见分光光度计

可见分光光度计（图1-11）是实验结果分析中经常会使用到的仪器，其工作原理是物质吸收了入射光中的某些特定波长的光能量，相应地发生了能级跃迁，由于各种物质吸收光能量情况不同，特定波长的吸光度也不同，从而可以根据吸光度测定该物质的含量。

1. 使用方法

（1）接通电源，使仪器预热20min。

（2）用<MODE>键选择测试方式：透射比（T），吸光度（A），已知标准样品浓度值方式（C）和已知标准样品斜率（F）方式。

（3）用波长选择旋扭设置所需的分析波长。

（4）将参比样品溶液和被测样品溶液分别倒入比色杯中，打开样品室盖，将盛有溶液的比色杯分别插入比色槽中，盖上样品室盖。一般情况下，参比样品放在第一个

图 1-11　可见分光光度计

槽位中，被测样品放在第二个槽位中。

（5）将 0%T 校具（黑体）置入光路中，在 T 方式下按"0%T"键，此时显示器显示"000.0"。

（6）将参比样品拉入光路中，按"100"%T 或"0"A，此时显示器显示的"BLA"直至显示"100.0"%T 或"0.000"A 为止。

（7）当仪器显示器显示出"100.0"%T 或"0.000"A 后，将被测样品拉入光路，这时，便可从显示器上读出被测样品的透射比或吸光度值。

（8）使用结束后，先关闭仪器开关，再拔下电源。

（9）蒸馏水冲洗比色杯，倒扣在比色杯架或平皿的滤纸上。

2. 注意事项

（1）使用仪器需要预热 20min，以稳定波长。若大幅度改变测试波长，需稍等片刻，等灯热平衡后，重新调零及满度后，再测量。

（2）预热和等待时，应保持打开比色皿暗箱的箱盖。为避免光电管因受光连续照射而疲劳，只有在测量时才将比色皿暗箱的箱盖放下。

（3）使用比色皿时要注意其方向性，并应配套使用，以延长其使用寿命。测试时，保证比色皿不倾斜（因为倾斜，就会使参比样品与待测样品的吸收光径长度不一致，还有可能使入射光不能全部通过样品池，导致测试准确度不符合要求）。

（4）使用完毕后，请立即用蒸馏水冲洗干净（测定有色溶液后，应先用稀酸进行浸泡，浸泡时间不宜过长，再用蒸馏水冲洗干净）。

第三节　微生物学实验常用器皿的洗涤和包扎技术

微生物学实验常用器皿，大多需要经过消毒灭菌后再使用。因此，实验器皿的清洗、包扎和灭菌是实验前的一项重要准备工作，对其洗涤和包扎方法均有一定要求。洗涤是否干净，灭菌是否彻底，直接影响实验结果。微生物学实验常用器皿及其用途见表1-1，部分常用器皿见图1-12。

表1-1 微生物学实验常用器皿及用途

名称	常用规格	用途
玻璃试管	18×180mm；15×150mm；12×120mm	制备琼脂斜面和用于血清学试验
德汉氏小套管	试管中倒置的小试管	观察细菌糖发酵的产气情况
塑料离心管	10ml，50ml	离心法测定菌丝浓度
eppendorf管	0.5ml，1.5ml	离心和PCR扩增等
微量移液器	0.1~2.5μl，2~20μl，20~200μl等	吸取试剂及菌液
tip头	10μl，20μl，100μl	吸取试剂及菌液
微量进样器	20μl等	聚丙烯凝胶电泳等取样、上样
注射器	2ml，5ml，10ml，25ml等	注射抗原或静脉取样（层析电泳等）
接种环	长25~30cm，不锈钢制作	接种菌种
接种针	长25~30cm，用铂、镍或竹制作	接种菌种
量筒	100ml，250ml，500ml等	配制培养基及其他试剂
瓷缸	500ml，1000ml等	配制培养基及其他试剂
烧杯	50ml，100ml，500ml等	配制培养基及其他试剂
容量瓶	50ml，100ml，250ml等	准确定量溶液
培养皿	直径75mm，90mm	菌种的分离、计数和生物效价测定
移液管	0.5ml，1ml，5ml，10ml，20ml	定量移取菌液或液体试药
载玻片	25.4mm×76.2mm×2mm	微生物涂片和形态观察
盖玻片	18mm×18mm×0.1mm	微生物形态观察
玻璃架	18.5×45mm×20mm	放置载玻片进行涂片染色
凹玻片	较厚玻片中间有一个圆形凹窝	悬滴培养活菌
三角爬	直径2mm、长300mm	琼脂平板上涂布分离微生物
滴管	长200mm	吸取菌液或液体试药
双层瓶	外层放二甲苯，内层放香柏油	使用油镜头观察微生物
滴瓶	60ml	盛装染色液、乳酸、石炭酸等
三角烧瓶	100ml，250ml，500ml等	盛装培养基等和振荡培养微生物

一、器皿的洗涤

1. 新购置的玻璃器皿的洗涤 新购置的玻璃器皿含游离碱较多，使用前先用洗衣粉或自来水洗净，然后在稀盐酸溶液中浸泡数小时，取出后再用清水冲洗干净。

2. 使用过的玻璃器皿的洗涤 微生物实验用过的玻璃器皿，如试管、培养皿等，应先经高压蒸汽灭菌或用清水煮沸，倒掉培养基，用热的肥皂水洗刷，然后再用清水冲洗干净。

洗涤移液管时，可先将移液管尖端与装在水龙头上的胶管连接，用水将棉花冲出

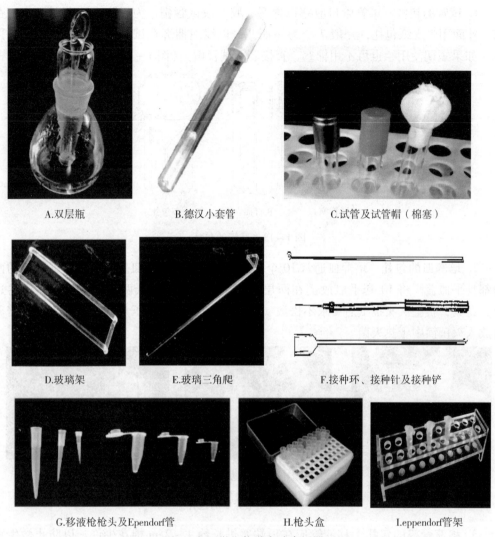

图1-12 微生物学实验常用器皿

后（取出棉花，以免堵塞水池），再冲洗片刻。若移液管吸过微生物培养物，应首先置于3%的来苏尔液中浸泡1h，再按常规洗刷。

3. 载玻片和盖玻片的洗涤 新的载玻片或盖玻片，先用肥皂水清洗、2%的稀盐酸溶液浸泡，再用水冲洗，最后用蒸馏水冲洗干净后，浸入95%的乙醇中，使用时，在火焰上烧去乙醇即可。

使用过的载玻片，可先用擦镜纸擦去镜头上的香柏油，再用另一块擦镜纸小心沾取二甲苯，擦去残存的油迹，然后将载玻片煮沸5min或浸于0.25%的新洁尔灭溶液中消毒，洗净后浸在95%乙醇中备用。

二、器皿的包扎

为了保证无菌效果，玻璃器皿使用前通常要用旧报纸或牛皮纸包扎好，然后干热或高压蒸汽灭菌。不同的器皿采用不同的包扎方法。

1. 试管的包扎 试管管口应塞以棉塞、胶塞或试管帽。为了防止灭菌时进水，在塞子外面用牛皮纸包扎，一般 7 个为一捆儿，以线绳捆绑在试管壁上，湿热或干热灭菌（如果短期使用，也可不用棉塞，直接套上试管帽）（图 1-13）。

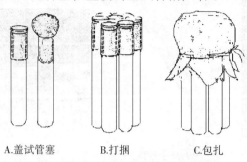

A.盖试管塞　　B.打捆　　C.包扎

图 1-13　试管的包扎

2. 培养皿的包扎 培养皿通常 10 个一组，最下面一套平皿反着放，保证平皿的两边都是平皿盖。将 10 套平皿放倒在两层旧报纸的一边，报纸两头折起压好，边卷边压，卷紧，折叠两头报纸后要求不松散（图 1-14），采用高压蒸汽灭菌，也可将培养皿放入灭菌筒中干热灭菌。

A.报纸两头折起压好　　　　B.边卷边压好两边报纸　　　　C.包扎完成

图 1-14　培养皿的包扎

3. 移液管等的包扎 移液管应在顶端塞进长约 1～2cm 棉花少许，以防止微生物液吸入口中或口中杂菌吹入移液管内，棉花的松紧程度以吹气时通气流畅而不下滑为准。准备 40～50cm 宽的长条纸，左边折回 6cm 的双层，将移液管尖端放在纸条的最左端，约与纸条成 45°，折叠纸条，包住移液管尖端，然后将移液管紧紧卷入纸条内，末端剩余纸条折叠打结（图 1-15）。如果一次使用很多根移液管，可将移液管塞好棉花后，很多根一起放在灭菌袋里（注意移液管头放在里边），再用报纸包好，扎紧。

采用相同方法对接种环、接种铲、滴管等进行包扎。要求包扎美观、紧密、使用方便，不同物品包扎方法略有不同。

4. 三角瓶等的包扎 三角瓶、茄形瓶等的包扎与试管的包扎相同，即以胶塞、棉塞或纱布盖（视培养物的需氧情况选择）塞口（图 1-16），再用牛皮纸包住，线绳扎紧。

5. 棉塞的制作 培养瓶（管）口上的棉塞，可过滤空气，防止杂菌侵入。它的质量直接影响无菌操作效果和通气状况，所以正确制备棉塞，是培养基制备的重要一环（图 1-17）。制作棉塞所用的棉花为市售的普通棉花，方法是按器皿口径大小，取适量

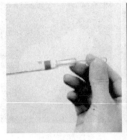   

A.塞棉花　　　　　B.纸条顶端对折　　　C.卷滚二圈后紧密卷滚　　　D.包扎完成

图 1-15　移液管的包扎

A.纱布盖对折　　　　　　　B.塞入瓶中　　　　　　　C.纱布盖塞口

D.棉塞塞口　　　　E.胶塞塞口　　　　F.牛皮纸包住　　　　G.线绳扎紧

图 1-16　三角瓶的包扎

棉花铺成正方形，横向折叠 2/5 长度，然后纵向紧密卷起来，拧紧，放在正方形纱布里塞入试管口，系好纱布，形状和大小均应与瓶（管）口完全配合，松紧适度。棉塞过紧会影响空气流通，过松起不到空气过滤除菌的作用，棉塞过小容易掉进试管或三角瓶中，过大容易脱落造成染菌。正确的棉塞的头部较大，长度约比容器口径大一半（如口径 20mm 则棉塞插入部分长度应为 30mm），有 2/3 在管内，1/3 在管外，外露部分要短些，略粗些，结实些。

6. 孢子打碎与过滤装置的包扎　打碎装置是由三角瓶和装入瓶中约 Φ4mm 的玻璃珠组成，100ml 三角瓶大约装入 80 个玻璃珠，或者刚好盖住三角瓶底部的量就可以，

A.棉花铺成正方形　　B.紧密卷起来　　C.塞入试管口　　D.制作完成

图1-17　棉塞的制作方法

然后以胶塞或棉塞封口，再用牛皮纸包住，线绳扎紧（图1-18）。

孢子过滤装置包括三角瓶或试管、漏斗、滤纸或脱脂棉、小平皿盖，接口处缠上纱布条防止染菌。再用牛皮纸包住，线绳扎紧（注意线绳要扎到三角瓶上，防止三角瓶脱落）。

A.孢子打碎装置　　B.缠纱布条　　C.过滤装置　　D.加皿盖　　E.线绳扎紧

图1-18　孢子打碎与过滤装置

第四节　微生物实验室常用物品的消毒灭菌和无菌操作技术

实验室常用的消毒灭菌方法有高压蒸汽灭菌、高温干热灭菌及化学消毒灭菌等等，具体采用哪种方法，应具体分析所要消毒的对象和消毒的目的等，从而有针对性地使用消毒和灭菌手段，选择简便、有效、不损坏物品的消毒灭菌方法。

一、微生物学实验室常用物品的消毒灭菌

1. 实验室、培养室和摇床间等房间的消毒　实验室、培养室和摇床间是观察、培养微生物的专用场所。为确保室内洁净，要做到定期消毒，主要是通过药物熏蒸、喷洒和紫外线灯照射等手段相结合来实现。定期消毒可采用5%来苏尔或0.1%新洁尔灭喷洒全室，擦拭工作台面、地面等，然后开紫外灯照射3~4h即可。

2. 仪器设备的消毒　培养箱和冰箱等仪器的消毒一般可采用5%来苏尔或0.1%新洁尔灭擦洗箱内，再用手提紫外灯照射灭菌。如果发生经常染菌情况时，可以先用75%来苏尔或0.1%新洁尔灭擦拭内壁，然后用甲醛+高锰酸钾熏蒸1h，再通风两整天，即可使用。

3. 玻璃器皿的灭菌　玻璃器皿的灭菌多采用高温干热灭菌，也可以采用高压蒸汽灭菌。高温干热灭菌通常用烘箱于160～170℃灭菌2h。注意灭菌前器皿必须是干燥的，避免升温引起玻璃的破碎。灭菌后，温度降到60℃以下时方可打开门，否则玻璃可能因突然遇冷而破碎。高压蒸汽灭菌是将物品放在密闭的高压蒸汽灭菌锅中，通常以121℃灭菌20～30min。

4. 塑料器皿的灭菌　塑料器皿和不能加热灭菌的材料一般采用辐射灭菌和化学药品灭菌。目前应用的杀菌射线有紫外线和$^{60}Co-\gamma$射线。波长在200～300nm的紫外线，容易被细胞中的核酸吸收，可以用于杀菌，但紫外线的穿透力弱，因此，只适用于器皿表层杀菌。$^{60}Co-\gamma$射线的穿透力强，可在包装完好的情况下灭菌。实验室中不能遇热的器皿消毒，也可采用化学药品灭菌，常用的化学药品有：0.25%新洁尔灭、3%～5%的甲醛溶液、75%乙醇等。

5. 接种工具等　接种环、接种铲和某些金属用具可以用高压蒸汽灭菌法，也可以用火焰烧灼灭菌法，将接种环或接种铲在火焰上烧红数分钟，冷后使用。

二、微生物学实验无菌操作技术

1. 无菌室工作规程

（1）每次使用无菌室前，先开启房间和场景工作台的紫外线灯，照射30min以上。之后关掉紫外灯，开启超净工作台的风机，平衡无菌空气15min。

（2）操作者用肥皂洗手，换上工作服、工作帽，戴好口罩。把所需器材放入无菌间。

（3）操作前，用75%乙醇棉球擦手和超净工作台台面。进行无菌操作时，动作要轻缓，不要讲话，尽量减少空气波动和地面扬尘。手持部位要远离瓶（管）口，即手持三角摇瓶（管）的底部，接种时接种针等不要碰到瓶（管）口。为达到纯种培养目的，要求动作规范，做到轻、准、稳、快、静、洁。

（4）操作过程中，如遇含菌培养物洒落或打碎盛菌容器时，应用浸润5%石炭酸的抹布包裹后，并用浸润5%石炭酸的抹布擦拭台面或地面，用乙醇棉球擦手后再继续操作。

（5）工作结束，立即将台面收拾干净，将不应在无菌室存放的物品和弃物全部带出无菌室，用75%乙醇棉球擦超净工作台台面，清理地面之后，开紫外线灯照射30min。

2. 无菌分装培养基、摆试管斜面操作　将灭菌后三角瓶装培养基温度降到60℃左右时，在超净工作台中保证无菌操作的状态下，将培养基转移到无菌试管中，每只15mm×150mm试管装量大约为5ml（注意转移培养基时，三角瓶口不要碰到试管口），趁着培养基还有一定温度时迅速在超净工作台里摆成斜面，这样可以减少凝集水出现。斜面由管底的上角开始，斜面培养基上部应距棉塞5cm以上，不应与棉塞接触。待其完全凝固后，可以垂直或水平放置，将试管斜面于37℃恒温箱内培养24～48h，检查灭菌的彻底性，如无菌生长则为合格，有杂菌生长则应废弃。为避免干燥和污染杂菌，一般将斜面培养基保存于冷暗处，保存期不应超过1个月，否则不宜使用。

（刘晓辉）

第五节　实验报告的书写要求

实验报告是实验者对其完成的实验工作进行的扼要的文字总结，是综合评定实验课成绩的重要依据之一。学生每完成一次实验都应提交相应的实验报告。实验报告通常的格式如下：

1. 实验目的　本实验最主要的目的。

2. 实验原理　与实验直接相关的基本理论依据。

3. 实验材料　包括主要仪器、药品与试剂、菌种、培养基及其他相关实验材料等。当使用的仪器、试剂或菌种与实验指导不一致时，一定要完整记录实际使用的仪器、试剂和菌种等。

4. 实验方法与步骤　实验方法与步骤可进行简要描述。

5. 实验原始记录　应详细记录原始数据及时间，根据原始资料，真实准确地记录所观察到的实验现象，记录的实验数据一定要整洁，不得涂改、删除数据，也不能插入数据。

6. 实验结果　将原始数据整理，以图或表格的形式表达实验结果，绘制表格时，应绘制三线表，表内布局合理，标题在表格的上方。

7. 讨论　讨论是根据已知的理论知识和已有的文献资料对实验结果的解释与分析，要根据自己的实验结果讨论，判断所得到的结果是否为预期结果，对于非预期结果要分析可能的原因。

第六节　验证性、综合性、设计性实验概述

一、什么是验证性、综合性、设计性实验

实验是教学的重要组成部分，学生通过实验能加深对理论知识的理解，并能掌握一定的操作技能，养成严谨、认真和实事求是的科学态度。实验可分为三类，分别是验证性实验、综合性实验和设计性实验。教育部在《普通高等学校本科教学水平评估方案（试行）》中对这三类实验做了明确注释。

1. 验证性实验　验证性实验是验证者对已知的实验结果进行的以验证实验结果、巩固和加强有关知识内容，培养实验操作能力，掌握实验原理为目的的重复性实验活动。是一种再现式的以书本知识为主，以传授理论知识和实验技能为主要特征的教学模式。

2. 综合性实验　综合性实验是指在学生具有一定实验基础知识和基本操作技能的基础上，运用某一课程或多门课程的综合知识，对学生实验技能和实验方法进行综合训练的一种复合性实验。综合性实验目的在于锻炼学生对所学知识的综合应用能力，培养学生分析和解决复杂问题的能力，培养学生数据处理以及查阅中外文资料的能力。

3. 设计性实验　设计性实验是学生在教师的指导下，根据给定的实验目的和实验条件，自行设计实验方案，选择实验器材，拟定实验程序，并加以实现的实验。它不

但要求学生综合多科知识和多种实验原理来设计实验方案，还要求学生能运用已有知识去发现、分析和解决问题。设计性实验目的在于培养学生掌握设计实验的一般方法和步骤，进行探究性学习，激发学生学习的主动性、培养学生的创新思维和创新能力。

二、综合性、设计性实验的基本步骤

1. 选题 选题以教师为主，学生参与，师生共同讨论确定。这一阶段培养学生选择研究课题及检索文献资料的能力。

基本方法：

（1）教师根据教学科研的经验及学生的实际情况，提出 2~4 个实验题目。

实验题目需满足如下几点：①综合性强；②灵活性大，有发挥的余地；③实验的材料、仪器设备容易得到。

（2）学生可以根据自己的兴趣并通过查阅文献与工具书，选定教师推荐的题目或自选符合要求的题目。

2. 实验方案设计 在查阅资料的基础上，应用所学知识拟定和完善实验方案，论证方案的可行性，分析实验中可能出现的各种问题。

此项工作以学生为主体，然后师生共同讨论，确定出最合理的可行方案。

具体的内容和格式要求如下：

（1）实验题目

（2）立题依据（研究目的、意义，以及打算解决的问题和国内外现状）。

（3）实验器材与药品（器材名称、型号、规格和数量）药品或试剂名称、培养基名称及配制规格、实验用菌种等。

（4）实验方法与操作步骤（包括实验的技术路线、实验的时间安排、具体实验操作过程、设立的观测指标和指标检测手段）。

（5）预期结果

3. 实验准备 根据实验的设计方案、准备实验所需要的材料和实验菌种。

4. 预实验 按照实验设计方案和操作步骤进行预实验。在预实验过程中，做好实验原始记录。实验结束后，及时整理实验数据，发现和分析预实验中存在的问题，确定需要改进、调整的内容。

5. 正式实验 按照修改后的实验设计方案和操作步骤进行正式实验，记录好原始记录。

教师的主要作用是引导学生独立思考，协助学生解决在实验中遇到的问题，进行宏观调控，把握学生实验进度，保证按时完成。

6. 实验结果与讨论 学生对实验数据进行归纳整理，并加以分析。

在整个实验过程中，学生是主体，教师的任务是对学生在查阅资料、方案设计等各环节给予方法上的指导。对学生的设计和实验不宜过多干涉，要注意培养学生的独立思考能力和创新能力，鼓励学生大胆使用新方法，提出新观点和新思路，充分调动学生的积极性。在教学过程中，学生会遇到许许多多的问题，教师要善于引导学生分析问题，找出解决问题的办法。

（徐 威）

第二章 微生物学基础性实验

Chapter 2 Basic Experiments in Microbiology

第一节 显微镜和显微技术

微生物学的研究对象大多是肉眼无法直接分辨的微小生物，正是由于微生物的这个特性，使得显微镜成为这个学科中最为重要的工具之一。本章从最常用的普通光学显微镜（即明视野显微镜）入手，介绍光学显微镜的构造、原理及其使用方法，这其中也涵盖了相差显微镜和荧光显微镜。在本章的最后将介绍如何利用显微镜进行细菌运动性观察和微生物大小的测量。

Section 1 Microscope and Microscopic Techniques

Microbiology is the study of organisms too small to be seen distinctly with the unaided eye. Because of this characteristic of microbes, the microscope is an essential tool for this discipline. Thus, to understand how the microscope works and the way to prepare the specimens are of great importance. This part focuses on the former and begins with the most commonly used bright – field light microscope with its structure, principle and use. Some other types of light microscopes are also included as phase – contrast microscope and fluorescence microscope. This section closes with bacterial motility and microscopic measurement of microbes.

实验一 普通光学显微镜的结构、使用与维护

【目的】

1. 掌握普通光学显微镜（以下称"明视野显微镜"）的使用，尤其是油镜的使用和保养方法。
2. 了解明视野显微镜的构造和工作原理。

【基本原理】

光学显微镜有多种分类方法：按使用目镜的数目可分为双目和单目显微镜；按图像是否有立体感可分为立体视觉和非立体视觉显微镜；按观察对象可分为生物和金相

显微镜等；按光源类型可分为普通光、荧光、红外光和激光显微镜等。

明视野显微镜，又称明视野光学显微镜是最常用的光学显微镜。之所以叫它"明视野"，是因为它就像你现在所看到的本页书上的文字一样，能够在较亮的背景下形成一个比较暗的图像。

（一）明视野显微镜的结构

明视野显微镜的基本结构如图 1-1 所示，分光学系统和机械系统两部分。

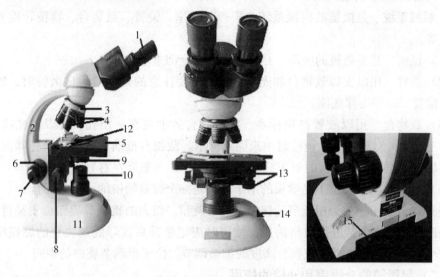

图 1-1　明视野显微镜

Figure 1-1　Bright-field microscope

1. 目镜　2. 镜臂　3. 物镜转换器　4. 物镜　5. 载物台　6. 粗调节旋钮　7. 细调节旋钮
8. 聚光器调节旋钮　9. 聚光器（含光圈）　10. 光源　11. 镜座　12. 玻片夹
13. 载物台推动旋钮　14. 光源调节旋钮　15. 电源开关
1. Ocular (eyepiece)　2. Arm　3. Nosepiece　4. Objective lens　5. Stage
6. Coarse adjustment knob　7. Fine adjustment knob　8. Condenser knob
9. Condenser (including diaphragm)　10. Light source　11. Base　12. Stage clips
13. Stage adjustment knobs　14. Brightness control　15. Switch

1. 光学系统　显微镜的光学系统主要包括物镜、目镜、聚光器和光源等。

（1）物镜　装在转换器上，是显微镜质量和性能最关键的部位。转换器上共有4个物镜，分别是4倍的搜索物镜（红圈），10倍的低倍镜（黄圈），40倍的高倍镜（蓝圈）和100倍的油镜（白圈）。需要注意的是这些物镜之所以用颜色进行标记是为了更容易进行辨识。在必要情况下镜头只能用擦镜纸进行清洁。

物镜根据使用条件不同可分为干燥系物镜和油浸系物镜。前者以空气为介质，包括搜索物镜、低倍镜和高倍镜等。后者指物镜与标本之间的介质是一种与玻璃折光率相近的香柏油，因此这种物镜也称为油镜。

（2）目镜　其作用是把物镜放大了的实像进行第二次放大，并把物像映入观察者的眼中。目镜由两组透镜组成，可在其上安装目镜测微计（见实验5）。实验室常用的目镜放大倍数为 5×、10×、15×、20×。显微镜的放大倍数由目镜和物镜的乘积决定。

（3）聚光器　位于载物台下方，其功能是将穿过标本的光线进行汇聚，以得到最强的照明，使物像获得明亮清晰的效果。聚光器下面还有彩虹光圈，用于调节穿过标本的光线强弱。

（4）光源　光源位于显微镜的基座上，由开关控制并对其亮度进行调节。显微镜不使用的情况下应将光源关掉。

2. 机械系统　显微镜的机械系统主要包括镜座、镜臂、载物台、物镜转换器和调节旋钮等。

（1）镜座　是显微镜的底座，是支撑显微镜全部重量的部件。

（2）镜臂　用以支撑载物台和光学镜筒。需要注意的是在搬运显微镜时，需要一手握住镜臂，一手支撑底座。

（3）载物台　用以放置被检标本，在载物台的中央有一小孔，可以使光线穿过。载物台上装有一对玻片夹，镜检时用来固定标本。载物台推动旋钮位于聚光器的一侧，分上下旋钮，作用是在镜检时对于载玻片的水平位置（前后左右）进行调节。

（4）物镜转换器　位于光学头的下半部，是一个带有物镜的可旋转圆盘。

（5）调节旋钮　分为粗调节旋钮和细调节旋钮。较大的粗调节旋钮位于镜臂两侧，用于快速升降载物台，进行粗略调焦，使用时应记住升高或降低载物台的旋钮旋转方向。较小的细调节旋钮用于载物台高度的细微调节，位于粗调节旋钮的中间。

（二）显微镜的分辨率和油镜的使用

显微镜所观察到的应是放大且清晰的图像。其清晰度由分辨率来决定。分辨率（R）是指能够将非常靠近的两点进行区分的能力。分辨率越小，其分辨能力越高。下面的方程式显示了决定分辨率的主要参数：

$$R = 0.61 \frac{\lambda}{NA}$$

式中，λ 是所用的光线波长，NA 叫作数值孔径。由方程可见，所用波长越小，呈像越清晰，电子显微镜所具有的高分辨率就与此有关。分辨率的另一个决定因素是数值孔径。数值孔径是用来描述镜头改变光线方向的相对有效性的数学常数。数值孔径的大小如下面方程显示：

$$NA = n \times \sin\theta$$

式中，n 代表物镜与样本间介质的折射率，θ 则定义为穿过样本进入镜头的光锥角度的 1/2，即镜口角（图 1-2）。

由于数值孔径越大，分辨率效果越好。所以要提高数值孔径，可以通过提高 n 和 $\sin\theta$ 来达到。空气的折射率是 1，并且由于 $\sin\theta$ 不会大于 1，所以任何透镜在空气中的数值孔径都不会大于 1。要使数值孔径大于 1 的唯一可行方法就是通过使用香柏油来提高折射率，香柏油是一种无色液体，与载玻片具有相同的折射率。当用香柏油代替空气时，入射光线就不会发生反射和折射。因此，就能够得到较大的数值孔径和更好的分辨效果（图 1-3）。

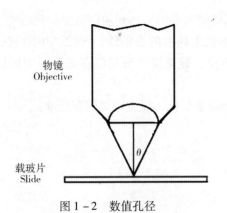

图1-2 数值孔径

Figure 1-2　Numerical apertures

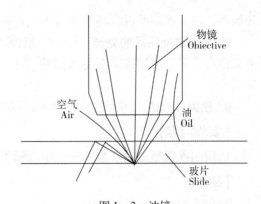

图1-3 油镜

Figure 1-3　the oil immersion objective

【仪器与材料】

1. 菌种标本

（1）细菌基本形态标本　金黄色葡萄球菌（*Staphyloccocus aureus*）、大肠埃希菌（*Escherichia coli*）。

（2）细菌特殊结构标本　芽孢（枯草芽孢杆菌 *Bacillus subtilis*）、鞭毛（变形杆菌 *Bacillus proteus*）、荚膜（肺炎双球菌 *Diplococcus pneumoniae*）。

（3）真菌基本形态标本　酿酒酵母（*Saccharomyces cerevisiae*）。

2. 培养基与试剂　双层瓶（内装香柏油和二甲苯）。

3. 设备　明视野显微镜。

4. 其他　擦镜纸。

【方法与步骤】

1. 取显微镜　一手握住镜臂，一手托住镜座，取出显微镜置于实验台上，镜座距离实验台边缘约3~4cm。为减少疲劳，镜检者应姿势端正，两眼睁开。

2. 放置标本　将标本用玻片夹固定在载物台上，通过调节标本推动旋钮将涂有标本的位置置于载物台光源通过孔的正上方。

3. 调焦　适当调节光圈、聚光器和光亮度，使视野得到均匀照明。

4. 低倍镜观察　将载物台上升至最高点，然后用粗调节旋钮缓慢下降载物台，同时从目镜中观察，直至视野中出现图像。当视野中出现图像时，改用细调节螺旋将图像调至清晰。注意：细调节旋钮不可沿同一方向过度调节。

5. 高倍镜观察　此时你的标本仍应在视野中只不过比原来放大了4倍。用细调节旋钮将物像调至清晰。真菌标本用高倍镜即可观察清楚。

6. 油镜观察　降低载物台，在载玻片上光源透过的位置加一滴香柏油，然后将100倍物镜（即油镜）转入光路，用粗调节旋钮升高载物台，并从侧面观察，使镜头浸入油中并轻轻与载玻片接触。注意：不可用力过猛，以免压碎玻片，损坏镜头。然后从目镜中进行观察，同时将光亮度升至最大。用粗调节旋钮缓缓下降载物台至出现物像时，改用细调节旋钮调整至物像清晰。需要注意的是：当使用油镜时，一旦镜头接触香柏油，就不要再将高倍镜转入，否则高倍镜头会因为沾到油而需要彻底清洗。

7. 显微镜使用后的处理　用擦镜纸擦除油镜上残留的香柏油，再用二甲苯进一步清理镜头，最后要用擦镜纸将残留的二甲苯擦掉。显微镜镜头只能在必要时用擦镜纸清洁。

8. 显微镜归位　将4倍物镜转入，载物台降至最低，显微镜送回存放处。

【实验内容】

使用明视野显微镜观察不同微生物标本。

【结果】

分别绘制镜下各标本形态。

【注意事项】

1. 显微镜应该放置在通风干燥的地方，避免阳光直射或暴晒，通常罩起来放入箱内。

2. 目镜和物镜必须保持清洁，如有灰尘只能用擦镜纸擦拭，不得用其他物品擦拭。

3. 油镜观察后，二甲苯用量不能太多，因为物镜中的透镜是用树胶粘合在一起的，一旦树胶溶解，透镜会脱落。

4. 二甲苯为有毒的有机物质，可以用乙醚乙醇混合液代替（无水乙醚：无水乙醇 = 7∶3，V/V）。

【思考题】

1. 使用油镜时应注意哪些问题？
2. 用明视野显微镜能否观察病毒？为什么？

Experiment 1　Bright – Field Light Microscope

Objectives

1. Grasp how to use the bright – field light microscope (abbreviated as "bright – field microscope") correctly, especially the oil immersion lens.

2. Understand the structure and principles of the bright – field microscope.

Principles

There are different ways to classify microscopes according to the number of ocular eyepiece, presenting stereo image or not, object of observation and the type of light source, etc.

The bright – field microscope is the most commonly used type of light microscope. It is named "bright – field" as it forms a dark image against a brighter background just like the words on this page.

Ⅰ. The structure of the bright – field microscope

The fundamental structure of a bright – field microscope including optical system and mechanical system is illustrated in Figure 1 – 1.

1. Optical system Mainly includes objective lens, ocular, condenser and light source.

(1) Objective lens The objective lens are located on the nosepiece. A 4 × scan objective or lens (red band), a 10 × low-power objective (yellow), a 40 × high-power objective (blue), and a 100 × oil immersion lens (white) may be found. Note that these lenses are color-coded, and remember which color is which for easy identification. Only lens paper should ever be used on the lenses if necessary. The objectives could be classified as dry- and oil immersion lens according to the working condition. The oil immersion lens is so called because cedar wood oil is applied between the objective and the specimen. While, in the dry objectives air is presented.

(2) Ocular The objective lens forms an enlarged real image within the microscope, and the eyepiece lens (ocular) further magnifies the primary image. The total magnification of any set of lenses is determined by multiplying the magnification of the objective by the magnification of the ocular. Ocular micrometer (in Experiment 5) can be placed in ocular tube for microscopic measurement. 5 ×, 10 ×, 15 × and 20 × oculars are commonly used.

(3) Condenser and iris diaphragm They are located under the stage. The condenser focuses the light going through the specimen and the iris diaphragm is used to regulate the amount of light passing through the specimen.

(4) Light source It is mounted on the base of the microscope, and is controlled by an on-off switch which can also adjust the brightness. Keep the light off when not in use.

2. Mechanical system Mechanical system mainly includes base, arm, stage, nosepiece and adjustment knobs.

(1) Base is the main support for the microscope.

(2) Arm supports the stage and the optical head. Note that the proper way to carry a microscope is to grasp the arm with one hand and support the base with the other hand.

(3) Stage is the flat area upon which the specimen is placed. It has a hole in the center through which light may pass. Mounted on the stage is a pair of stage clips. This is designed to keep the slide in place from moving while you are looking at it. The stage adjustment knobs can move the slide horizontally when searching under the microscope.

(4) Nosepiece is a revolving plate on the lower side of the optical head which holds the objective lenses.

(5) Adjustment knobs A pair of large knobs, called the coarse adjustment knobs, which permit rapid rising or lowering of the stage, are located on both sides of the arm. Memorize the direction to rotate this knob to lower or raise the stage. Pair of smaller knobs, called the fine adjustment knobs, which permit smaller adjustments in the stage height, are located "inside" the coarse adjustment knobs.

II. Microscope resolution (resolving power) and the oil immersion objective

Microscope is used to provide not just a magnified image but a clear one. Resolution (R) is essential for this. Resolution is the capacity of an optical system to distinguish or separate two adjacent objects or points from one to another. The smaller the resolution is, the better the resolving ability is. The following equation expresses the main determining factors in resolution:

$$R = 0.61 \frac{\lambda}{NA}$$

Lambda (λ) is the wavelength of light used to illuminate the specimen and the *NA* is the numerical aperture. From the equation, it indicates that a major factor in resolution is the wavelength of light used. The shorter the wavelength is, the clearer the image will be. This will help to understand the high resolution of the electron microscope. Another factor in resolution is *NA*. *NA* is a mathematical constant that describes the relative efficiency of a lens in bending light rays. It could be expressed as:

$$NA = n \times \sin\theta$$

In the equation, *n* represents the refractive index of the medium between the objective and the specimen. θ is defined as 1/2 the angle of the cone of light that enters a lens from a specimen. (Figure 1 - 2)

Remember that a higher *NA* can provide better resolution. There are two ways to increase the *NA* value which are higher *n* and $\sin\theta$. The refractive index for air is 1.00. Since $\sin\theta$ cannot be greater than 1.00, no lens working in air can have a *NA* higher than 1.00. The only practical way to increase the *NA* to above 1.00 is to increase the refractive index with immersion oil, a colorless liquid with same refractive index as the slide. When air is replaced with immersion oil, reflection and refraction of the light from the condenser to the objective will be avoided effectively. Thus, an increase in numerical aperture and resolution is achieved (Figure 1 - 3).

Apparatus and Materials

1. Specimens (on glass slides)

(1) The basic morphology of bacteria – *Staphylococcus aureus* and *Escherichia coli*.

(2) The special structures of bacteria – endospore (*Bacillus subtilis*), flagella (*Bacillus proteus*) and capsule (*Diplococcus pneumoniae*).

(3) The basic morphology of fungi – *Saccharomyces cerevisiae*.

2. Cultures and reagents   Dual – bottle (with immersion oil and xylene).

3. Apparatus   Bright – field microscope.

4. Others   Lens paper.

Methods and Procedures

1. Always carry the arm of the microscope with one hand and support the base with the other hand. Place it on the desk 3 to 4cm away from the desk edge. In order to reduce fatigue, viewers should posture properly and open both eyes.

2. Place the provided slide on the stage, and secure it firmly using the stage clips. Try to guesstimate where the specimen is located on the slide, and place it in the center of the hole

allowing light through the stage.

3. Adjust the iris diaphragm, condenser, and brightness of light carefully to get suitable illumination.

4. With the low-power objective (10 × objective) in position, raise the stage to the top position. Then lower the stage slowly using the coarse adjustment knobs while looking at the microscope until the object comes into view. This can avoid striking the objective to the slide. Once the object is in view, switch over to the fine adjustment knobs to focus the desired image. Note: Do not overuse the fine adjustment knobs in one direction.

5. Rotate the high-power objective (40 × objective) in position. Your specimen should still be seen in the field of vision, but 4 times larger now. Use your fine adjustment knobs to clarify the image. Fungi specimen can be observed clearly with the high-power objective.

6. Use the oil immersion lens to examine the stained bacteria samples provided. Lower the stage and place a single drop of immersion oil on the slide right over where the light is coming through the stage. Then rotate the 100 × objective (oil immersion lens) into position. Raise the stage with the coarse adjustment knob while looking at the objective from the side until the lens immerses in oil and just touches the slide. Note: Do this gently to avoid breaking the slide and the objective. Now look through the oculars, increasing your light for maximum illumination. Lower the stage slowly with the coarse adjustment knobs. Once the object is in view, use the fine adjustment knobs to focus clearly. Note: Once you have gone into immersion oil, do not go back to the 40 × objectives. The objective will get oil on it, and you will have to really clean it to get the oil off.

7. After you are finished with the microscope, use the lens paper to wipe the oil from the 100 × objective lens. Xylene could be used to clean the oil if necessary, but remember to remove xylene with the lens paper as well. Clean all the microscope's lenses only with lens paper if necessary.

8. Rotate the 4 × objective lens into place. Use the coarse adjustment knobs to lower the stage to the bottom. Finally, return the microscope to the appropriate storage area.

Experiment contents

Observe different microbial specimens under bright-field microscope.

Results

Draw the typical morphology of microorganisms you observed through the bright-field microscope.

Notes

1. Microscopes should be placed in the appropriate area avoiding sunlight.

2. Only lens paper should ever be used on the lenses if necessary.

3. Xylene applied should not be much, or the gum in the objective lens will be dissolved.

4. Xylene could be replaced by diethyl ether and ethanol solution (7∶3, V/V) because of its toxicity.

Questions

1. What should be paid attention to when use the oil-immersion lens?
2. Can you observe virus with the bright-field microscope? Why?

实验二 相差显微镜的结构与使用

【目的】

了解相差光学显微镜的基本原理并学会如何正确使用。

【基本原理】

使用明视野显微镜并不能够清楚观察未染色的活细胞及其内部结构，这是因为细胞与水环境或细胞内部结构之间的对比度太小。不过细胞的不同结构之间确实存在密度上的细微差别，从而能够将通过其中的光线有所改变。即光线通过透明的活细胞后，由于细胞各部分密度的差异（或折射率不同），而使光波的相位发生变化，形成相位差。

相差光学显微镜（以下简称"相差显微镜"）可以很容易地将这种细微的改变转化成为可检测出的光强变化。因此这种显微镜非常适用于观察活细胞。

与明视野显微镜相比，相差显微镜有另外两个部件：一个是位于物镜后聚焦面的相板；另外一个是与相板相匹配的位于聚光器上的环形光阑，它是一个黑色圆盘，上面带有一个透明圈。相板上有一环状区域，经过其中的光线不会发生改变，而经过环状区域以外的其他区域的光线则会被延迟1/4波长。

带有环状光阑的聚光器可以产生中空的光锥。当光锥通过细胞的时候，部分光线会由于标本中各结构密度与折射率的不同发生偏斜，并被推迟 $1/4\lambda$（波长），由这些光线汇聚形成了备检物体的一个图像。未发生偏斜的直射光照射到位于相板上的环形区域，而衍射光则会错过这一区域而从相板上的其他区域穿过。由于相板可以使通过的衍射光再被推迟 $1/4\lambda$，直射光与衍射光之间就会存在 $1/2$ 的相位差，当两者汇聚在一起时，这部分就会相互抵消而呈现出较暗的图像（图2-1）。

【仪器与材料】

1. 菌种 枯草芽孢杆菌（*Bacillus subtilis*）24h液体培养物、酿酒酵母（*Saccharomyces cerevisiae*）48h液体培养物。

2. 培养基与试剂 无菌水、双层瓶（内装香柏油和二甲苯）。

3. 仪器 相差显微镜。

4. 其他 载玻片、盖玻片、擦镜纸、吸水纸。

【方法与步骤】

1. 制片 制备枯草芽孢杆菌水浸片。

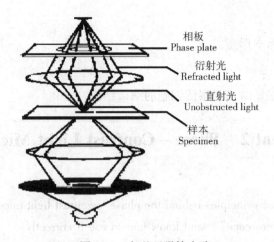

图 2-1 相差显微镜光路

Figure 2-1 Light pathway of a phase-contrast microscope

2. 放置标本 将标本置于载物台上，用玻片夹固定好。

3. 低倍镜调节 先转至 10 倍物镜，同时将环形光阑也调至 10 倍位置，用粗调节旋钮和细调节旋钮将物像调至清晰。然后将目镜取下，换上合轴调整望远镜，调整至亮环与暗环都清晰（图 2-2A），固定合轴调整望远镜。再调节聚光器至两环相互重叠（图 2-2B）。调整完毕后，将目镜重新安上，用细调节旋钮将图像调节清晰并观察。

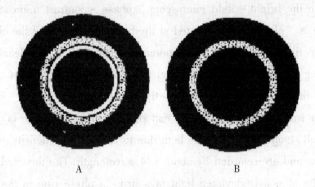

图 2-2 相差显微镜的调节

Figure 2-2 Adjustment of the phase-contrast microscope condenser

A. 调节亮环和暗环清晰 B. 两环重合

A. Adjust the dark ring and the bright ring clearly

B. The dark ring and the bright ring superimposition

4. 高倍镜及油镜调节 按同样的方法分别用 40 倍物镜和油镜观察，同时要将环形光阑也调至相应位置。记录观察结果。油镜使用完毕后需清洁镜头。

5. 酿酒酵母观察 按照相同的步骤观察酿酒酵母水浸片，并记录观察结果。由于酿酒酵母是真菌，所以不必使用油镜观察。

【实验内容】

使用相差显微镜观察枯草芽孢杆菌和酿酒酵母。

【结果】

分别绘制镜下各标本形态。

【思考题】

为什么可以用相差显微镜观察活细胞的细微结构？

Experiment 2　Phase – Contrast Light Microscope

Objectives

Understand the basic principles behind the phase – contrast light microscope (abbreviated as "phase – contrast microscope") and know how to use it correctly.

Principles

Unstained living cells and their internal components are not clearly visible in the bright – field microscope because there is little difference in contrast between cell and water or the internal structures. But cell structures do differ slightly in density which can alter the light that passes through them in subtle ways.

A phase – contrast microscope can converts the slight differences into easily detected variations in light intensity. Thus this microscopy is an excellent way to observe living cells.

Compared with the bright – field microscope, phase – contrast microscope has two additional components: a "phase plate" located at the back focal plane of the objective lens and a matching "annular diaphragm" in the condenser consisting of a clear ring on a black field. There is a ring – shaped area which is on the phase plate. Light can be retarded exactly 1/4 wavelength by the area except the ring on the phase plate.

The condenser with annular diaphragm can produce a hollow cone of light. As this cone passes through a cell, some light rays are bent due to variations in density and refractive index within the specimen and are retarded by about 1/4 wavelength. The deviated light is focused to form an image of the object. Undeviated light rays strike a phase ring in the phase plate while the deviated rays miss the ring and pass through the rest of the plate. Because the phase plate is constructed in such a way that the deviated light passing through it is advanced by 1/4 wavelength, the deviated and undeviated waves are now about 1/2 wavelength out of phase and will cancel each other when they come together to form an image. (Figure 2 – 1)

Apparatus and Materials

1. Specimens　24h broth culture of *Bacillus subtilis* and 48h broth culture of *Saccharomyces cerevisiae*.

2. Cultures and reagents　Dual – bottle (with immersion oil and xylene), sterile water.

3. Apparatus　Phase – contrast microscope.

4. Others　Glass slide, cover slip, lens paper, blotting paper.

Methods and Procedures

1. Make a wet – mount slide of *B. subtilis*.

2. Place the slide on the stage and secure with stage clips.

3. Rotate the 10 × objective into place as well as the annular diaphragm that corresponds to the 10 × objective. Adjust the coarse and fine adjustment knobs to focus the image. Replace the ocular lens with a phase centering telescope. Adjust the telescope until the dark ring and bright ring are clear (Figure 2 – 2A). Fix the telescope. Adjust the condenser to make the two rings coincide (Figure 2 – 2B). Put the ocular lens back. Use the fine adjustment knobs to clarify the image and observe.

4. Do the same with the 40 × objective and the oil immersion lens. Remember to adjust the corresponding diaphragm in position. Record the results. Clean the oil immersion lens after use.

5. Do the same with *S. cerevisiae* under 10 × and 40 × objectives. Oil immersion lens is not applied for fungi because of the lager size of cells.

Experiment contents

Observe *B. subtilis* and *S. cerevisiae* under the phase – contrast microscope.

Result

Draw the typical morphology of *B. subtilis* and *S. cenevisiae* observed through the phase – contrast microscope.

Discussion

Why can the phase – contrast microscope be used to observe the microstructure of living cells?

实验三　荧光显微镜的结构与使用

【目的】

1. 熟悉荧光显微镜的使用方法。

2. 了解荧光显微镜的构造和基本原理。

【基本原理】

有些化合物（荧光素）可以吸收紫外线并放出一部分光波较长的可见光，这种现象称为荧光。因此在紫外线的照射下，发荧光的物体会在黑暗的背景下表现为光亮的有色物体，这就是荧光显微技术的原理。荧光显微镜是利用紫外光（不可见光）的照射，使标本内的荧光物质转化为各种不同颜色的荧光（可见光）后，来观察和分辨标本内某些物质的性质与存在位置。

在明视野显微镜中，我们看到的是标本的直接本色，因此光源所起的作用仅是照明。而在荧光显微镜中，我们看到的不是标本的本色而是它的荧光，也就是说这里的光源起到的不是照明作用，而是一种激发作用。由此可知，这种光源必须提供使标本中的荧光物质能够被激发的能量。荧光显微镜的光源是超高压汞灯，使用时要预热

15min 才能达到最亮点。荧光显微镜有透射式和落射式两种。所谓落射式是指激发光自上而下穿过物镜，继而照射在标本上的激发方式。这种激发方式可以得到较清晰的荧光图像。与之相对应的是透射式荧光显微镜，由于图像效果不如落射式，现已很少使用。本实验将重点介绍落射式荧光显微镜的结构与原理。

与明视野显微镜相比，由于荧光显微镜的光源为紫外光，由实验一中的原理可知荧光显微镜的分辨率将优于普通光学显微镜。此外，荧光显微镜有两个特殊的滤光片，光源前激发滤片用以滤除可见光，目镜和物镜之间的阻断滤片用于滤除紫外线，用以保护人眼（图3-1）。荧光显微镜通常采用暗视野聚光器，这样会与物体产生的荧光形成鲜明对比。

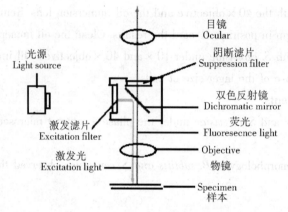

图 3-1 荧光显微镜原理示意图
Figure 3-1 Fluorescence microscope

目前荧光显微镜已广泛用于微生物检验及免疫学方面的研究，还可以用来在显微镜下区分死菌与活菌等，特别是用于抗酸细菌（如结核分枝杆菌）的观察。在荧光显微检验中，常用的荧光染料有金胺、中性红、品红、硫代黄素、樱草素等。其中用金胺可检查抗酸细菌。

【仪器与材料】

1. 菌种 结核分枝杆菌（*Mycobacterium tuberculosis*）。

2. 培养基与试剂 金胺（0.01g 溶于 95% 乙醇 10ml 中，后加 5% 石炭酸至 100ml）、3% 盐酸乙醇、0.5% 高锰酸钾水溶液、双层瓶（内装香柏油和二甲苯）。

3. 仪器 荧光显微镜。

4. 其他 擦镜纸、接种环、载玻片等。

【方法与步骤】

1. 制片 制备好观察用的玻片（方法见实验6）。

2. 初染 金胺溶液染色 10min 后水洗。

3. 脱色 以盐酸乙醇脱色至看不到黄色后水洗。

4. 复染 高锰酸钾溶液染色 2min，水洗，吸干。

5. 镜检　加香柏油，在荧光显微镜下观察。观察完需清洁油镜镜头。

【实验内容】

使用荧光显微镜观察结核分枝杆菌。

【结果】

菌体呈黄色荧光。

【注意事项】

1. 荧光镜检应在暗室进行。

2. 高压汞灯启动后需等15min才能达到稳定，亮度达到最大，方可使用。关闭后3min内不得再启动。若开启次数多，时间短，则会大大缩短汞灯寿命。

3. 因荧光物质受紫外光照射时随时间的延长而使荧光逐渐减弱，因此镜检时应经常变换视野。

4. 观察与摄影尽量在短时间内完成。

5. 紫外线对眼睛有伤害，不能直视激发光。

6. 光源附近因温度高不可放置易燃品。

Experiment 3　Fluorescence Microscope

Objectives

1. Be familiar with how to use the fluorescence microscope correctly.

2. Understand the principles and the structures behind the fluorescence microscope.

Principles

When some molecules absorb radiant energy (UV), they become excited and later release much of their trapped energy as light. Any light emitted by an excited molecule will have a longer wavelength than the radiation originally absorbed. Fluorescent light is emitted very quickly by the excited molecule as it gives up its trapped energy (UV) and returns to a more stable state. This is the basis of fluorescence microscopy.

In bright-field microscope, light source is used for illumination. However, in fluorescence microscope, light source is used to make the specimen excited. The fluorescence microscope exposes a specimen to UV light and forms an image of the object with the resulting fluorescent light. A mercury vapor arc lamp produces an intense beam and needs 15min preheating for maximum brightness. There are two types of fluorescence microscope: Transmission fluorescence and epifluorescence microscope. The latter one can produce a better image so will be the focus of this experiment.

According to the principles in experiment one, because of the use of UV light in fluorescence microscope, it has a better resolution than bright-field one. The light passes through an exciter filter that transmits only the desired wavelength. A dark-field condenser provides a black background against which the fluorescent objects glow. Usually the specimens have been stained with dye molecules, called fluorochromes that fluoresce brightly upon exposure to light of a specific wavelength but some microorganisms are autofluorescing. The microscope forms an image of the fluorochrome – labeled microorganisms from the light emitted when they fluoresce. A barrier filter positioned after the objective lenses removes any remaining UV light, which could damage the viewer's eyes (Figure 3 – 1).

The fluorescence microscope has become an essential tool in microbiology and immunology. It also can be used to differentiate between live and dead microbes. Bacterial pathogens (e.g., *Mycobacterium tuberculosis*) can be identified after staining them with fluorochromes. Auramine, neutral red, pinkish red, sulfo – flavin and primin are the commonly used fluorochromes.

Materials and Apparatus

1. Specimens *Mycobacterium tuberculosis*.

2. Cultures and reagents Auramine (Dissolved 0.01g auramine in 10ml of 95% ethanol, then add 5% phenol to make the volume to 100ml), 3% hydrochloric acid ethanol, 0.5% potassium permanganate solution, dual bottle (with immersion oil and xylene).

3. Apparatus Fluorescence microscope.

4. Others Lens paper, inoculation loop, glass slide.

Methods and Procedures

1. Prepare the glass slide for observation. (Ref to Experiment 6)

2. Use auramine solution for primary stain for 10min and then rinse.

3. Use hydrochloric acid ethanol for decolorization until colorless and then rinse.

4. Use potassium permanganate solution for counter stain for 2min. Rinse again and blot dry.

5. Observe under the fluorescence microscope with the oil immersion lens. Clean the lens after use.

Experiment content

Observe *Mycobacterium tuberculosis* with the fluorescence microscope.

Results

The microbes show yellow fluorescence.

Notes

1. Fluorescence microscope should be used in dark room.

2. 15min preheating for mercury vapor arc lamp is necessary for maximum brightness. The lamp should not be switched on again within 3 minutes after turning off. This will extend the working life of the lamp.

3. Move the slide frequently to obtain good image.

4. Observation should be finished as soon as possible.

5. Always keep eyes away from UV.

6. Keep in flammable things away from light source because of high temperature.

实验四 细菌运动性观察

【目的】

了解如何利用压滴法和悬滴法观察细菌的运动性。

【基本原理】

细菌是否具有鞭毛,以及鞭毛的着生方式是细菌分类鉴定的重要特征之一。使用压滴法(水封片法)或悬滴法直接在光学显微镜下检查活菌是否具有运动能力是判断菌体是否具有鞭毛的快速方法。

悬滴法就是将菌液滴加在洁净的盖玻片中央,在其周边涂上凡士林,然后将其倒盖在有凹槽的载玻片中央,即可放在明视野显微镜下观察。压滴法是将菌液滴在普通载玻片上,然后盖上盖玻片镜检。观察时宜选用幼龄菌体。

【仪器与材料】

1. 菌种 变形杆菌(*Bacillus proteus*)、枯草芽孢杆菌(*Bacillus subtilis*)8~12h液体培养物。

2. 培养基与试剂 凡士林。

3. 仪器 明视野显微镜。

4. 其他 载玻片、盖玻片、接种环、镊子、凹玻片、滴管、擦镜纸等。

【方法与步骤】

1. 压滴法

(1) 取菌 用滴管取菌液1滴,放于清洁的载玻片中央,注意液滴不可太大。

(2) 盖盖玻片 用镊子夹住盖玻片覆盖于菌液上。放置时先使盖玻片一边接触菌液,再将另一端松开放下,以玻片间不产生气泡为佳。

(3) 镜检 先用低倍镜找到标本,再用高倍镜观察(或用暗视野显微镜观察)。

(4) 标本处理 观察完毕用镊子取下盖玻片和载玻片分别投入消毒缸内。

2. 悬滴法

(1) 涂凡士林 取洁净的盖玻片一张,用火柴杆在四周涂凡士林少许。

(2) 取菌 用滴管加1滴菌液至盖玻片中央。

(3) 加盖玻片 将凹玻片的凹槽对准盖玻片中央的菌液,轻轻盖在盖玻片上,使两者粘在一起,然后迅速翻转盖玻片,使菌液正好悬在凹槽的中央,然后再轻轻按下,使玻片四周边缘闭合,以防菌液干燥或溢出(图4-1)。

(4) 镜检 先用低倍镜找到悬滴边缘,再将液滴移至视野中央,然后转换至高倍镜观察。由于样本未经染色,镜检时光线宜稍弱,可将光圈适当缩小,聚光器下降。镜检时要仔细辨别是细菌的运动还是分子的布朗运动,前者在视野下可见细菌由一处游动至另一处,而后者只是在原地左右摆动。细菌的运动速度和方式因菌种不同而异,应仔细观察。

(5) 标本处理 观察完毕,用镊子取下盖玻片,将载玻片和盖玻片分别投入消毒

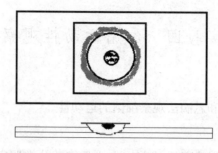

图 4-1 悬滴法标本（上：正面；下：侧面）

Figure 4-1 Hanging drop preparation

(Upwards: front view; Down: side view)

缸内。本法较压滴法复杂，但由于使用了凡士林，标本便于较长时间观察。

【实验内容】

使用明视野显微镜观察变形杆菌和枯草芽孢杆菌的运动性。

【结果】

分别绘出所观察到的细菌形态图，并标出运动方向。

【思考题】

1. 如何观察活菌，有何意义？
2. 检测细菌是否具有运动性的方法有哪些？

Experiment 4   Bacterial Motility Observation

Objectives

Understand how to observe bacterial motility with wet mount and hanging drop methods.

Principles

The existence of flagella and the arrangement of flagella are important in bacterial classification. Wet mount and hanging drop methods are commonly used to quickly identify the existence of flagella in bacteria.

Hanging drop means adding a drop of bacterial inoculum to a coverslip with a ring of Vaseline around the edge of the coverslip and then being covered by a depression slide. Turn over the slide and observe under the microscope. Wet mount means adding bacterial inoculum onto a glass slide and then covering the coverslip for microscopic observation. Young age cultures are preferred.

Apparatus and Materials

1. Specimens   8~12h broth cultures of *Bacillus proteus* and *Bacillus subtilis*.
2. Cultures and reagents   Vaseline.
3. Apparatus   Bright-field microscope.
4. Others   Glass slide, coverslip, inoculation loop, forceps, depression slide, drop-

per, lens papers.

Methods and Procedures

1. Wet – Mount Techniques

(1) Transfer a small drop of bacterial inoculum to the center of a clean glass slide.

(2) Use forceps to handle the coverslip carefully by one edge, and place it on the drop. Gently press on the coverslip to avoid bubbles.

(3) Place the slide on the microscope stage and observe it with low power. Then switch to high power and record the representative fields. (Dark – field microscope is optional)

(4) Discard the coverslips and any contaminated slides in a container for disinfection.

2. Hanging drop techniques

(1) With a matchstick, spread a small ring of Vaseline around a coverslip.

(2) Add a drop of bacterial inoculum to the center of the coverslip.

(3) Lower the depression slide, with the concavity facing down, onto the coverslip so that the drop protrudes into the center of the concavity of the slide. Press gently to form a seal. Quickly invert it so the drop is suspended. (Figure 4 – 1)

(4) Examine the drop under low power by locating the edge of the drop. Then move the drop to the center of the field. Switch to high power. As the specimen has not been stained, reduce the brightness with the iris diaphragm, and focus. Make sure the bacterial movement is due to Brownian movement or true motility. Brownian movement is caused by the molecules in the liquid striking an object and causing the object to shake or bounce. In Brownian movement, the microorganisms all vibrate at about the same rate and maintain their relative positions. Motile microorganisms move from one position to another. Their movement appears more directed than Brownian movement, and occasionally the cells may spin or roll according to different microbes. Record the representative observations.

(5) After observation, discard the coverslips and any contaminated slides in a container for disinfection. This method is more complicated than wet mount but suitable for long – time observation due to the use of Vaseline.

Experiment content

Observe the motility of *Bacillus proteus* and *Bacillus subtilis* with bright – field microscope.

Results

Draw a representative field for each bacterial specimen and point out the direction of motion.

Questions

1. How to observe living bacteria?
2. How to determine the motility of bacteria?

实验五　微生物测微技术

【目的】
1. 掌握用显微测微计测定微生物细胞大小的方法。
2. 了解目镜测微计和镜台测微计的构造及使用原理。

【基本原理】

在一定条件下，微生物细胞的大小是重要的形态特征之一，也是分类鉴定的依据之一。微生物细胞微小，只能在显微镜下测量。通常，用于测量微生物细胞大小的工具叫作显微测微计，由镜台测微计和目镜测微计两个部件组成。

目镜测微计（图5-1）是一块可放在目镜内的圆形玻片，其中央一般有100等分的小格，每小格代表的长度随目镜、物镜放大倍数的改变而改变，必须以镜台测微计来校准，从而计算出在一定的物镜和目镜的光学系统中，目镜测微计每格的实际代表长度，最后方可利用目镜测微计直接测量被测对象的长度和宽度。

镜台测微计（图5-2）是中央部分刻有精度等分线的专用载玻片。一般将1mm等分为100格，每格长度为10μm，专门用于校正目镜测微

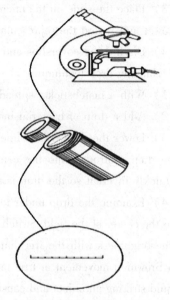

图5-1　目镜测微计
Figure 5-1　Ocular micrometer

计的。校正时，将镜台测微计放在载物台上，由于镜台测微计与细胞标本处于同一位置，都需要经过物镜和目镜的两次放大成像进入视野，因此从镜台测微计上得到的读数就是细胞的真实大小。即用镜台测微计的已知长度在一定放大倍数下校正目镜测微计，即可求出目镜测微计每格所代表的实际长度，然后移去镜台测微计，换上待测标本，用校正好的目镜测微计在同样放大倍数下测量微生物细胞大小。

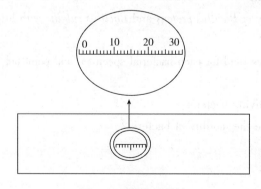

图5-2　镜台测微计
Figure 5-2　Stage micrometer

【仪器与材料】

1. 菌种　酿酒酵母（*Saccharomyces cerevisiae*）24h 液体培养物。
2. 仪器　明视野显微镜、镜台测微计、目镜测微计。
3. 其他　盖玻片、载玻片、滴管、擦镜纸等。

【方法与步骤】

1. 目镜测微计的校正

（1）放置目镜测微计　取出目镜，旋开接目透镜（即上透镜），将目镜测微计放在目镜的隔板上（有刻度的一面向下），然后旋上接目镜，最后将此目镜插入目镜镜筒内（图 5 - 1）。

（2）放置镜台测微计　把镜台测微计放在显微镜载物台上，刻度朝上。

（3）校正目镜测微计　用低倍镜观察，对准焦距，通过调焦能看清镜台测微计的刻度，移动镜台测微计和转动目镜测微计使两者刻度平行。调整载物台推进旋钮从而使两测微计某段起、止线完全重合，数出两条重合线之间目镜测微计和镜台测微计的格数。因为镜台测微计的刻度每格长 10μm，所以由下列公式可以算出目镜测微计每格所代表的实际长度（图 5 - 3）。

图 5 - 3　目镜测微计的校正

Figure 5 - 3　Calibration of ocular micrometer

$$目镜测微计每格长度（\mu m） = \frac{两重合线间镜台测微计的格数 \times 10}{两重合线间目镜测微计的格数}$$

例如：测得某光学显微镜的目镜测微计 15 格相当于镜台测微计 2 格，则目镜测微计每格代表的长度为：$2 \times 10/15 = 1.33$ μm/格。

用同法分别校正在高倍镜和油镜下目镜测微计每小格所代表的长度。由于不同显微镜及附件的放大倍数不同，因此校正目镜测微计必须针对特定的显微镜和附件（物镜、目镜）进行，而且只能在该显微镜上使用，当更换不同显微镜目镜或物镜时，必须重新校正目镜测微计。

2. 测定酿酒酵母细胞的大小

（1）取下镜台测微计，将用压滴法制成的酿酒酵母水浸片放在显微镜载物台上。

（2）先在低倍镜下找到菌体，然后转至高倍镜观察，调至物像清晰后，转动目镜测微计，测量酵母菌细胞的长度和宽度分别占有几个格数（不足一格的部分估计到小数点后一位数），再将测得的格数乘以目镜测微计每格的校正值即可求出酵母菌细胞的大小。

（3）在同一张玻片上测定10～20个酵母细胞并求出平均值，这样才能代表酵母细胞的平均大小。待测微生物需用培养至对数生长期的菌体进行测量。

（4）最后取出目镜测微计，将目镜测微计和镜台测微计分别用擦镜纸擦拭后放回干燥处保存。

【实验内容】

在明视野显微镜下测定酵母菌细胞大小。

【结果】

表5-1 目镜测微计的校正结果

物镜放大倍数	目镜放大倍数	镜台测微计格数	目镜测微计格数	目镜测微计校正值（μm/格）
10×				
40×				

表5-2 高倍镜下酵母菌细胞大小结果

酵母细胞	长		宽		菌体大小（长×宽）
	目镜测微计格数	菌体长度（μm）	目镜测微计格数	菌体宽度（μm）	
1					
2					
3					
4					
5					
6					
7					
8					
9					
10					
平均值					

【思考题】

目镜测微计每格代表的长度是否是固定的，为什么？

Experiment 5　Microscopic Measurement of Microbes

Objectives

1. Grasp how to measure the microorganisms under microscope.

2. Understand the structures and principles of ocular micrometer and stage micrometer.

Principles

Size determinations are often indispensable in the identification of a microbial unknown. The size of microbes is generally determined by the use of a microscope equipped with an ocular micrometer and a stage micrometer.

An ocular micrometer (Figure 5 – 1) is a small glass disk on which uniformly spaced lines of unknown distance, ranging from 1 to 100, are etched. The ocular micrometer is inserted into the ocular of the microscope and then calibrated against a stage micrometer (Figure 5 – 2), which has uniformly spaced lines of known distance etched on it. The stage micrometer is usually divided into 10μm graduations. The ocular micrometer is calibrated using the stage micrometer and then can be used to measure microbes.

Apparatus and Materials

1. Specimens   24h broth culture of *Saccharomyces cerevisiae*.
2. Apparatus   Bright – field microscope, ocular micrometer, stage micrometer.
3. Others   Coverslip, glass slide, dropper, lens paper.

Methods and Procedures

1. Calibrating an ocular micrometer

(1) Put ocular micrometer in the body tube of ocular eyepiece (Figure 5 – 1).

(2) Put stage micrometer in place with the graduations upwards.

(3) Use low power objective first. When in place, turn the ocular until the lines of the ocular micrometer are parallel with those of the stage micrometer. Match the lines of the two micrometers in certain segment by moving the stage micrometer.

Calculate the actual distance between the lines of the ocular micrometer by observing how many spaces of the stage micrometer are included within a given number of spaces on the ocular micrometer (according to the doublication lines).

Because the smallest space on the stage micrometer equals 10μm, you can calibrate the ocular micrometer using the following formula (Figure 5 – 3).

1 space on the ocular micrometer (μm) = spaces on the stage micrometer × 10/spaces on the ocular micrometer

For example, if 15 spaces on the ocular micrometer = 2 spaces on the stage micrometer, then 1 ocular space = 2 × 10/15 = 1.33μm.

Calibrate for each of the objectives on the microscope. Note that the numerical value holds only for the specific objective – ocular lens combination used and may vary with different microscopes.

2. Microscopic measurements of yeast cells

(1) Replace stage micrometer with the prepared wet mount slides of *Saccharomyces cerevisiae*.

(2) Focus the cells under low power then high power. Determine the spaces of the length and width of the cells (to 0.1 spaces) and then calculate the dimensions of yeast

cells.

(3) Measure 10~20 yeast cells on the same slide and get the average. The culture should be in log phase for representative morphology.

(4) Take off ocular micrometer and clean both micrometers thoroughly.

Experiment content

Measure the size of yeast cells under bright-field microscope.

Results

Table 5-1 Calibration of ocular micrometer

Objective lens	Ocular eyepiece	Spaces on stage micrometer	Spaces on ocular micrometer	Calibrated value of ocular micrometer (μm/space)
10×				
40×				

Table 5-2 Dimensions of yeast cells under high power

Yeast cell	Length		Width		Dimensions
	Space number	Length (μm)	Space number	Width (μm)	(length × width)
1					
2					
3					
4					
5					
6					
7					
8					
9					
10					
Average					

Questions

Is the actual distance of an ocular micrometer constant? Why?

(马晓楠)

第二节　微生物染色、形态结构与菌落特征观察

微生物体积小且透明，直接观察形态很不清楚，更谈不上识别细菌结构。而经过染色后，借助普通光学显微镜，除可清楚地观察其形态外，还可观察菌体表面及内部

着色结构。因此,微生物染色技术是观察微生物形态结构的重要手段。

Section 2　Microorganisms Stain, Morphology and Colony Characteristics

It is difficult to observe the structures of microbes directly, because they are tiny and transparent. After staining, the shape, surface, and even some structures could be observed by a light microscope. Therefore, staining is an important technique for experimental microbiology.

实验六　细菌的单染色法与细菌菌体形态观察

【目的】

1. 掌握细菌的单染色制片方法。
2. 掌握显微镜(油镜)的使用方法。

【基本原理】

细菌体积微小且透明,与周围的水或其他介质环境的光学性质相近,在普通显微镜下难以直接观察,经过染色后,可与背景形成明显反差,有助于细菌标本的观察。

单染色法即用单纯一种染料进行染色。常见染料可分为带有阴离子发色基团的酸性染料和带有阳离子发色基团的碱性染料。考虑到在一般的生理条件下(pH7.4左右)细菌菌体都带有负电荷,所以常选用易于菌体结合的阳离子碱性染料,如美兰、结晶紫和稀释复红等。此染色方法只能显示细菌的形态及排列方式,尚不能辨别其结构。

【仪器与材料】

1. 菌种　大肠埃希菌(*Escherichia coli*)、金黄色葡萄球菌(*Staphyloccocus aureus* 209P)琼脂斜面 18~24h 培养物(培养基见附录一,2)。

2. 试剂　蒸馏水、碱性美兰染液(附录三,1)、稀释复红染液(附录三,2)、香柏油、二甲苯。

3. 仪器　显微镜。

4. 其他　接种环、玻璃棒、载玻片、纱布、擦镜纸、酒精灯、滤纸本、双层瓶(内装香柏油和二甲苯)。

【方法与步骤】

1. 涂片　在洁净的载玻片中央滴加一滴蒸馏水,再以无菌操作,用接种环取少量细菌与水混匀并涂成直径 1cm 左右的薄层。(如样本为液体培养物,可取一环直接涂于载玻片上即可)。然后将接种环置于火焰上灼烧,杀死残留的细菌。

2. 干燥　将载玻片置于空气中自然干燥,若想加速干燥,可将玻片远离火焰烘干,但切忌直火加热,避免细菌变形。

3. 固定　将涂菌面向上,在火焰上来回烘烤 2~3 次,待冷却后再加染料。固定的目的是杀死细菌,使菌体与载玻片粘附牢固,在染色时不易被染液和水冲掉,使细菌易被着色。

4. 染色 滴加稀释复红或碱性美兰染液1~2滴,使染液盖满涂菌面,染色时间约1~2min。

5. 水洗 倾去染色液,用细水流将玻片洗净,至无残余染料为止。

6. 吸干 自然干燥或用滤纸吸干。

7. 镜检 用油浸镜观察。

以上步骤参见图6-1。

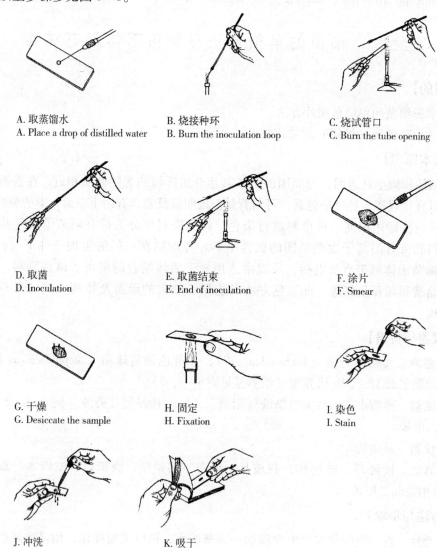

A. 取蒸馏水
A. Place a drop of distilled water

B. 烧接种环
B. Burn the inoculation loop

C. 烧试管口
C. Burn the tube opening

D. 取菌
D. Inoculation

E. 取菌结束
E. End of inoculation

F. 涂片
F. Smear

G. 干燥
G. Desiccate the sample

H. 固定
H. Fixation

I. 染色
I. Stain

J. 冲洗
J. Rinse

K. 吸干
K. Sip the residue solution

图6-1 细菌单染色法步骤
Figure 6-1 Procedures for bacteria simple stain

【结果】

绘出上述两种细菌经单染色后的形态图。

【思考题】

1. 为什么细菌染色时所用的染料多属于碱性染料？
2. 根据实验体会，你认为制备染色标本时应注意哪些事项？
3. 单染色法在微生物学上有何实际意义？

Experiment 6　　Simple Staining

Objective

1. Be familiar with the simple stain of bacteria and aseptic manipulation.
2. Be familiar with the use of oil immersion.

Principle

Microorganisms are very small, which could hardly be observed under a microscope. However, it could be easily visualized after the simple stain, which could distinguish the microbes out from the background.

Only one dye is used in simple staining, such as methylene blue and crystal violet. Common dyes could be divided into acid and alkaline dyes with cationic and anionic chromophores. Under normal physiological conditions (pH 7.4), bacteria cells carry negative charges. Cationic alkaline dyes such as methylene blue, crystal violet and safranine are usually used. Bacteria are dyed to be the same color to indicate their morphology and arrangement, but failed to explore their structures.

Apparatus and Materials

1. Specimens　18～24h nutrient agar slant culture of *Escherichia coli* and *Staphylococcus aureus* 209P (Appendix I, 2).

2. Reagents　Distilled water, Methylene blue (Appendix Ⅲ, 1), Safranine (Appendix Ⅲ, 2), cedar wood oil, xylene.

3. Apparatus　Microscope.

4. Others　Inoculation loop, glass spreader, slide, gauze swab, lens paper, blotting paperr, dual-bottle (with immersion oil and xylene), lamp or burner.

Methods and Procedures

1. Smear　Place a drop of distilled water on a clean glass slide. Transfer some bacteria into the water drop with the sterile inoculation loop; smear it to form a bacteria membrane about 1cm diameter. Flame the inoculation loop.

2. Dry　Let the smear to air dry at room temperature. If fast dry was needed, keep the slide warmed above the flame with the temperature tolerated by your own hands.

3. Fixation　Holding the glass slide at one end, pass the smear through the flame of

the burner for 3~4 times. Fixation of the bacteria on the surface of the glass slide is necessary for sample dying, washing, and keeping the arrangement of bacteria while cell been stained.

4. Staining Cover the smear with methylene blue for 1~2 min.

5. Wash Gently wash the smear with tap water to remove excess stain.

6. Water residue absorption Air dry or use the blotting paper to make the smear totally dry.

7. Observation Examine the stained slides under the oil immersion. The above procedures were summarized in Figure 6-1.

Results

Draw a representative field for each kind of bacteria after simple staining.

Questions

1. Why the dyes used in the staining process belong to the alkaline dyes?

2. What steps should be noticed in the manipulation of staining samples?

3. What kinds of applications could the simple staining process used in the microbiology?

实验七 细菌的革兰染色

【目的】

1. 熟悉革兰染色的原理并掌握革兰染色的方法。
2. 熟悉显微镜（油镜）的使用及保养方法。

【基本原理】

革兰染色法是细菌学中使用最广泛的一种鉴别染色法。1884年由丹麦医师（Christan Gram）创立。该方法首先用结晶紫对细菌染色，添加媒染剂（增加染料和细菌的亲和力）后，用脱色剂（乙醇或丙酮）脱色，再用复染剂染色。如果细菌不被脱色而保存原染液颜色者为革兰阳性菌（G^+）；如被脱色，而染上复染液的颜色者为革兰阴性菌（G^-）。此染色法可将所有具有细胞壁的细菌分为两大类：革兰阳性菌和革兰阴性菌。

一般认为细菌对革兰染色的不同反应主要是由革兰阳性菌和阴性菌的细胞壁结构与化学组成不同而导致的（表7-1）。革兰阳性菌的细胞壁较厚，肽聚糖含量多，且交联度大，脂类含量少，经95%乙醇脱色时，肽聚糖层的孔径变小，通透性降低，与细胞结合的结晶紫与碘的复合物不易被脱掉，因此细胞仍保留初染时的颜色。而革兰阴性菌的细胞壁较薄，肽聚糖的含量较少，而其外膜层含有较多的脂类成分，在乙醇脱色时易被溶解，增加细胞壁的通透性，使初染的结晶紫和碘的复合物易于渗出，结果细菌被脱色，经复染后，又染上复染液的颜色。

表 7-1　革兰阳性菌与革兰阴性菌细胞壁结构比较

细胞壁结构	革兰阳性菌	革兰阴性菌
厚度	厚，15~80nm	薄，10~15nm
肽聚糖含量	多，占细胞壁干重50%~80%	少，占细胞壁干重10%左右
脂类含量	少，约1%~4%	多，约11%~22%
磷壁酸	有	无
外膜	无	有
脂蛋白	无	有
脂多糖	无	有

【仪器与材料】

1. 菌种　金黄色葡萄球菌（*Staphylococcus aureus* 209P）、大肠埃希菌（*Escherichia coli*）琼脂斜面 18~24h 培养物。

2. 试剂　蒸馏水、结晶紫染液（附录三，3）、稀释复红染液（附录三，2）或 0.5% 番红花红液（附录三，5）、卢戈（Lugol）碘液（附录三，4）、95% 乙醇、香柏油、二甲苯。

3. 仪器　显微镜。

4. 其他　接种环、玻璃棒、载玻片、纱布、擦镜纸、滤纸本、酒精灯、双层瓶（内装香柏油和二甲苯）。

【方法与步骤】

1. 涂片、干燥、固定　同实验六中单染色法的对应步骤。

2. 初染　在已固定的标本上滴加结晶紫染液，覆盖涂片区域，染色 1~2min，细水冲洗。

3. 媒染　加碘液 1min 左右（在此其间更换碘液 2~3 次），使碘液与涂片区域充分接触。细水冲洗。

4. 脱色　加 95% 乙醇数滴，轻轻晃动玻片，然后斜持玻片，使乙醇流去；如此反复至流下的乙醇无色或稍呈淡紫色为止，细水冲洗。

5. 复染　滴加稀释复红液或番红花红染液染色 1min 左右，水洗，滤纸本吸干残留水分。

6. 镜检　用油浸镜观察。

以上步骤可参见图 7-1。

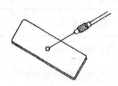
A. 取蒸馏水
A. Place a drop of distilled water

B. 烧接种环
B. Burn the inoculation loop

C. 烧试管口
C. Burn the tube opening

D. 取菌
D. Inoculation

E. 取菌结束
E. End of inoculation

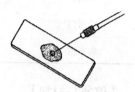
F. 涂片
F. Smear

G. 干燥
G. Desiccate the sample

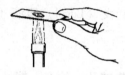
H. 固定
H. Fixation

I. 初染
I. Primary stain

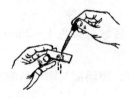
J. 冲洗
J. Rinse

K. 媒染
K. Mordent stain

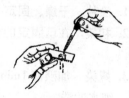
L. 冲洗
L. Rinse

M. 脱色
M. Decolorization

N. 冲洗
N. Rinse

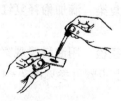
O. 复染
O. Counter stain

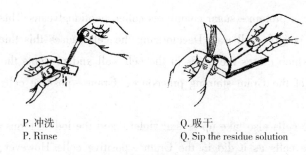

P. 冲洗　　　　　　　　　　　Q. 吸干
P. Rinse　　　　　　　　　　　Q. Sip the residue solution

图 7 – 1　细菌革兰染色法步骤
Figure 7 – 1　Procedures for bacteria gram stain

【注意事项】

1. 乙醇脱色是革兰染色的重要环节，一定要控制好。因细菌细胞壁的渗透性是相对的，所以如脱色过度，阳性菌可被误染为阴性菌。如脱色不够，则阴性菌可能被误染为阳性菌。脱色时间的长短应由涂片的厚薄来决定。一般涂片时，取菌要少，涂片薄而均匀为好。

2. 被检菌的培养条件、培养基成分、菌龄大小等因素均会影响染色结果。如革兰阳性菌的陈旧培养物也有出现阴性的情况，所以被检查的材料、菌龄最好处在对数生长期内。

【结果】

注意样本中两种细菌的形态、大小、排列方式，以及染色结果。若细菌被染成紫色则为革兰阳性菌（G^+），若细菌被染成（粉）红色则为革兰阴性菌（G^-）。绘图说明革兰染色结果，分别绘出两种细菌的形态、革兰染色性质、颜色，并标注出细菌菌名。

【思考题】

1. 简述革兰染色法的全过程及其实验结果。
2. 革兰染色过程中，哪些操作步骤是成败关键，为什么？

Experiment 7　Gram Staining

Objectives

1. Be familiar with Gram's stain principles and procedures.
2. Be familiar with the application and maintenance of oil immersion.

Principles

The Gram stain is the most widely used staining procedure in microbiology. It was first established in 1884 by a Danish scientist named Chrstian Gram, which was used to differentiate between Gram – positive organisms and Gram – negative organisms.

Gram – positive cells take up the crystal violet, which is then fixed in the cell with the iodine mordant. This forms a crystal – violet iodine complex which remains in the cell even after decolorizing. It is thought that this happens because the cell walls of Gram – positive organisms include a thick layer of protein – sugar complexes called peptidoglycans. This layer makes up to 90% of the Gram – positive cell wall. Decolorizing the cell causes this thick cell wall to dehydrate and shrink which closes the pores in the cell wall and prevents the stain from exiting the cell. At the end of the Gram staining procedure, Gram – positive cells will be stained a purplish – blue color.

Gram – negative cells also take up crystal violet, and the iodine forms a crystal violet – iodine complex in the cells as it did in the Gram – positive cells. However, the cell walls of Gram – negative organisms do not retain this complex when decolorized. Peptidoglycans are present in the cell walls of Gram – negative organisms, but they only comprise 10% ~ 20% of the cell wall. Gram – negative cells also have an outer layer, which Gram – positive organisms do not have; this layer is made up of lipids, polysaccharides, and proteins.

Exposing Gram – negative cells to the decolorizer (the alcohol); dissolves the lipids in the cell walls, which allows the crystal violet – iodine complex to leach out of the cells. This allows the cells to subsequently be stained with safranin. At the end of the Gram staining procedure, Gram – negative cells will be stained a reddish – pink color.

Apparatus and Materials

1. Specimens   18 ~ 24h nutrient agar slant culture of *Escherichia coli* and *Staphylococcus aureus* 209P.

2. Reagents   Distilled water, Crystal violet (Appendix III, 3), Safranine (Appendix III, 2), Lugol's iodine (Appendix III, 4), 95% ethanol, Cedar wood oil, xylene.

3. Apparatus   Microscope.

4. Others   Inoculation loop, glass spreader, slide, blotting paper, burner, dual – bottle (with immersion oil and xylene).

Methods and Procedures

1. Smear, Dry and Fixation   Steps are the same as in Experiment 6.

2. Primary stain   Cover the smear with crystal violet for 1 to 2 min, and gently wash off the crystal violet with water by squirting the water so it runs through the smear.

3. Mordant   Add Lugol's iodine to the smear for 1 min (20s x 3), and gently wash the smear with tap water from one side of the slide to remove excess iodine.

4. Decolorization   Add 95% ethanol to the smear, decolorize the smear for 30s, and gently wash the smear with tap water from one side of the slide to remove ethanol.

5. Counterstain   Cover the smear with safranine for 1 min, and gently wash the smear with tap water from one side of the slide to remove excess dyes. Air dry or absorb water with paper without wiping the slide.

6. Examination   Examine all stained slides under oil immersion. (Figure 7 – 1)

Notes

1. Decolorization is an essential step in gram stain. Excessive decolorization could render gram positive bacteria to gram negative result, and vice versa. The period of decolorization

depends on the square of the sample smear.

2. Culture conditions, media ingredients, age of bacteria etc. would have contributions to the gram stain results. The bacteria samples should be selected from their log phase in growth.

Results

Draw a representative field for Gram positive and Gram negative bacteria, indicating their colors, Gram properties and bacteria names.

Question

1. Please summarize the procedures for Gram stain and the results.
2. Which is the essential step in Gram stain? Why?

实验八　细菌的芽孢染色法

【目的】

1. 掌握细菌芽孢染色的方法。
2. 熟悉细菌芽孢染色的原理。

【基本原理】

芽孢具有厚而致密的多层壁膜结构，折光性强，通透性低，不易着色，用一般染色法只能使菌体着色而芽孢不着色。芽孢染色法就是根据芽孢具有既难以染色而一旦染上色后又不易脱色的特点而设计的。所有的芽孢染色法都基于以下原则，即除了用着色力强的染料外，还需要加热或延长染色时间等，促进标本着色。然后使菌体脱色而芽孢上的染料仍保留。经复染后，菌体和芽孢呈不同的颜色。芽孢染色的方法有很多，这里仅介绍孔雀绿染色法。

【仪器与材料】

1. 菌种　枯草芽孢杆菌（*Bacillus subtilis*）肉汤琼脂斜面24h培养物。

2. 试剂　蒸馏水、7.6%孔雀绿水溶液（附录三，6）、稀释复红染液（附录三，2）或0.5%番红花红液（附录三，5）、香柏油、二甲苯。

3. 仪器　显微镜。

4. 其他　接种环、玻璃棒、载玻片、纱布、擦镜纸、滤纸本、酒精灯。

【方法与步骤】

1. 涂片、干燥、固定　同实验六中单染色法的对应步骤。

2. 初染　在已固定的标本上滴加孔雀绿染液，覆盖涂片区域，染色15~20 min，细水冲洗。

3. 复染　加稀释复红染液或番红花红染液染色5 min左右。细水冲洗，滤纸本吸干残留水分。

4. 镜检　用油浸镜观察。

以上步骤可参见图8-1。

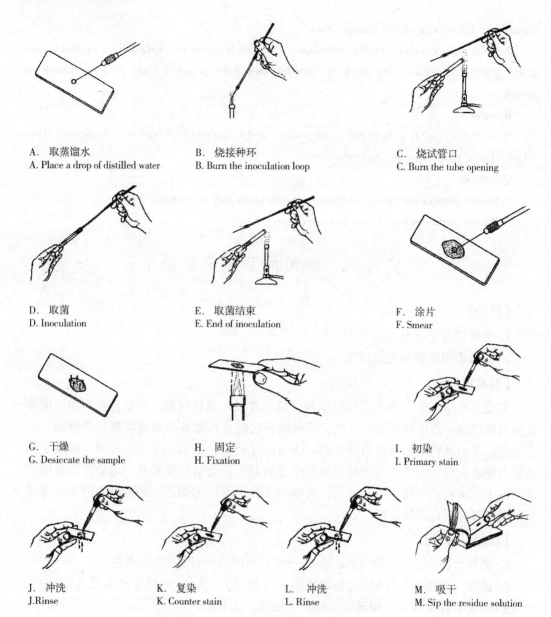

图 8-1 细菌芽孢染色步骤

Figure 8-1 Procedures for bacteria spore stain

【注意事项】

染色过程中染料不能完全干燥。

【结果】

绘图说明芽孢染色结果，绘出枯草芽孢杆菌菌体形状、芽孢着生位置及形状，判断是否有游离的芽孢存在。

【思考题】

为什么芽孢与营养细胞能染成不同的颜色？

Experiment 8   Spore Staining

Objectives
1. Be familiar with the principle of the spore stain.
2. Be familiar with the procedure of the spore stain.

Principles
Spores have thick and durable multilayer structures, with outer coats of dense, strong refraction, low permeability and difficult to be dyed by general staining methods. Long time staining of malachite green renders both the vegetative cell and spores into green, with heating or extending the staining time. Then the bacteria were rinsed and counterstained by safranine to make the vegetative cell into red, which then may differentiate the spores and vegetative cell clearly.

Apparatus and Materials
1. Specimens   24h nutrient agar slant culture of *Bacillus subtilis*.
2. Reagents   Distilled water, 7.6% Malachite green (Appendix Ⅲ, 6), Safranine (Appendix Ⅲ, 2), cedar wood oil, xylene.
3. Apparatus   Microscope.
4. Others   Inoculation loop, glass spreader, slide, paper, lamp, dual-bottle (with immersion oil and xylene).

Methods and Procedures
1. Smear, Dry and Fixation   These steps were the same as in Experiment 6.
2. Primary stain   Stain the smear with malachite green for 15~20 min, and gently wash the smear with tap water from one side of the slide to remove excess dyes.
3. Counterstain   Stain the smear with Safranine for 5 min, and gently wash the smear with tap water from one side of the slide to remove excess dyes. Air dry or absorb water with paper without wiping the slide.
4. Examination   Examine all stained slide under oil immersion Figure 8-1.

Notes
Themalachite green solution could not be totally dried during the staining process.

Results
Draw a representative field for spore stain, indicating the position, shape and color of the spores and vegetative cells.

Question
Why the spore and vegetative cell could be stained in different colors?

实验九　细菌荚膜、鞭毛的染色观察

【目的】

1. 掌握细菌荚膜染色和鞭毛染色的方法。
2. 熟悉细菌荚膜染色和鞭毛染色的原理。

【基本原理】

1. 荚膜染色原理　荚膜是某些细菌在生活过程中，向细胞壁外分泌的一层疏松、透明的黏液状物质。它是由多糖类衍生物或多肽所聚集而成。荚膜的折光性低，与染料的亲和力弱，往往是通过将背景染色，从而使不能着色的荚膜衬托出来，在显微镜下呈现透明的荚膜层，又称为负染色法。常用的有黑斯（Hiss）法。

2. 鞭毛染色原理　细菌鞭毛极细，直径一般为 10~20 nm，需要用电子显微镜进行观察。如果采用特殊染色法，则在普通光学显微镜下也能看到。鞭毛染色方法很多，但基本原理相同，即都在染色前先进行预处理，让媒染剂沉积在鞭毛上，使鞭毛加粗，然后进行染色。常用的媒染剂由丹宁酸和钾明矾等配制而成。本实验介绍众多鞭毛染色方法中的一种——户田法。

【仪器与材料】

1. 菌种　肺炎双球菌（*Diplococcus pneumoniae*）、变形杆菌（*Proteus vulgaris*）肉汤琼脂斜面 15~18h 培养物。

2. 试剂　蒸馏水、荚膜染色液（附录三，7）、5% 结晶紫染色液（附录三，3）、20% 硫酸铜溶液、鞭毛染色液（附录三，8，9，10）、香柏油、二甲苯。

3. 仪器　显微镜。

4. 其他　接种环、玻璃棒、载玻片、纱布、擦镜纸、滤纸本、酒精灯。

【方法与步骤】

（一）荚膜染色法

1. 取肺炎双球菌培养液注射于小白鼠腹腔中（约 0.5ml）。
2. 小白鼠死亡后，立即取腹腔液。
3. 将腹腔液涂布于载玻片上，自然干燥、固定。
4. 滴 5% 龙胆紫染色液于涂片上，在弱火上略加热，使冒蒸气为止。
5. 用 20% 硫酸铜液将涂片上的染液洗去，此时切勿再用水洗。以滤纸吸干后镜检。（菌体呈紫色，菌体周围有一圈淡紫色或无色的荚膜）

（二）鞭毛染色法

1. 菌液的制备及涂片　将变形杆菌预先连续移接 5~7 代。染色前将其转种于新鲜的琼脂斜面上，于 37℃ 培养 15~18h 后，用接种环挑取斜面底部的菌苔数环，轻轻移入盛有 1ml 37℃ 无菌水中，不要振动，让有活动能力的菌游入水中，至轻度浑浊。在

37℃下保温10min，然后从上层液面挑数环菌液，置于洁净玻片的一端，稍稍倾斜玻片，使菌液缓缓地流向另一端。置空气中自然干燥。

2. 染色　滴加染色液2～3min，用水轻轻冲洗，然后自然干燥、镜检（菌体和鞭毛呈红色或紫色）。

【注意事项】

用于鞭毛染色的玻片必须十分清洁。

【结果】

绘图说明荚膜与鞭毛染色结果，绘出肺炎双球菌菌体及荚膜位置，绘出变形杆菌鞭毛着生部位及数量。

【思考题】

1. 为什么用于鞭毛染色的玻片必须十分清洁？
2. 鞭毛染色时为什么要连续多次地进行传代？

Experiment 9　Capsule and Flagella Stain

Objectives

1. Be familiar with the stain principles of the capsule and flagella.
2. Be familiar with the stain methods of the capsule and flagella.

Principles

Capsule is an outer layer of loose, transparent mucus – like gelatinous substance surrounded the bacteria. The composition of capsule are composed of polysaccharide and polypeptides. Capsule is difficult to be stained and negative stain is usually required. With a low refractive property, capsule is nonionic, which renders weak affinity to bind with the dyes. The common strategy is to stain the background to provide enough contrast between the background and capsule to make them visualized under the microscope.

Flagella are fine, thread like organelles of locomotion, with a diameter of 10～20 nm, which could be observed under the electronic microscope. However, there are some practical methods could be used to stain the flagella, which may made them observed under light microscope. The principles for flagella stain are to make some pretreatments which render the mordents deposited on and thicken the flagella before stain process. The common flagella staining solutions were prepared with tannic acid and potassium alum.

Apparatus and Materials

1. Strains Specimens　15～18h nutrient agar slant culture of *Diplococcus pneumoniae* and *Proteus vulgaris*.

2. Reagents　Distilled water, capsule stain (AppendixⅢ, 7), crystal violet (AppendixⅢ, 3), 20% copper sulfate, flagella stain (AppendixⅢ, 8, 9, 10), ceda wood oil, xylene.

3. Apparatus Microscope, lamp.

4. Others Inoculation loop, glass spreader, slide, water-absorption paper, immersion oil, xylene.

Methods and Procedures

1. Capsule staining

(1) Inject 0.5ml of *Diplococcus pneumoniae* cultures into the mice abdomen.

(2) Take out the peritoneal fluid after the mice decease.

(3) Smear the peritoneal fluid on the slide, dry and fixation.

(4) Stain the smear with crystal violet, gently heat the slide.

(5) Rinse the smear with 20% copper sulfate; use the water-absorption paper to make the smear totally dry.

2. Flagella staining

(1) Culture preparation Inoculate the *Proteus vulgaris* for successive 5~7 generations. Transfer the cultures onto a fresh agar slant, culture at 37℃ for 15~18h. Take some lawn at the bottom of slant into 1ml 37℃ sterilized water. Keep the cube steady to free the motivate bacteria into the water. Incubate at 37℃ for 10 min, take some cultures onto one end of the slide, tilt the slide gently to make the cultures flow to the other end of the slide. Dry the smear in the air.

(2) Stain Stain the smear for 2~3 min, rinse, dry and examine.

Notes

The slides used for flagella stain should be very clean and clear.

Results

Draw a representative field for each capsule and flagella stain, indicating the position, shape and color of the capsules and flagella.

Question

1. Why should the slides used in the flagella staining be very clean and clear?

2. Why it should be inoculated for several generations for the flagella stain?

实验十 放线菌形态观察

【目的】

1. 掌握观察放线菌孢子丝形态特征的培养方法。
2. 熟悉放线菌的各种形态特性。

【基本原理】

多数放线菌的菌落在培养基上着生牢固，不易被接种针挑取，由于孢子的存在，常使菌落表面呈粉末状。放线菌由纤细的丝状菌丝细胞组成菌丝体，菌丝内无隔膜。

菌丝体分为两部分,即潜入培养基中的营养菌丝和由营养菌丝向上生长的气生菌丝。气生菌丝上部分化成孢子丝,可呈直形、弧形或螺旋状等,其着生形式有丛生、互生和轮生等。放线菌菌丝有各种颜色,有的还能分泌脂溶性色素或水溶性色素,常常使菌丝或培养基呈现一定颜色。孢子圆或椭圆,表面光滑或粗糙,有各种颜色,这些形态特点都是鉴定放线菌的重要依据。

【仪器与材料】

1. 菌种 链霉菌（*Streptomyces* 1787-3，*Streptomyces* 12-21，*Streptomyces* 4.794）、诺卡氏菌（*Nocardia* 71-N）和小单孢菌（*Micromonospora* m-220）。

2. 培养基与试剂 蒸馏水、美蓝染液、高氏一号培养基（附录一,5）、马铃薯琼脂培养基（附录一,6）、葡萄糖天门冬素培养基（附录一,9）。

3. 仪器 显微镜。

4. 其他 接种环、L型玻棒、载玻片、纱布、擦镜纸、滤纸本、酒精灯、盖玻片。

【方法与步骤】

（一）插片法

1. 取一接种环小单孢菌m-220,涂抹在预先制好的葡萄糖天门冬素琼脂平板培养基上,以无菌L形玻棒将菌均匀涂布在整个平板表面。将无菌盖玻片以45°角插入平板内的培养基中（注意：盖玻片勿插入到培养基底部）,盖好皿盖（图10-1）,倒置于37℃恒温箱中培养7~8d,待长出孢子（培养基表面转黑褐色）,取出培养皿,用小镊子夹取一片盖玻片,用乙醇棉球擦去一侧培养物,将有菌的一面向下放到载玻片上,用低倍镜和高倍镜观察。注意菌丝直径、孢子着生位置、孢子的形状和大小等。

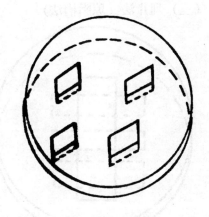

图10-1 插片法

Figure 10-1 Coverslip Insertion method

2. 将链霉菌1787-3和12-21分别划线接种在两个高氏一号平板上,采用插片法（盖玻片插入方向要与接种划线方向相垂直）插片,倒置于28℃恒温箱中培养5~7d后,取出培养皿,用小镊子夹取一片盖玻片,放在载玻片上,先用低倍镜观察,然后用高倍镜观察菌丝、孢子丝和孢子的自然形态（图10-2）。用于插片法的平板培养基不宜太薄,每皿应在20ml左右。

（二）平板划线培养基直接观察法

将链霉菌4.794以连续划线法接种于高氏一号平板上,各条线之间留一定空隙。接种完毕,倒置于28℃恒温箱中培养7~10d,待菌长好后,打开皿盖,将培养物直接放在低倍或高倍镜下观察,在菌落边缘或划线的空隙处,可观察到菌丝和孢子丝（该菌为二级轮生）的形态特点。

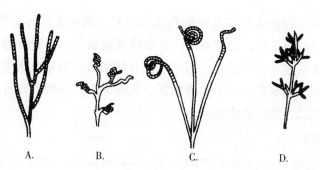

图 10－2　常见放线菌的孢子丝形态
Figure 10－2　Morphology of actinomycetes fibrillae of spores
A. 直单股分枝　B. 紧螺旋　C. 顶端大螺旋　D. 二级轮生
A. Straight　B. Flexuous　C. Open spirals　D. Biverticiliate

如平板培养物的菌丝体和培养基太厚，可用小铲刀切一小方块培养物放置载玻片上，再用刀片切除下层厚的培养基部分，把带菌丝体的上部薄层留在载玻片上，菌面朝上在显微镜下进行观察。

（三）埋片法（琼脂槽法）

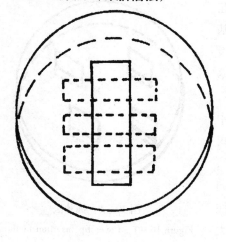

图 10－3　埋片法
Figure 10－3　Agar notch method

取一接种环诺卡氏菌 71－N，于马铃薯平板培养基表面，用无菌 L 形玻棒将菌均匀地涂满整个平板。再用无菌铲在平板上挖 2～3 条平行等距离的沟槽（每条 1cm×5cm 左右），造成通气条件，将无菌载玻片放在与沟槽相垂直方向的培养基上面，轻轻按一下玻片，使玻片与平板贴紧（图 10－3），盖好皿盖倒置于 28℃ 恒温箱中培养。可如此做成若干这样的培养物，于 6h、10h、15h 等不同培养时间分别取出载玻片，使菌面朝上，在低倍镜和高倍镜下观察已长在玻片上的菌丝在不同时间的生长状态。同样，用埋片法也可在高氏一号培养基上观察链霉菌的菌丝和孢子丝形态。

（四）印片法

取洁净的盖玻片一块，在培养好的菌落上面轻轻地压一下，然后将有压痕的一面朝下，放在有一滴美兰染液的载玻片上，并将印迹部分浸在染液内，用油浸镜观察孢子丝的形状。

【结果】

绘图说明各样本中放线菌的形态特征。

【思考题】

为什么放线菌属于原核微生物?

Experiment 10　Actinomyces Morphology

Objectives

1. Be familiar with the morphologic characteristics of the Actinomyces.

2. Be familiar with culture methods of the mycelia of Actinomyces.

Principles

Colonies of Actinomyces are steadily affiliated to the culture media, which could hardly be picked up by the inoculation loop. Spores may render the colony surface as powdered looking. Actinomyces are prokaryotic microorganisms, which were consisted of substrate filament, aerial filament and sporebearing filament. The aerial filaments differentiate to sporebearing filaments. The shapes of sporebearing filament are sphere, ellipse and rod, etc. The substrate and aerial filaments may contain pigments, which may render colony and media different colors. These morphological characteristic could be used to identify the Actinomycetes.

Apparatus and Materials

1. Specimens　*Streptomyces* 1787 - 3, *Streptomyces* 12 - 21, *Streptomyces* 4.794, *Nocardia* 71 - N, *Micromonospora* m 220.

2. Reagents　Distilled water, Methylene blue, Gause's Synthetic Agar Medium (Appendix I, 5), Potato medium (Appendix I, 6), Glucose asparagines Agar Medium (Appendix I, 9).

3. Apparatus　Microscope.

4. Others　Inoculation loop, L - shaped spreader, slide, water - absorption paper, lamp, coverslip.

Methods and Procedures

Ⅰ. Coverslip Insertion method

1. Take one loop of *Micromonospora* m 220 culture, spread the culture on the prepared glucose asparagines agar media plate with the L - shaped spreader. Insert an aspectic coverslip into the agar with 45 degree (DO NOT insert the cover slip to the bottom of petri dish.) (Figure 10 - 1), incubate the plate upside down for 7 ~ 8 days at 37℃. When the spores were generated (the surface of Petri dish changed into dark brwon), take out the coverslip, wipe off the culture media with alcohol cotton, place the coverslip on a slide with the spores attached side down. Observe the sample under microscope. Note the diameter of filaments, generation sites, shape and sized of spores.

2. Streak inoculated *Streptomyces* 1787 – 3 and *Streptomyces* 12 – 21 on two Gause's Synthetic Agar Medium plates. Insert coverslips into the media as mentioned above. Incubate the plates at 28℃ for 5 ~ 7 days. Take out the coverslip and observe under microscope.

II. Direct observation with streak inoculation

Continuous streak inoculated *Streptomyces* 4.794 on a Gause's Synthetic Agar Medium plate. Incubate the plate at 28℃ for 7 ~ 10 days. Once the fungi grew mature. Observe the Petri dish directly under the microscope.

If the plate media or the mycelia is too thick, take out one clot of culture and cut off the media. Observe the mycelia under the microscope.

III. Agar notch method

Take one loop of *Nocardia* 71 – N1 culture; spread the culture on the prepared potato agar media plate with the L – shaped spreader. Averagely dig several notches in the plate (1 cm × 5 cm each). Gently press a aseptic slide cross upon those notches. Incubate the plate at 28℃. Take out the slide at 6h, 10h, 15h to observe under the microscope. Similarly, this method could be applied to observe the filaments and spores of Streptomyces.

IV. Stamp method

Gently press a coverslip on the colony surface of Actinomyces. Place the coverslip on a drop of methylene blue on a clean slide, with the spores attached side down.

Results

Draw a representative field for each strain of Actinomyces.

Question

Why the Actinomyces were classified as the prokaryotic microorganisms?

实验十一 酵母菌形态观察

【目的】

1. 掌握观察酵母菌形态的方法。
2. 熟悉酵母菌的单染色法。
3. 了解酵母菌子囊孢子的观察方法。

【基本原理】

酵母菌是圆形、卵圆形、不运动的单细胞真核微生物。菌体比细菌大，繁殖方式也比较复杂，无性繁殖主要是出芽生殖，如酿酒酵母菌；还有一些以二分裂法方式繁殖，如裂殖酵母属；有性繁殖是通过单倍体细胞接合产生子囊孢子。

有些酵母如假丝酵母，常形成发达的假菌丝。幼龄期的酵母菌具有较均匀的细胞质，当细胞变老时，细胞内便出现颗粒和液泡，在高倍镜下观察，可见到肝糖、脂肪滴及异染颗粒等。

美兰是一种无毒性染料，其氧化型呈蓝色，还原型为无色。由于酵母菌活细胞的新陈代谢作用，细胞内具有较强的还原能力，经染色的酵母菌活细胞能使美兰从蓝色的氧化型变为无色的还原型而呈无色，而死细胞或代谢缓慢的细胞则无还原能力或还原能力弱，从而被染成蓝色或淡蓝色。

【仪器与材料】

1. 菌种 酿酒酵母（*Saccharomyces cerevisiac*）、白假丝酵母菌（*Candida albus*）。

2. 培养基与试剂 蒸馏水、0.1%碱性美兰染液（附录三，1）、石炭酸复红染液（附录三，2）、3%酸性乙醇、孔雀绿芽孢染液、碘液、乳酸石炭酸棉蓝染色液（附录三，11）、沙氏培养基（附录一，7）、葡萄糖-醋酸盐培养基（附录一，52）。

3. 仪器 显微镜。

4. 其他 接种环、玻璃棒、载玻片、纱布、擦镜纸、滤纸本、酒精灯、盖玻片。

【方法与步骤】

1. 酿酒酵母美兰染色法

（1）在载玻片中央滴加一滴0.1%碱性美兰染液，无菌操作用接种环取少许酿酒酵母，于美兰染液中混匀。

（2）用镊子取一片盖玻片，使其一侧与液滴边缘接触，然后将整个盖玻片慢慢放在液滴上，避免气泡产生，标本静置3min以上。

（3）先用低倍镜观察，然后用高倍镜观察，注意细胞的大小、形状、芽殖情况。根据细胞是否被染色来区别活细胞和死细胞。

2. 肝糖颗粒的观察 在载玻片中央加一滴碘液，然后用接种环取少量24酿酒酵母斜面培养物与碘液混匀，同上法盖上盖玻片，在高倍镜上能看到液泡内呈褐色的肝糖颗粒。

3. 异染颗粒观察 加一滴美兰于载玻片中央，取72h酿酒酵母斜面培养物与美兰混匀，加上盖玻片，在高倍镜下观察，可见细胞内有深蓝色颗粒，即为异染颗粒。

4. 假菌丝及厚膜孢子观察 将白假丝酵母菌接种于玉米斜面培养基上，37℃培养7d。取一张洁净载玻片，滴加乳酸石炭酸棉蓝染色液一滴，用接种环取少量菌混匀，盖上盖玻片，在低倍镜及高倍镜下观察，厚膜孢子呈蓝色，假菌丝和正常细胞无色。

5. 子囊孢子的观察

（1）用接种环取葡萄糖-醋酸盐培养基上的酿酒酵母于载玻片上，涂片，干燥，固定。

（2）加石炭酸复红染液，在火焰上加温染色5~10min（不要沸腾）。

（3）3%酸性乙醇洗30~60s，细水冲洗。

（4）碱性美兰染液复染1min，细水冲洗。

（5）油镜观察，酵母菌子囊孢子为红色，菌体为青色。如果用孔雀绿芽孢染液进行单染色（不加热），油镜观察可见绿色的子囊孢子。

【结果】

1. 绘图说明酿酒酵母的菌体形态及出芽方式。

2. 绘图说明酿酒酵母子囊和子囊孢子。

3. 绘图说明酿酒酵母肝糖颗粒和异染颗粒及其存在位置。

4. 绘图说明白假丝酵母菌厚膜孢子和假菌丝形态。

【思考题】

1. 什么叫子囊孢子？

2. 什么叫异染颗粒和肝糖颗粒？

3. 酵母细胞和细菌细胞在大小、细胞结构上有何区别？

Experiment 11　Morphology of Yeast

Objectives

1. Be familiar with the simple stain of yeast.

2. Observe the morphology of yeast.

3. Be familiar with the observation methods for yeast ascospore.

Principles

Yeasts are round, egg shaped static eukaryotic organisms, which is bigger than bacteria. Yeasts reproduce asexually mainly by budding (parent cells forms a bud that eventually breaks off, such as *Saccharomyces cerevisiac*) and sometimes by fission (divide evenly to produce two new cells, such as *Schizosaccharomyces pombe*) as well. Yeasts may reproduce sexually by the conjugation of haploids to produce ascospores.

When buds are produced and fail to detach, they form a chain of cells called pseudohypha (*Candida albicans* for example). Homologous cytoplasm could be observed in nascent and young yeast cells. When the cell grows older, granules and vacuoles could be produced and observed under the microscope.

Methylene blue is a non-toxic dye, which appears blue in its oxidative form and colorless in its reduced form. The metabolism in living yeast cells may renders strong reductive ability. The stained yeast cell may change their color from blue in oxidative form to colorless in reduced form. However, the old and dead yeast cells have only weak reductive ability, which could be stained in blue.

Apparatus and Materials

1. Specimens　*Saccharomyces cerevisiac* and *Candida albus*.

2. Reagents　Distilled water, Methylene blue (Appendix Ⅲ, 1), Safranine (Appendix Ⅲ, 2), 3% acid alcohol, malachite green, iodine, Lactic Medan staining solution (Appendix Ⅲ, 11), Sabouraud media (Appendix Ⅰ, 7), Glucose-acetate media (Appendix Ⅰ, 52).

3. Apparatus　Microscope.

4. Others　Inoculation loop, glass spreader, slide, water-absorption paper, lamp, coverslip.

Methods and Procedures

1. Methylene blue staining method

(1) Place a drop of methylene blue solution on a slide. Inoculate a loop of yeast into the solution and spread.

(2) Take a coverslip upon the drop. Wait for at least 3 min.

(3) Observe the sample under the microscope. Differentiate the living and dead yeast cell according to their colors.

2. Observe the glycogen granules   Place a drop of iodine solution on a slide. Inoculate a loop of *Saccharomyces cerevisiac* 24h culture into the solution and spread. Take a coverslip upon the drop. Observe the sample under the microscope. Find the glycogen granules in brown color.

3. Observe the metachromatic granules   Place a drop of methylene blue solution on a slide. Inoculate a loop of *Saccharomyces cerevisiac* 72h culture into the solution and spread. Take a coverslip upon the drop. Observe the sample under the microscope. Find the metachromatic granules in dark blue color.

4. Observe the pseudohypha and chlamydospores   Inoculate the *Candida albus* on the corn media agar slant, incubate at 37℃ for 7 days. Place a drop of Lactic Medan staining solution on a slide. Inoculate a loop of *Candida albus* culture into the solution and spread. Take a coverslip upon the drop. Observe the sample under the microscope. Find the chlamydospores in blue color, the vegetative cell and pseudohypha colorless.

5. Observe the ascospores

(1) Take a loop of *Saccharomyces cerevisiac* culture from the glucose – acetate media onto a slide. Smear, dry and fix.

(2) Stain with the safranine, heat on the flame for 5 ~ 10min. (Do NOT boil.)

(3) Decolorization with 3% acid alcohol for 30 ~ 60s. Rinse with tap water.

(4) Counterstain with methylene blue for 1min. Rinse with tap water.

(5) Observe the sample under the microscope. Find the yeasts ascospores in red color and vegetative body in green color. If malachite green is used (without heating), green ascospores could be observed.

Results

1. Draw a representative field for shape and budding methods of *Saccharomyces cerevisiac*.

2. Draw a representative field for ascus and ascospores of *Saccharomyces cerevisiac*.

3. Draw a representative field for glycogen granules and metachromatic granules in *Saccharomyces cerevisiac*.

4. Draw a representative field for clamydospores and pseudohypha of *Candida albus*.

Questions

1. What is the definition of ascospores?

2. What is the definition of glycogen granules and metachromatic granules?

3. What kind of shape and structure differences could be found between the yeasts and bacteria?

实验十二 霉菌形态及观察

【目的】
1. 掌握观察霉菌的基本方法。
2. 熟悉霉菌的基本形态。

【基本原理】
霉菌是指真菌中一些由单细胞或多细胞菌丝组成的具有较复杂结构的丝状真菌。霉菌的菌丝粗壮，在固体培养基上长成颜色不同的绒毛状或棉絮状；繁殖方式也多种多样，各类霉菌都具有其特殊的繁殖器官，这些都为霉菌的分类鉴定工作提供了依据。

由于霉菌菌丝较粗，细胞容易收缩变形，而且孢子很容易飞散，所以制标本时常用乳酸石炭酸棉蓝染色液。此染液制成的霉菌标本的特点是：细胞不变形；具有杀菌防腐作用；不易干燥，能保持较长时间；溶液本身呈蓝色，有一定染色效果。

【仪器与材料】
1. 菌种 青霉菌（*Penicillium sp.*）、曲霉菌（*Aspergillus sp.*）和根霉菌（*Rhizopus sp.*）。

2. 试剂 乳酸石炭酸棉蓝染色液（附录三，11）。

3. 仪器 显微镜。

4. 其他 接种环、玻璃棒、载玻片、纱布、擦镜纸、滤纸本、酒精灯。

【方法与步骤】
1. 青霉、曲霉水浸标本制备及观察

（1）于载玻片上滴一滴乳酸石炭酸棉蓝染色液，用小镊子从霉菌菌落边缘处取不过于老化的带有孢子的菌丝体一小块，孢子面向上放置于载玻片的液滴内，再小心用小镊子将菌丝轻轻拨散开（注意不是切碎），然后加盖盖玻片，避免气泡产生。

（2）先用低倍镜观察，再换高倍镜进行观察。观察青霉菌孢子穗的帚状分枝、梗基、小梗及分生孢子排列方式；观察曲霉菌孢子穗的顶囊、梗基、小梗及分生孢子的排列方式。找出其各部位，并比较一下青霉、曲霉的分生孢子穗有什么区别（图12-1，图12-2）。

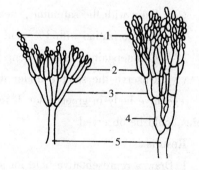

图 12-1 青霉分生孢子穗结构

Figure 12-1 Conidiophore structure of *Penicillium*

1. 分子孢子 2. 小梗 3. 梗基 4. 副枝 5. 分生孢子梗

1. Conidia 2. Sterigmata 3. Metulae 4. Branch 5. Conidiophore

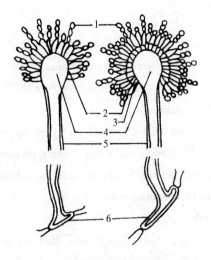

图 12－2 曲霉分生孢子穗结构
Figure 12－2　Conidiophore structure of *Aspergillus*
1. 分生孢子　2. 小梗　3. 梗基　4. 顶囊　5. 分生孢子梗　6. 足细胞
1. Conidia　2. Terigmata　3. Metulae　4. Vesicles　5. Conidiophore　6. Stolon

2. 根霉形态观察　将黑根霉接种在无菌沙氏培养基平板上，28℃倒置培养至菌丝长满平皿空间为止，轻轻取下平皿盖，换配上一洁净平皿底，将平皿直接用低倍镜观察，不断调节粗细准焦螺旋，可观察到在自然条件下生长的黑根霉的匍匐菌丝、假根、孢子囊梗和孢子囊。并注意观察菌丝横隔的有无及孢子囊梗与假根的相对位置（图12－3）。

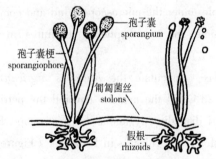

图 12－3　黑根霉基本结构
Figure 12－3　Morphology of *Rhizopus oryzae*

【结果】

绘图说明青霉、曲霉和根霉的基本结构。

Experiment 12　Morphology of Molds

Objectives

1. Be familiar with the basic methods for molds observation.

2. Be familiar with the fundamental shapes of molds.

Principles

Molds are filamentous fungi consisted of single or multiple cell filaments. Molds have strong filaments, which may display colorful colony shapes on the culture media. Different molds have different reproduction means and reproductive organs. These characteristics could be used to identify the molds classification.

The molds filaments are strong easy to shrink and their spores are easy to scatter. Lactic Medan staining solution is usually used to make the staining process, which could maintain the cell shape, has antiseptic effect, uneasy to dry to maintain a longer staining period, and has some staining effect due to its blue natural color.

Apparatus and Materials

1. Specimens   *Penicillium sp*, *Aspergillus sp* and *Rhizopus sp*.

2. Reagents   Lactic Medan staining solution (Appendix Ⅲ, 11).

3. Apparatus   Microscope.

4. Others   Inoculation loop, glass spreader, slide, water-absorption paper, lamp, coverslip.

Methods and Procedures

1. Observe the *Penicillium* and *Aspergillus*

(1) Place a drop of Lactic Medan staining solution on a clean glass slide. Take a clot of mycelia from the margin of the colonies with forceps. Spread the mycelia in that drop (spread not cut/ stir). Lay a coverslip.

(2) Observe the sample under the microscope. Find and compare the conidia, sterigmata, metulae, branch, vesicle and conidiophores structures in *Penicilium* and *Aspergillus* (Figure 12-1, Figure 12-2).

2. Observe the *Rhizopus*   Inoculate the *Rhizopus nigricans* on the Sabouraud media plate. Incubate the plate at 28℃ till the mycelia covered the petri dish. Change a clean petri dish cover. Observe the petri dish directly under the microscope. Find the sporangium, columella, sporangiophore, stolon, and rhizoids in Rhizopus (Figure 12-3).

Results

Draw a representative field for fundamental structures of *Penicillium*, *Aspergillus* and *Rhizopus*.

实验十三　细菌、放线菌、酵母菌、霉菌菌落的特征观察

【目的】

熟悉细菌、放线菌、酵母菌和霉菌代表种类的菌落形态特征。

【基本原理】

将某种微生物接种在一定的培养基上，经过一定时间培养后，由单个菌体形成的

肉眼可见的群体形态,称为菌落(colony)。细菌、放线菌、酵母菌和霉菌每一类微生物在一定培养条件下形成菌落各具有某些相对的特征,如菌落的大小、形状、表面状态与颜色都不尽相同。观察这些特征,可初步识别和区分各类微生物。图 13-1 是平板上常见菌落的一些基本特征。

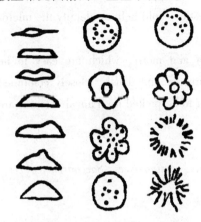

图 13-1 常见菌落的基本特征
Figure 13-1 Characteristics of microorganism colonies

一般来说,细菌和酵母菌的菌落是比较光滑湿润的。用接种环极易将菌体挑起。放线菌的菌落硬度较大,干燥致密,且与培养基结合紧密,不易被接种针挑起。霉菌菌落长成了绒毛状或棉絮状。

【仪器与材料】

1. **菌种** 大肠埃希菌(*Escherichia coli*)、金黄色葡萄球菌(*Staphylococcus aureus*)、放线菌(*Actinomycetes sp.*)、酿酒酵母菌(*Saccharomyces cerevisiac*)、青霉菌(*Penicillium sp.*)和曲霉菌(*Aspergillus sp.*)的单个菌落平板。

2. **仪器** 显微镜。

3. **其他** 接种环。

【方法与步骤】

肉眼观察菌落特征。

【结果】

观察细菌、放线菌、酵母菌和霉菌的菌落特征,并完成表 13-1。

表 13-1 细菌、放线菌、酵母菌和霉菌的菌落特征比较

微生物类型	大小	形状	表面状态	菌落形态边缘	颜色	组成	与培养基结合紧密程度
细菌							
酵母菌							
放线菌							
霉菌							

【思考题】

如何区别细菌、放线菌、酵母菌和霉菌的菌落?

(徐慰倬)

Experiment 13 Colonies Characteristics of Bacteria, Actinomyces, Yeasts and Molds

Objective

Be familiar with the colony characteristics of bacteria, actinomyces, yeasts and molds.

Principles

After inoculated on agar media and cultured for some time, microorganisms grow to form congregate from single cell which could be seen by naked eye. It is called colonies. Different microorganisms have different cultural characteristics under some conditions, such as the shape, surface and color of the colonies. These characteristics could help to identify the microorganism (Figure 13-1).

Usually, colonies of bacteriaand yeasts are smooth and moist, which are easy to be picked up by inoculation loop. Colonies of actinomyces are dry and dense, closely combined with the media with a higher hardness, which are difficult to be picked up. Fungal colonies are often grown into fluffy or cotton shape.

Apparatus and Materials

1. Specimens *Escherichia coli*, *Staphylococcus aureus*, *Actinomyces*, *Saccharomyces cerevisiac*, *Aspergillus*, *Penicillium*.

2. Apparatus Microscope.

3. Others Miscellaneous, inoculation loop.

Methods and Procedures

Observe these colonies with naked eyes.

Result

Observe these colonies, to fulfill their characteristics in Table 13-1.

Question

How to differentiate the colonies from bacteria, actinomyces, yeasts and molds?

第三节 培养基的制备、灭菌和除菌技术

Section 3 Media Preparation, Sterilization and Disinfection Techniques

实验十四 微生物常用培养基的配制

【目的】

1. 掌握培养基的配制原则。
2. 熟悉不同种类微生物的常用培养基配制方法。

【基本原理】

微生物的生长依赖于可利用的营养物质和适宜的生长环境。在实验室研究微生物时，首先需要制备出供微生物生长的培养基。培养基是人工配制的用于培养、转移、

贮存微生物的固体、半固体或液体营养基质。培养基中必须含有微生物生长所需的全部营养物质，包括碳源、氮源、能源（有时可能与碳源相同）、无机盐、水分和生长因子。适宜的 pH 对微生物的生长也是必不可少的。

培养基的物理状态有三种：液体、半固体和固体。这些培养基的主要区别是在固体和半固体培养基中加有固化剂——琼脂。琼脂是从某种海藻中得到的结构复杂的长链多糖，具有很多优良性质。当加入溶液中时，琼脂要在98℃才能融化形成有一定黏性的液体，并在42℃凝固。凝固之后要重新加热至98℃才能再次融化。琼脂具有的其他特性包括：不易被微生物降解；接近无色，因此易于观察菌落等。对于固体培养基需要加入琼脂粉的量约为1.4%~1.6%，半固体培养基需加入约0.6%~0.8%。

根据培养基成分的来源不同，可将培养基分为两种：一种叫合成培养基，由化学成分完全了解的物质配制而成，只能满足特定微生物的基本生长需求。另一种叫天然培养基，由化学成分不清楚或不恒定的物质组成，例如肉的水解产物（蛋白胨），能够充分满足不同微生物的生长需要。天然培养基如营养肉汤培养基在实验室中通常用于培养细菌，沙氏培养基用于培养真菌。合成培养基如高氏一号合成培养基则通常用于培养放线菌。本实验将介绍不同种类微生物培养时常用的培养基配方及其配制方法。

培养基在配制过程中会由于接触到容器、药品、称量纸或其他表面而被污染，因此必须进行灭菌，同时不能将培养基中的营养物质破坏。关于培养基的灭菌方法将在实验十五中介绍。

【仪器与材料】

1. 培养基与试剂 见附录一相应配方。

2. 仪器 天平、高压蒸汽灭菌锅、电炉等。

3. 其他 药匙、称量纸、量筒、烧杯、pH 试纸、三角瓶、硫酸纸等。

【方法与步骤】

1. 营养肉汤琼脂培养基（肉汤琼脂培养基）的配制（用于细菌的培养） 配方见附录一，2。

（1）计算、称量：按照配方准确计算并称量每一种药品的实际用量。牛肉膏因为其黏性高需用硫酸纸称取。

（2）溶解：用量筒称取1000ml 自来水倒入烧杯。然后将各药品（除琼脂外）加入烧杯（将称好的牛肉膏与硫酸纸一并放入烧杯中，待溶解后将硫酸纸取出），加热并搅拌至溶解。

（3）调 pH：用 1mol/L HCl 或 1mol/L NaOH 调节 pH 至 7.2~7.6。

（4）分装：将配好的培养基分装到三角瓶中至不超过1/2高度处。如果配制固体（琼脂1.4%~1.6%）或半固体（0.6%~0.8%）培养基，将称好的琼脂直接放入三角瓶中（琼脂的用量要按照三角瓶中培养基的实际装入量来称取）。盖好胶塞。若配制斜面，则需先将液体培养基煮沸，再把琼脂放入，继续加热至琼脂完全融化。在加热过程中应注意不断搅拌，以防琼脂沉淀在瓶底，也要控制火力，以免培养基因沸腾而溢出容器。待琼脂完全融化后，用热水补足因蒸发损失的水分。凝固前分装至试管中，以试管高度 1/5~1/4 为宜。注意分装时避免培养基挂在瓶口或管口上引起杂菌污染。

（5）灭菌：用牛皮纸包装好三角瓶（试管）并注明培养基名称、学生名字和日期。然后将三角瓶（试管）放入高压蒸汽灭菌锅中灭菌 30min。

（6）摆放斜面：将灭菌的试管培养基在凝固前倾斜摆放，摆放的斜面长度以培养基不超过试管总长的一半为宜。

（7）无菌检查：见实验十五。

（8）保存：如果不立即使用，应将灭菌后的培养基置于 4℃ 保存。

2. 高氏一号合成培养基的配制（用于放线菌的培养） 配方见附录一，5。

（1）计算、称量：用量筒称量 900ml 自来水至烧杯中，在煤气灯上加热。称取可溶性淀粉至小烧杯中，加入 50~100ml 自来水调成糊状，再加至上述热水中搅拌混匀。

（2）溶解：分别称取药品并加入烧杯中至完全溶解，最后补足水分。

（3）调 pH：待溶液冷却至室温时，调 pH 至 7.2~7.4。

（4）分装：配置固体培养基时，按前面叙述方法加入琼脂并制备斜面。

（5）包扎、标记、灭菌及无菌检查。

3. 沙氏培养基的配制（用于真菌的培养） 配方见附录一，7。

配制步骤仍按照：计算→称量→溶解→调 pH→分装→包扎→标记→灭菌等程序进行。

4. 生化培养基的配制（用于细菌生理生化反应的检测）

（1）单糖发酵培养基（单糖发酵试验） 配方见附录一，10。

配好后分装到带一倒立小试管（Durham 管）的中试管中，每管装量约为 4~5ml（以小倒管完全被浸没为标准），培养基应呈绿色。包扎，灭菌。

（2）蛋白胨水培养基（吲哚试验） 配方见附录一，12。

配好后分装至小试管中至 1/4 高度。包扎，灭菌。

（3）葡萄糖蛋白胨水培养基（甲基红试验和 V-P 试验） 配方见附录一，13。

甲基红试验时，配好后分装至小试管至 1/4 高度。V-P 试验时，每管定量 2ml。包扎，灭菌。

【实验内容】

配制营养肉汤培养基、高氏一号合成培养基和沙氏培养基。

【注意事项】

1. 配制培养基时使用的容器不宜用铜、铁器皿。因为培养基中含铜量超过 0.30mg/L 或含铁超过 0.14mg/L 就可能影响微生物的正常生长。

2. 培养基放置一段时间后，再次使用前应做无菌检查。

3. 灭菌后的培养基不宜保存过久，以免营养成分起化学变化。培养基在贮存期间，因能吸收空气中的 CO_2，而变为酸性。所以用贮存过久的糖发酵培养基做生理生化试验就可能不出现正确结果。

【思考题】
在配制培养细菌的培养基时为什么要将 pH 调至中性?

Experiment 14   Preparation of Commonly Used Microbial Media

Objectives
1. Grasp the principles of media preparation.
2. Be familiar with the preparation methods of commonly used media for different microbes.

Principles
The growth of microorganisms depends on available nutrients and a favorable growth environment. Microbial laboratory works require the culture media to grow the microorganisms. A culture medium is a solid, semisolid or liquid preparation used to grow, transport and store microorganisms. The medium must contain all the nutrients the microorganisms require for growth, including carbon source, nitrogen source, energy source (can be the same as carbon source), inorganic salts, water and growth factors. A suitable pH is also needed for the microbial growth.

Three physical states media are used: liquid media, semisolid media, and solid media. The major difference among these media is that solid and semisolid media contain a solidifying agent - agar. Agar is a complex, long chain polysaccharide derived from certain marine algae and has several useful properties. When added to a solution, it melts at 98℃ forming a slightly viscous liquid that solidifies at 42℃. After solidification, the agar will not melt unless the temperature is again raised to 98℃. Some other useful properties of agar include its resistance to microbial degradation and its translucence for easy viewing of colonies. When a solid medium is necessary, a 1.4% ~ 1.6% concentration of agar is typical. For semi - solid medium around 0.6% ~ 0.8% agar is employed.

According to the sources of media ingredients, two different types of media are designed. A defined or synthetic medium contains known amounts of pure chemicals which are usually the basic requirements for the growth of particular microbes. A complex medium contains undefined ingredients such asprotidtemns of meat (peptones) which provide enough nutrients to sustain the growth for different microbes. Complex media such as nutrient broth is routinely used for cultivation of bacteria in the laboratory works as well as Sabouraud's (SAB's) medium for fungi and Gause's synthetic medium for Actinomycetes. All these media mentioned above will be included in this experiment.

Culture media must be made sterile without inactivating nutrients necessary for growth of the microorganisms due to the contamination of containers, media ingredients, weighing papers, or other surfaces that come in contact with the medium during preparation. This involves the use of an autoclave and will be introduced in the Experiment 15.

Apparatus and Materials

1. Cultures and reagents   Relevant recipes in Appendix I.

2. Apparatus   Balance, autoclave, electric stove.

3. Others   Spoon, weighing paper, cylinder, beaker, pH test strips, triangular flask, vegetable parchment.

Methods and Procedures

1. Preparation of nutrient agar broth (for bacteria, Appendix I, 2)

(1) Calculation and weighing: Calculate and weigh out each component required for the nutrient broth agar medium. For the beef extract, weigh it on a vegetable parchment because of its viscous property.

(2) Dissolving: Measure 1000ml tap water using cylinder and pour it into the beaker. Add all the components to the beaker except the agar. Put the beef extract with the vegetable parchment into the beaker together and take the parchment out after beef extract dissolving. Stir and heat until dissolved.

(3) Adjusting pH: Using 1mol/L HCl or 1mol/L NaOH to adjust the pH to 7.2~7.6.

(4) Distribution: pour the medium to the flasks less than half height of them. When solid (1.4%~1.6% concentration of agar) or semisolid (0.6%~0.8% concentration of agar) medium is needed, add agar into the flask directly (the amount of agar should be according to the real mount of media in the flask). Put on the plug. When preparing agar slant, the solution should be heated to boil first and then add agar, heating until agar dissolved completely. Stir the solution during heating process to avoid sedimentation of agar. Do not overheat to prevent overflowing. When agar dissolved completely, add water to volume of the original amount because of the loss of water during heating. Distribution should be done before solidification to 1/5~1/4 height of the tube.

(5) Sterilization: Wrapping the flask (tube) with brown paper labeled with media name, student name and date. Now the flask (tube) can be put into the autoclave for sterilization for 30min.

(6) Making agar slant: After sterilization, incline the tube before solidification. The length of the slant surface should be less than half of the total length of the tube.

(7) Sterility examination: Reference to Experiment 15.

(8) Preservation: Sterilized media should be preserved at 4℃ if they are not used right now.

2. Preparation of Gause's synthetic medium (for Actinomycetes, Appendix I, 5)

(1) Calculation and weighing: Add 900ml of tap water into a beaker and heat. Weigh out soluble starch and mix with 50~100ml tap water to make paste. Then add the paste into the 900ml hot water and stir.

(2) Dissolving: Add all the other ingredients to the beaker and stir to dissolve. Make the media volume to 1000ml.

(3) Adjusting pH: When the media cools to room temperature, adjust the pH to 7.2~7.4.

(4) Distribution: When solid media is needed, agar should be added directly to the container. Agar slant can be made according to the procedure in nutrient agar broth preparation.

(5) Wrapping, labeling, sterilization and sterility test.

3. Preparation of SAB's media (for fungi, Appendix I, 7)   The procedure is still following: calculation→weighing out→dissolution→adjusting pH→distribution→wrapping→labeling→sterilization.

4. Preparation of biochemical media (for physiological and biochemical tests)

(1) Monosaccharide fermentation broth (for monosaccharide fermentation test, Appendix I, 10): Distribute the prepared media (green color) into test tubes with inverted Durham tubes inside. 4~5ml is appropriate for each tube (the media should immerse the Durham tube). Wrap, label and sterilize.

(2) Peptone broth (for Indole test, Appendix I, 12): Distribute the prepared media to small tubes to 1/4 height. Wrap, label and sterilize.

(3) MR – VP broth (for methyl – red test and V – P test, Appendix I, 13)

Methyl – red test: distribute the media into small tubes to 1/4 height. Wrap, label and sterilize.

V – P test: distribute the media to test tubes with exact 2ml in each tube. Wrap, label and sterilize.

Experiment content

Prepare nutrient broth medium, Gause's synthetic medium and SAB's medium.

Notes

1. Copper and iron containers should not be used to avoid growth inhibition from dissolved copper or iron.

2. Media should be tested again for sterility before use longer than one week storage.

3. The sterile media should not be preserved for long time to avoid any chemical changes in nutrients. Because of the absorption of CO_2 in the air, preserved media could become acidic which will result in mistake in monosaccharide fermentation test.

Question

Why the media pH needs to be adjusted to neutral for bacterial cultivation?

实验十五 培养基的灭菌和灭菌验证

【目的】

1. 掌握高压蒸汽灭菌法的原理、高压蒸汽灭菌锅的使用及注意事项。
2. 熟悉巴氏消毒、紫外杀菌和干热灭菌的原理和步骤。

3. 验证细菌芽孢的耐热性。

【基本原理】

由于培养基在配置过程中并非是无菌操作，其接触的药品、容器等也不是无菌的，所以在配制后要求立即灭菌，这样做的目的是防止配制过程中混入的杂菌利用培养基中的营养物质进行繁殖从而破坏培养基的性能。实验室最常用的灭菌方法就是利用高温处理达到杀菌效果，即热力灭菌法。因为高温能够使蛋白质和核酸变性，破坏细胞膜结构，最终达到杀菌目的。热力灭菌法分为湿热灭菌法和干热灭菌法两大类。湿热灭菌法主要包括高压蒸汽灭菌法、巴氏消毒法和煮沸法等；干热灭菌主要包括火焰灼烧和干烤法等。除热力灭菌外，紫外杀菌和过滤除菌也是常用的消毒灭菌方法。每种方法的使用特点与适用范围将通过杀菌效果展示出来。本实验中将逐一介绍。

1. 高压蒸汽灭菌法及灭菌验证

（1）高压蒸汽灭菌法：是微生物实验室应用最广、效果最好的湿热灭菌方法。其原理是：将待灭菌的物品放置在盛有适量水的高压蒸汽灭菌锅内，然后开始加热煮沸，除尽锅内原有的冷空气后将锅密闭。再继续加热就会使锅内的蒸汽压逐渐上升，温度也随之升到100℃以上。为达到良好的灭菌效果（杀灭包括芽孢在内的一切微生物），一般要求温度达到121℃（压力为0.1MPa），时间维持15～30min。当培养基中含有糖时，也可在较低的温度（如，113℃，55.21kPa）下维持30min。此法适用于微生物学实验室对培养基等的灭菌处理。

使用高压蒸汽灭菌法进行灭菌处理时，锅内冷空气的排尽与否至关重要，否则在同一压力下，锅内所能达到的实际温度与理想温度就会有差距（表15-1）。

表15-1 高压蒸汽灭菌锅内温度与空气排除量的关系

压力表读数（kPa）	灭菌锅内温度（℃）				
	空气完全排除	空气2/3排除	空气1/2排除	空气1/3排除	空气全未排除
34.52	108.4	100	94	90	72
68.94	115.2	109	105	100	90
103.46	121.3	115	112	109	100
137.88	126.2	121	118	115	109

高压蒸汽灭菌器（锅），有直立式（图15-1）、横卧式和手提式三种。结构上都有坚固的双层金属壁和严密的盖。此外，都必须具有压力表及温度计，表示锅内的压力和温度；锅上有放气阀，用于加热时排出冷空气；另外还装有安全阀，当压力超过规定限度时，安全阀可以自动排气。

（2）灭菌验证：培养基配制完成后，必须对灭菌效果进行验证，这将直接关系到实验结果正确与否。灭菌验证方法很多，在此介绍如何利用嗜热脂肪芽孢杆菌纸片来进行灭菌验证的方法。

将嗜热脂肪芽孢杆菌纸片（以下简称菌片）用无菌镊子放入密封试管中。分别放

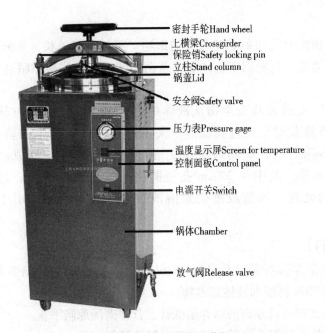

图 15 - 1 高压蒸汽灭菌锅

Figure 15 - 1　Autoclave

置在高压蒸汽灭菌锅的蒸汽口处、底部排气口处及底部出水口处等不同位置，然后进行灭菌处理。灭菌后的菌片以无菌操作放入灭菌后的溴甲酚紫胨水培养基内56℃培养48h，观察颜色变化。如培养基变为黄色，说明菌片中的嗜热脂肪芽孢杆菌没有被完全杀灭，从而在培养基中生长，分解葡萄糖产酸变为黄色。如培养基颜色不变化仍为紫色，则说明芽孢菌已全部灭活。同时要用未经灭菌的菌片放入培养基内作为阳性对照，不加菌片的空白培养基作为阴性对照。

2. 巴氏消毒法　是用于对牛奶或其他液体进行加热以破坏腐败或病原微生物的消毒措施。操作温度通常在63～72℃，因此只能杀死营养细胞，对芽孢无效。

3. 干热灭菌法　许多物品需要在干燥条件下进行灭菌，即干热灭菌。干热灭菌有火焰灼烧法和烘箱法。前者指将待灭菌对象直接用火焰灼烧来灭菌的方法，因其操作温度高，所以灭菌效果彻底，但破坏力强，因此只适用于接种环、三角瓶口（试管口）等的灭菌处理。烘箱法是指将待灭菌物品置于烘箱中160～170℃灭菌1～2h，这样能够保证彻底杀灭包括芽孢在内的一切微生物。烘箱温度不能超过180℃，否则棉花和包扎纸会被烧焦。此法适于玻璃仪器、金属仪器和粉末等水蒸汽无法有效穿入并且在高温下不会被破坏的物品的灭菌。

4. 紫外线杀菌法　紫外线的波长在100～400nm，其中260nm的紫外线具有最强杀菌性，因为DNA分子的最大吸收峰在此。当微生物暴露在紫外光下时，DNA会吸收紫外线，并使其中相邻的胸腺嘧啶形成二聚体。DNA复制时，无法对此二聚体进行配对，引起DNA链复制错误。换句话说，紫外照射能够引起突变并导致蛋白质合成发生错误。如果突变过多，细菌的代谢就会被阻断从而引起微生物死亡。由此可知，紫外照射不能用于杀灭产芽孢的细菌，因为处于芽孢这个休眠期间是没有DNA复制过程发生的。

紫外线的杀菌能力受照射时间长短的影响：照射时间越长，杀菌效果越好。紫外照射由于其穿透力差，只能用于物体表面杀菌。同时紫外线对眼睛和皮肤的细胞也会产生伤害。

5. 过滤除菌 是通过将微生物从液体中移走而不是杀死的方法来达到消毒灭菌效果的。将待除菌对象经过膜滤器，大于滤膜孔径的微生物就会被截留下来，与滤液分开。所用的微孔滤膜通常由硝酸纤维或醋酸纤维制成，其孔径大小从 0.025~25μm 不等，其中 0.22μm 孔径的细菌滤膜较为常用。过滤除菌适用于对热敏感液体的处理，如酶或维生素溶液、血制品等，还可用于啤酒生产中代替巴氏消毒。

【仪器与材料】

1. 菌种 枯草芽孢杆菌（*Bacillus subtilis*）24h 斜面和液体培养物、大肠埃希菌（*Escherichia coli*）20h 斜面和液体培养物。

2. 培养基与试剂 营养肉汤培养基试管、营养肉汤琼脂平板。

3. 仪器 温控水浴锅、烘箱、紫外灯、培养箱等。

4. 其他 接种环、无菌滴管等。

【方法与步骤】

1. 高压蒸汽灭菌法（对配制后的培养基进行灭菌）

（1）加水 向锅内加入足够的水。

（2）摆放培养基 放入配好的培养基，注意不要装得太挤，关闭锅盖，按下开始按钮。

（3）排除冷空气 打开排汽阀，待冷空气完全排尽，开始持续冒出白色水蒸气后，关上排汽阀，蒸汽由夹套进入内层，于是温度开始上升。

（4）灭菌计时 当达到 121℃ 时，计时器开始计时 15~30min（视所灭菌的培养基种类而定）。灭菌后，锅内的液体需缓慢冷却方可打开锅盖，以防止由于压力骤降引起液体喷溅。

（5）无菌验证 参照本实验中原理部分。

2. 巴氏消毒法

（1）取样 分别接种枯草芽孢杆菌和大肠埃希菌到试管内的营养肉汤培养基中，再用另一管未接种的培养基作为空白对照。

（2）消毒 将 3 个试管共同置于 63℃ 恒温水浴 30min 进行消毒。

（3）消毒验证 然后将试管置于 37℃ 恒温培养 24h，观察细菌在消毒后的生长状况。

3. 紫外消毒法

（1）预热 打开紫外灯预热 10min。

（2）取样 用无菌滴管分别滴加 3 滴枯草芽孢杆菌和大肠埃希菌液体培养物至两个琼脂平板表面，并涂布均匀。（2）~（4）步都必须无菌操作。

(3) 加牛皮纸　将镊子灭菌取一张无菌的星形牛皮纸，放至平板中央并将边缘按压。

(4) 照射紫外线　移去平板盖，在距离紫外灯20cm左右距离照射30min。时刻提醒不要将皮肤或眼睛直接暴露在紫外灯下以免受伤。

(5) 取牛皮纸　照射后，用镊子通过无菌操作将牛皮纸取下，重新盖好在37℃倒置培养24h。

(6) 记录杀菌效果　观察微生物的生长情况。

4. 烘箱灭菌

(1) 物品摆放　将待灭菌物品（平皿、试管、吸管等）用报纸包好后放入电烘箱内，注意物品不要摆得太挤。

(2) 加热灭菌　打开烘箱开关，直至达到160～170℃，开始计时2h。

(3) 冷却　待电烘箱内温度降到70℃以下取出灭菌物品。

【实验内容】

1. 利用高压蒸汽灭菌器对培养基进行灭菌并验证灭菌效果。
2. 检验巴氏法的消毒效果。
3. 检验紫外消毒法的效果和特点。

【结果】

1. 观察试管中培养状况（澄清或浑浊），比较巴氏消毒法对大肠埃希菌和枯草芽孢杆菌的消毒效果。
2. 观察牛皮纸下与牛皮纸外菌体生长状况，比较紫外照射法对大肠埃希菌和枯草芽孢杆菌的消毒灭菌效果及紫外杀菌的特点。

【注意事项】

1. 高压蒸汽灭菌时

(1) 冷空气一定要排尽。

(2) 灭菌物品不能堆得太满、太紧，以免影响温度均匀上升。

(3) 达到目标温度时才开始计时。

(4) 停止加热后，一定要自然冷却，切不可急于打开排气阀，这样会导致压力骤降使培养基剧烈沸腾喷出。

2. 干热灭菌时

(1) 灭菌物品不能堆得太满、太紧，以免影响温度均匀上升。

(2) 灭菌物品不能直接放在烘箱地板上，防止包装纸等烤焦。

(3) 灭菌温度若超过180℃，棉花、报纸会烧焦甚至燃烧。

(4) 温度自然降至70℃左右时才能打开箱门取出物品。

3. 在紫外消毒时

(1) 要避免光复活现象。

(2) 保护好眼睛和皮肤不被紫外线直接照射。

【思考题】

1. 干热灭菌的注意事项有哪些？
2. 怎样才能获得枯草芽孢杆菌的芽孢悬液？
3. 灭菌结束后，为什么要等到灭菌锅内压力降为常压时才能打开灭菌锅盖？

（马晓楠）

Experiment 15　Culture Media Sterilization and Sterility Test

Objectives

1. Grasp the principles, methods and precautions of autoclaving. Grasp how to use the autoclave correctly.

2. Be familiar with the principles and methods of Pasteurization, UV disinfection and dry – heat sterilization.

3. Verify the heat – resistance of bacterial endo – spore.

Principles

Culture media must be made sterile due to the contamination of containers, media ingredients, weighing papers, or other surfaces that come in contact with the medium during media preparation. The microbes coming from contamination will consume the nutrients in the media. Sterilization is the process of rendering a medium free of all forms of life. The most useful approach is heating. Heat is thought to kill by degrading nucleic acids and by denaturing enzymes and other essential proteins. It may also disrupt cell membranes. There are two ways for heat sterilization which are moist heat and dry heat sterilization. Moist heat includes autoclaving, Pasteurization and boiling water. Dry heat includes fire and oven. UV disinfection and filtration are also basic ways for sterilization for different items. All the characteristics and application scopes will be introduced in this experiment.

1. Autoclaving and verification of sterilizing effectiveness

（1）Autoclaving：It is the most useful approach of moist heat sterilization in microbial laboratory with best effectiveness. Items are sterilized by exposure to steam at 121℃ and 0.1MPa of pressure for 15 ~ 30min. Under these conditions, microorganisms, even endospores, will not survive longer than 15 minutes. This method is rapid and dependable. Autoclaves are designed to ensure that all of the air has been expelled and only steam is present in the autoclave chamber. They are carefully temperature controlled as well. When sugar is involved in the media, lower temperature (113℃ at 55kPa for 30min) is applied. Almost all media in microbial lab can be sterilized in this way.

The expelling of air in the autoclave chamber is crucial to sterilizing effectiveness which will result in different actual temperature away from the ideal one (Table 15 – 1).

Table15-1 Influence of expelled air on sterilization temperature in autoclave

Pressure (kPa)	Actual temperature in autoclave (℃)				
	All air expelled	2/3 of air expelled	1/2 of air expelled	1/3 of air expelled	No air expelled
34.52	108.4	100	94	90	72
68.94	115.2	109	105	100	90
103.46	121.3	115	112	109	100
137.88	126.2	121	118	115	109

In an autoclave (Figure 15-1), you can find double deck chamber, well sealed lid, pressure gage, thermometer, safety valve and release valve.

(2) Verification of sterilizing effectiveness: After sterilization, the sterilizing effect must be tested to guarantee the following experiment. Here will focus on the use of *Bacillus stearothemophilus* strip.

Put the strips into sealed test tubes and place them into the autoclave chamber to where steam enters, air releases and water releases, then start to sterilize. After sterilization, aseptically transfer the strips into bromcresol purple broth and incubate at 56℃ for 48h. If the broth turns yellow, it means *B. stearothemophilus* is still alive because of acid production during the bacterial growth. If the medium keeps purple, sterilization is successfully done. Positive control with unsterile strip inoculated and blank control without strip are also necessary.

2. Pasteurization It is a disinfection approach for preserving milk and reducing milk transmissible diseases. Milk, beer, and many other beverages are now pasteurized. Because of the manipulation temperature of 63~72 ℃, Pasteurization does not sterilize a beverage, but it does kill any pathogens present and drastically slows spoilage by reducing the level of non-pathogenic spoilage microorganisms. There is no effect on endospores.

3. Dry-heat sterilization Often, some items need to be sterilized under dry condition, known as dry heat sterilization. There are two basic ways including fire and oven. Fire means sterilize the items by firing it directly. It has better effectiveness because of the ultra-high temperature. However, high destructive power makes it only suitable for the sterilization of inoculation loop, opening of flask or tube. Glassware such as pipettes and petri plates must be sterilized with oven. The glassware is placed in an electric oven set to operate from 160° to 170°C. Since dry heat is not as effective as wet heat, the glassware must be kept at this temperature for about 1~2 hours or longer. The oven temperature must not rise above 180°C or any cotton or paper present will scorch. Besides glassware, oven is usually operated on metal ware and powder which resist to high temperature.

4. UV disinfection Radiation between 100 and 400 nm is called ultraviolet (UV). The most lethal UV radiation has a wavelength of 260 nm, the wavelength most effectively absorbed by DNA. The primary mechanism of UV damage is the formation of thymine dimers in DNA. When the microorganisms are exposed to UV, two adjacent thymines in a DNA strand

are covalently joined to inhibit DNA replication and function. With sufficient mutation, bacterial metabolism is blocked and the organism dies. Endospres are not sensitive to UV radiation because there is no DNA replication in the dormant period.

The microbicidal activity of ultraviolet (UV) light depends on the length of exposure: The longer the exposure the greater the microbicidal activity. An important consideration when using UV light is that it has very poor penetrating power. Because of this disadvantage, only microorganisms on the surface of a material that are exposed directly to the radiation are susceptible to destruction. UV light can also damage the eyes, cause burns, and cause mutation in cells of the skin.

5. Filtration   Filtration is an excellent way to disinfect or sterilize heat – sensitive materials, like solutions of enzymes and vitamins or blood products, sometimes can replace Pasteurization in beer industry. Rather than directly destroying contaminating microorganisms, the filter simply removes them. Porous membranes, made of cellulose acetate or cellulose nitrate are supplied. Although a wide variety of pore sizes are available (0.025 ~ 25μm), membranes with pores about 0.22μm in diameter are most commonly used.

Apparatus and Materials

1. Specimens   24h slant culture and broth culture of *Bacillus subtilis*, 20h slant culture and broth culture of *Escherichia coli*.

2. Cultures and reagents   Nutrient broth in test tubes, nutrient agar plates.

3. Apparatus   Water bath, drying oven, UV lamp, incubator.

4. Others   Inoculation loop, sterile droppers.

Methods and Procedures

1. Autoclaving (prepared media)

(1) Add enough water into the autoclave.

(2) Put the media into the chamber. Note: Do not pack too tightly. Lock the door and press the start button to begin sterilization.

(3) Open the release valve until the cool air is completely exhausted then close it. Steam enters into a jacket surrounding the chamber.

(4) The temperature in the chamber will go up to 121℃. At this point, the timer begins to count down 15 to 30 minutes depending on the type of media. After sterilization, autoclaved liquids must be cooled slowly to avoid splashing when the pressure is released.

(5) Verify the sterilizing effectiveness according to the principles in this experiment.

2. Pasteurization

(1) Inoculate *Bacillus subtilis* and *Escherichia coli* to nutrient broth tubes and take an un – inoculated tube as a blank control.

(2) Put these 3 tubes in a water bath at 63℃ for 30min for disinfection.

(3) After treatment, transfer the tubes to incubator at 37℃ for 24h and then observe the bacterial growth.

3. UV disinfection

(1) Warm up the UV lamp for 10min.

(2) Using sterile droppers add 3 drops of *Bacillus subtilis* and *Escherichia coli* broth cultures to the surfaces of 2 agar plates. Spread uniformly with a sterile swab to make a lawn. All these need to be done with aseptic techniques including the following steps.

(3) Sterilize the forceps and pick up a sterile brown paper in star shape. Put the brown paper on the center of the agar surfaces and press the edge.

(4) Remove the lid and expose the plate to UV light for 30min at 20cm distance to the lamp. Always remind yourself not to expose your skin or eyes to the UV light directly.

(5) After exposure, take out the brown paper with sterilized forceps and recover the plate. Incubate at 37℃ for 24h.

(6) Record the disinfection results and observe the growth of bacteria.

4. Sterilization with oven

(1) Put the items (petri plates, tubes and droppers) wrapped with newspapers into an electric oven. Note: Do not pack too tightly.

(2) Switch on the oven and time 2 hours until the temperature increases to 160~170℃.

(3) Take out the items until the temperature decreases to 70℃.

Experiment contents

1. Sterilize media with autoclave and verify the sterilizing effectiveness.

2. Examine the disinfection ability of Pasteurization.

3. Examine the disinfection effect and characteristics of UV radiation.

Results

1. Observe the incubation results in the test tubes (clear or turbid) and compare the disinfection effects of Pasteurization on *Bacillus subtilis* and *Escherichia coli*.

2. Observe the growth of bacteria under and outside the brown paper. Compare the disinfection effects of UV radiation on *Bacillus subtilis* and *Escherichia coli*. Verify the characteristics of UV radiation.

Notes

1. When autoclaving:

(1) Flush all air out of the chamber.

(2) The chamber should not be packed too tightly.

(3) Start to time when it reaches the target temperature.

(4) When autoclaving is finished, cool down naturally. Do not open the release valve immediately to avoid eruption of media.

2. When dry heating:

(1) The chamber should not be packed too tightly.

(2) Items should not be placed on the oven floor to keep wrapping paper from scorching.

(3) Operating temperature should not exceed 180℃ or the cotton or paper will scorch.

(4) Open the oven until temperature falls to around 70℃.

3. When UV radiation is performed:

(1) Avoid photo-reactivation.

(2) Protect eyes and skin from exposure to UV light directly.

Questions

1. What are the precautions for dry heat sterilization?

2. How to obtain suspension of endospore for *Bacillus subtilis*?

3. Why the autoclave can only be opened when the pressure inside falls to normal pressure after sterilization?

（马晓楠）

第四节　微生物的接种技术

在自然界中，各种微生物一般都是混杂生活在一起，即使取很少量的样品也是许多微生物共存的群体。为了从混杂的样品中获得所需要微生物纯种，或是在实验室中把受污染的菌种重新纯化，都离不开分离纯化技术、微生物接种技术和培养技术。

在从事微生物工作中，我们必须选择某些方法将纯培养的微生物（培养物）转移到另外的新鲜培养基中，同时要避免环境的污染。这种防止污染的技术被称为无菌技术。

无菌操作技术和微生物接种技术是微生物分离、纯化和培养的重要保证。

Section 4　Microbial Inoculation Technology

In natural environments, microorganisms usually exist as mixed populations, even for a small amount of samples. In order to get the pure microbes from the mixed samples, or purify the contaminated bacteria in the laboratory, the purification, inoculation and cultivation technology are all necessary.

In working with microorganisms, we must have a method of transferring growing organisms (called the inoculum) from a pure culture to a sterile medium without introducing any unwanted outside contaminants. This method of preventing unwanted microorganisms from gaining access is termed aseptic technique.

Aseptic technique and microbial inoculation technology are the basic guarantee for isolation, purification and cultivation of microbes.

实验十六　微生物的接种技术

【目的】

1. 掌握无菌操作技术在微生物接种过程中的重要性。

2. 掌握常用的微生物接种方法。

3. 熟悉常用接种工具的使用方法。

【基本原理】

将少量有菌的实验材料或纯的菌种转移到另一培养基上，并使之生长繁殖，这个过程就是接种。经过培养，在培养基中或表面的生长物即为培养物。有关培养基的概念已在第三章，实验14有详细介绍。

由于实验目的、培养基种类及实验器皿等不同，所用的接种方法不同。常用的微生物接种方法包括：斜面接种法、平板接种法、液体接种法和穿刺接种法。

本部分灭菌、无菌和纯培养技术显得尤为重要。接种必须在一个无菌的环境中进行，以免微生物污染。

微生物接种常用的工具如图16-1所示。接种工具由软硬适度，能经受火焰反复灼烧，又易冷却的铂金丝或镍铬丝（或细电炉丝）制成，如末端为环形，则为接种环；末端为针形，则为接种针；末端为铲形，则为接种铲、末端弯曲成约45°角，则为接种钩。

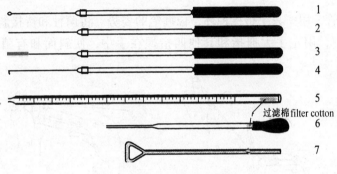

图16-1 接种工具

Figure 16-1 Inoculation Tools

1. 接种环 2. 接种针 3. 接种铲 4. 接种钩 5. 移液管 6. 滴管 7. 涂布棒

1. Inoculating loop 2. Inoculating needle 3. Inoculating shovel 4. Inoculation hook

5. Pipette 6. Dropper 7. Spreader

（1）接种环 是接种用的简单工具。其金属丝的直径要求以0.5mm为宜，长5～7cm，环的内径约2mm。接种环常用于细菌、酵母菌、产生较多孢子的放线菌的接种。

（2）接种针 接种针常用于半固体或明胶直立柱培养基的穿刺接种。

（3）接种钩和接种铲 对于产孢子很少或不产孢子的放线菌和真菌常使用接种钩或接种铲。

（4）涂布棒 涂布棒是将一段长约30cm、直径5～6mm的玻棒或不锈钢的一端烧红后压扁为"了"形，或将玻棒弯曲成"△"形，并使"△"的平面呈30°左右的角度。涂布棒接触平板的一侧，要求平直光滑。涂布棒常用于涂布接种。

此外，滴管和移液管常常用于液体接种法。

【仪器与材料】

（1）菌种 大肠埃希菌斜面培养物、枯草芽孢杆菌斜面培养物、链霉菌斜面培养物、黑曲霉斜面培养物、大肠埃希菌液体培养物、枯草芽孢杆菌液体培养物。

（2）培养基 肉汤琼脂培养基、高氏一号合成琼脂培养基。

(3) 仪器　超净工作台、恒温培养箱。

(4) 其他　接种环、接种钩、乙醇棉、记号笔等。

【方法与步骤】

(一) 斜面接种法

斜面接种是从已长好的菌种斜面或平板上挑取少量菌种移植到另一支新鲜斜面培养基上的一种接种方法。

在菌种扩大培养和菌种保藏时常采用斜面接种方法。斜面接种使用的接种工具为接种环和接种钩。

(1) 准备工作　消毒工作区；洗净双手并用乙醇棉消毒双手。

(2) 标记试管　接种前，在试管上标记培养基名称、菌名、接种人姓名、接种日期、组别等。

(3) 接种

1) 手持试管　用一只手持接种环，像握笔的姿势。将菌种和待接斜面等两支试管放在另一支手中，用左手大拇指和其他四指握在手中，使斜面和有菌种的一面向上 (图16-2)。

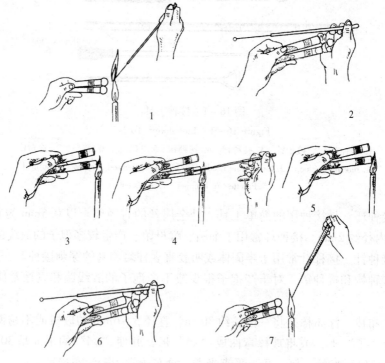

图 16-2　无菌操作斜面接种示意图

Figure 16-2　Inoculating procedures

1. 烧接种环　2. 拔试管帽　3. 试管口过火焰　4. 取菌种、接种

5. 试管口过火焰　6. 加盖试管帽　7. 灭菌接种环

1. Sterilize the inoculating loop　2. Remove the tube cap　3. Flame the mouth of the tube　4. Obtain inoculum

5. Flame the mouth of the tube　6. Replace the tube cap　7. Sterilize the inoculating loop

2) 转动棉塞或是试管帽,以备接种时易拔取。

3) 接种环灭菌　在煤气灯或酒精灯火焰的外焰将接种环一端烧红,然后将其斜持,沿环向上,将能深入试管的金属柄部分全部烧红。注意不要将灭菌后的接种环碰到试管壁,以免被污染,冷却接种环再取菌种。

4) 拔试管帽　用右手的无名指、小指和手掌取下菌种管和待接试管的试管帽,注意不要将试管帽任意放在桌上。

5) 将试管口灭菌,以防空气中的微生物污染接种物。

6) 取菌与接种　在火焰旁迅速将取过菌的接种环伸入待接斜面试管,由底部划"Z"线,直划到斜面上缘,注意不要划破培养基(图16-3)。

图16-3　斜面接种(Z字线)

Figure 16-3　Agar slant inoculation (Z)

7) 再次将试管口灭菌并盖上试管帽　接种完毕将接种环取出,灼烧试管3次;盖上试管帽。注意,盖试管帽时应保持试管不动,以免移动时不洁空气污染试管。

8) 接种环灭菌　接种环经灼烧冷却放回原处。

9) 培养　将接种后的细菌斜面放在37℃培养箱中培养,24h观察生长状况。酵母菌放在28~30℃培养箱中培养,48h后观察生长现象;放线菌和霉菌则放在28~30℃培养箱中培养,3~5d后观察生长现象。

如果是将分离纯化得到的单菌落接种至斜面,应先用灭菌接种环挑取已选好的菌落,注意切勿触及其他菌落。其他同上述方法接种。

(二) 液体接种法

1. 由斜面培养基接入液体培养基

(1) 如接种量小,可按斜面接种法取菌。

1) 取无菌液体培养基试管,用右手小手指开盖,切记不要将试管盖子放在实验台上。

2) 试管口灭菌。

3) 将接种环上的菌体培养物接种到液体培养基中(图16-4)并从试管中抽取接种环,注意不要将接种环放在实验台上。

4）再次灭菌试管口。

5）盖上试管帽。

6）将接种环灭菌放回原处。

7）将接种物放在适当温度下培养观察生长现象。

(2) 如接种量大，可用定量的无菌水注入斜面菌管中，用接种环把菌苔刮下研开，试管口在火焰上灭菌，把菌悬液在火焰旁倒入液体培养基中。盖好棉塞，在适宜条件下培养。

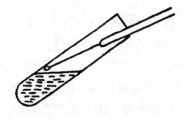

图 16-4　液体接种法

Figure 16-4　Liquid medium inoculation

2. 由液体培养物接种液体培养基

(1) 取移液管：取单支包装的无菌移液管时，可以将移液管的包装纸从中间撕开取出。

(2) 采用无菌操作技术，用右手打开盛菌液试管盖，管口灭菌并吸取所需菌液，再盖上盛菌液的试管帽。

(3) 接种：一只手持待接种的三角瓶或试管，注意勿使瓶口朝上，用另一只手小指和手掌边拔出瓶塞或管帽，瓶口缓慢过火。将移液管伸入三角瓶或试管，慢慢放出菌液，移出移液管，瓶口缓慢过火后，最后塞上瓶塞或试管帽（图 16-5）。

(4) 移液管处理　将用过的移液管放在指定地点，待灭菌处理。

(5) 将瓶口包扎，标明菌名、接种者和接种日期，放入特定温度培养箱培养。

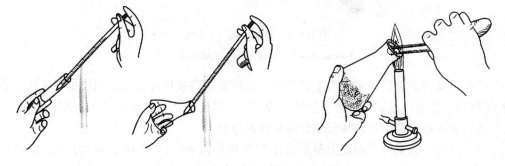

图 16-5　各种液体培养物接种示意图

Figure 16-5　Inoculating procedures

（三）穿刺接种法

此法多用于半固体培养基、明胶培养基等，可作为菌种保藏的一种形式，也可检查细菌运动能力或用作特殊试验等，适宜于细菌和酵母菌的接种培养。

方法：

(1) 穿刺接种法　常常使用接种工具为接种针。获得半固体穿刺培养菌体材料的方法见斜面接种法 (1) ~ (3)。

(2) 接种　迅速将沾有菌种的接种针伸入半固体培养基，由半固体培养基的中央处直刺至接近管底（注意不要穿透），然后沿着原路拔出接种针，试管口过火焰数次灭菌，盖好试管帽。注意接种时且勿搅动，以免接种不整齐而使结果不准确（图 16-6）；

接种针使用后灭菌、冷却以备他用。

（3）贴标签、培养　标明菌名、接种者和接种日期，放入特定温度培养。

（四）点种法

是指用接种针沾取少量霉菌孢子，在琼脂平板上用点种法接种，一般可在一个平板上点3点，经过经培养后形成3个菌落。其优点是在一个平皿上同种菌落有3个重复，因此，它是用于观察霉菌菌落特征的理想接种方法。此外，菌落彼此相近的边缘，菌丝生长稀疏，较透明，还分化出繁殖结构，因此可直接把培养皿放在低倍镜下观察，便于根据形态特征进行菌种的鉴定。

图16-6　穿刺接种法
Figure 16-6　Agar deep inoculation

点种法操作程序：

（1）制备平板　根据菌种选择适宜的培养基。

（2）在皿底贴标签　注明接种者姓名、组别，以及接种日期、菌名等。

（3）接种　将接种针在火焰上灭菌后冷却，再将接种针伸入菌种管，用针尖取少量霉菌孢子，以垂直法或水平法（图16-7）把接种针上沾着的孢子，点接到平板培养基表面相应的位置上。注意在点接时切勿刺破培养基。

（4）培养　将平板倒置于28℃恒温箱中培养，48h后开始观察生长情况（图16-8）。

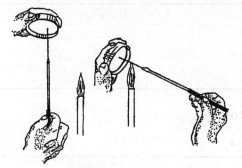

图16-7　点种法
Figure 16-7　Spot inoculation

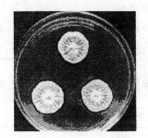

图16-8　点种结果
Figure 16-8　Result of spot inoculation

【实验内容】

1. 接种大肠埃希菌于牛肉膏、蛋白胨琼脂培养基斜面上，24h后观察生长现象。接种链霉菌于高氏一号合成琼脂培养基斜面上，5d后观察生长现象。

2. 用黑曲霉在PDA平板上做点种，并于培养72h后观察生长状况。

3. 将大肠埃希菌和枯草芽孢杆菌分别接种于营养肉汤培养基的试管中，并于培养24h后观察生长状况。

4. 将大肠埃希菌和枯草芽孢杆菌分别接种于营养琼脂半固体培养基中，并于培养24h后观察生长状况。

【结果】

1. 斜面培养 将生长状况填入表 16-1。

表 16-1

菌种	大肠埃希菌	链霉菌
生长（+或-）		
菌染（+或-）		
生长状况描述		

2. 点种 将生长状况填入表 16-2。

表 16-2

菌种	黑曲霉菌落1	黑曲霉菌落2	黑曲霉菌落3
生长（+或-）			
菌染（+或-）			
菌落状况描述			

3. 液体接种 将生长状况填入表 16-3。

表 16-3

菌种	大肠埃希菌	枯草芽孢杆菌
生长（+或-）		
菌染（+或-）		
生长状况描述		

4. 半固体穿刺接种 将生长状况填入表 16-4。

表 16-4

菌种	大肠埃希菌	枯草芽孢杆菌
生长（+或-）		
菌染（+或-）		
生长状况分布图		

【思考题】

1. 接种前和接种后均要烧接种环，目的是什么？
2. 为什么说无菌操作技术很重要？
3. 总结比较各种接种方法的特点。

Experiment 16 Microbial Inoculation Technology

Objectives

1. Grasp the importance of aseptic technique in microbial inoculation process.
2. Be familiar with the usage of inoculation tools.
3. Learn and grasp the common methods for microbial inoculation.

Principles

To cultivate, or culture, microorganisms, one introduces a tiny sample (inoculum) into a container of nutrient medium (pl. media), which provides an environment in which they multiply. This process is called inoculation. The observable growth that later appears in or on the medium is known as a culture. The important concept of media has been covered in more details in Chapter 3, Experiment 14.

The inoculation methods used depend on different experimental purposes, medium types and experimental containers. The inoculation methods commonly used include: agar slant inoculation, plate inoculation, liquid medium inoculation and agar deep inoculation.

Inherent in these practices are the concepts of sterile, aseptic and pure culture techniques. Contamination is a constant problem, so sterile techniques (media, transfer equipment) help ensure that only microbes that come from the sample are present.

The common tools for inoculation are shown in Figure 16-1. Aseptic transfer and inoculation are usually performed with a sterile, heat-resistant, no corroding Nichrome wire attached to an insulated handle. When the end of the wire is bent into a loop, it is called an inoculating loop; when straight, it is an inoculating needle, when the loop is bent at a 45° angle, it is an inoculating hook.

(1) Inoculating loop: Inoculating loop is a simple tool used for inoculation. The diameter of the wire is 0.5mm advisable, 5~7cm long, 2mm inner diameter. Inoculating loop is commonly used as inoculation tool for bacteria, yeasts, and actinomycetes with more spores.

(2) Inoculation needle: Inoculating needle is often used as agar deep inoculation in semi-solid puncture or gelatin upright medium.

(3) Inoculation hook and inoculation shovel: They are commonly used for the inoculation of actinomycetes or fungus, which produce little or no spores.

(4) Spreder: The spreader is a stainless steel or glass rod about 30 cm long, 5~6mm diameter, burn red the end of it and then squash it to the shape of Chinese character "了", or curve the glass rod into a "△" shape, and make the angle of "△" is about 30 degrees. Spreader require straight and smooth flat side to contact with. Spreaders are often used in spreading inoculation.

In addition, the dropper and pipette are often used for liquid inoculation.

Apparatus and Materials

(1) Specimens *Escherichia coli* agar slant cultures, *Bacillus subtilis* agar slant cultures, *Streptomyces* agar slant cultures, *Aspergillums* agar slant culture, *Escherichia coli* broth, *Bacillus subtilis* broth.

(2) Cultures Nutrient agar broth, Gause's, Gao – One synthetic agar medium.

(3) Apparatus Super clean workbench, constant temperature incubator.

(4) Others Inoculating loop, inoculating hook, alcohol cotton, marker, etc.

Methods and Procedures

Ⅰ. Agar slant inoculation

The agar slant inoculation is an inoculation method of picking off a few species from agar slant or plate to another fresh agar slant. It is usually used in large scale culture and culture collection. The tools used in the agar slant inoculation are inoculation loop and inoculation hook.

(1) Preparation: Disinfect your work area; Wash your hands with disinfectant soap and sanitize your hands with alcohol.

(2) Labeling tube : Before inoculation, label one tube of each medium with the name of the media, the name of microbe, your name, the date and your lab section.

(3) Inoculation

1) Hold the inoculating loop in your dominant hand hold as if it were a pencil, and the two agar slants (one is the pure culture slant, the other is uninoculated tube) in the other hand with the marker visible (Figure 16 – 2).

2) Turn the cotton plug or tube cap, so as to be easily pulled out when inoculation.

3) Sterilize your inoculating loop by holding it in the hottest part of the flame (at the edge of the inner blue area) until it is red – hot. Pass the inoculating loop at an angle through the flame until the entire length of the wire becomes orange from the heat. Never lay the loop down once it is sterilized or it may again become contaminated. Allow the loop to cool a few seconds to avoid killing the inoculum.

4) Remove the cap: Remove the cap of the pure culture slant and the microbe slant with the ring finger and little finger of your loop hand. Never lay the cap down or it may become contaminated.

5) Very briefly pass the mouths of the culture tubes through the flame three times before inserting the loop for an inoculum, which can reduce the chance of airborne microorganisms contaminating your cultures.

6) Keeping the culture tube at an angle, insert the inoculating loop and remove a loopful of inoculum, and then withdraw it from the tube. Inoculate the slant by streaking the loop back and forth across the agar surface from the bottom of the slant to the top with "Z" line (Figure 16 –3), being careful not to gouge the agar.

7) Again pass the mouth of the culture tube through the flame three times and replace the

cap. Note that the tube should be fixed when replacing the cap, in order to avoid the unclean air pollution.

8) Flame your loop and let it cool.

9) Incubate the inoculated slants at 37℃ (for bacteria) for 24 ~ 48h. Yeast should be incubated at 28 ~ 30℃ for 48h. Actinomyces and mold should be incubated at 28 ~ 30℃ for 3 ~ 5days.

If some isolated single colonies were chosen to be inoculated, you had better select the right colony, being careful not to touch the other colonies.

II. Liquid medium inoculation

1. Transferring the slant culture into a broth tube

(1) With less inoculation amount, you can get a loopful slant culture as the Agar slant inoculation method.

1) Pick up the sterile broth tube and remove the cap with the little finger of your loop hand. Do not set the cap down.

2) Brieflypass the lip of the broth tube through the flame.

3) Place a loopful of inoculum into the broth (Figure 16 - 4), and withdraw the loop. Do not lay the loop down!

4) Again pass the lip of the culture tube through the flame.

5) Replace the cap.

6) Resterilize the loop by placing it in the flame until it is hot - red. Now you may lay the loop down until it is needed again.

7) Incubate all cultures at 37℃.

(2) With more inoculation amount, you can get the inoculum as follows: Inject quantitative sterile water into the agar slant tube. Scrape the slant inoculum. Flame the mouth of the tube. Pour the bacteria suspension into the liquid medium beside flame. Replace the cap and incubate at appropriate temperature.

2. Transferring microorganisms into a broth tube

(1) Take a single packaging of sterilized pipette, tear off the middle part of the wrapping and remove the pipette.

(2) Remove the cap of the broth culture of some microbe with the little finger of your right hand, flame the mouth of the tube. Never lay the cap down or it may become contaminated. Aseptically transfer the broth culture of some microbe. Recap the tube by turning the tube into the cap.

(3) Inoculation Keeping the sterile culture tube or flask in your left hand at an angle. Remove the cap of the tube or flask with the little finger of your right hand. Flame the mouth of the tube or flask. Never lay the cap down or it may become contaminated. Stretch the pipette into the tube or flask; release the broth culture of some microbe slowly. Again pass the lip of the tube or flask through the flame, replace the cap (Figure 16 - 5)

(4) Put the pipette in the designated place to be sterilized.

(5) Wrap the mouth of the bottle. Label with the name of the culture, the name of microbe, your name, the date. Incubate at appropriate temperature.

III. Stab inoculation

This method is usually used for semi – solid medium, gelatin medium. It can also be one form of the preservation of pure cultures. Sometimes, this method can be used to determine the motility of bacteria and to localize a reaction at a speific site. Agar deep inoculation is suitable for the cultivation of bacteria and yeast.

Method:

(1) Obtain a nutrient semisolid agar deep, and using your inoculating needle, repeat steps I (1) ~ (3).

(2) Inoculate the semisolid agar deep by plunging the needle straight down the middle of the deep and then pulling it out through the same stab, as shown in Figure 16 – 6. Flame the mouth of the tube and replace the cap. Do not stir during inoculation to avoid irregular inaccurate results. Flame the needle and let it cool.

(3) Label with the name of the culture, the name of microbe, your name, the date and your lab section. Incubate at appropriate temperature.

IV. Spot Inoculation

Touch some spores with inoculation needle and then inoculate on the surface of the agar plate, usually 3 points each plate. By incubation for sometime, three colonies will form. It is used to be an ideal inoculation method for observing the colony characteristic, with its advantage that the same colony has three repetitions. In addition, thin and transparent mycelia form at the edge of the colony close to each other, sometimes, reproductive structures also can be differentiated. So, you can observe and identify the species by putting the petri dishes directly under low magnification.

Procedures for Bacterial Spot Inoculation:

(1) Prepare petri plate: Choose the appropriate medium according to the species.

(2) Label the bottom of the nutrient agar plate with your name and lab section, the date, and the source of the inoculum.

(3) Obtain a nutrient inoculum, and using your inoculating needle, repeat steps I(1) ~ (3). Inoculate the petri plate by gently touching some spores on the surface of the agar plate, in a vertical or horizontal method, usually 3 points each plate. As shown in Figure 16 – 7, flame the needle and let it cool. Be careful not to gouge the agar.

(4) Incubation Incubate the plates upside down (lid on the bottom) at 28℃, to prevent condensing water from falling down on the growing colonies and causing them to run together. Observe the growth after 48h, as shown in Figure 16 – 8:

Experiment contents

1. Inoculate *E. coli* on nutrient agar slant observe and describe the resulting growth on the slant after 24h of incubation. Inoculate *Streptomyces synthesis* on actinomycetes culture medium, observe and describe the resulting growth on the slant after five days of incubation.

2. Inoculate *Aspergillus niger* on PDA plate by spot inoculation, observe and describe the resulting growth on the slant after 72h of incubation.

3. Inoculate *E. coli* and *Bacillus subtilis* into the nutrient broth medium respectively, observe and describe the resulting growth on the slant after 24h of incubation.

4. Inoculate *E. coli* and *Bacillus subtilis* into semisolid agar medium; observe and describe the resulting growth on the slant after 24h of incubation.

Results

1. Nutrient Agar Slant   Record your results in Table 16 – 1.

Table 16 – 1

Specimens	*Escherichia coli*	*Bacillus subtilis*
Growth (+ or –)		
Contamination (+ or –)		
Pattern of growth		

2. Spot Inoculation   Record your results in Table 16 – 2.

Table 16 – 2

Specimens	*Aspergillus niger* 1	*Aspergillus niger* 2	*Aspergillus niger* 3
Growth (+ or –)			
Contamination (+ or –)			
Pattern of colony			

3. Liquid medium inoculation   Record your results in Table 16 – 3.

Table 16 – 3

Specimens	*Escherichia coli*	*Bacillus subtilis*
Grow (+ or –)		
Contamination (+ or –)		
Pattern of growth		

4. Agar deep inoculation   Record your results in Table 16 – 4.

Table 16-4

Specimens	*Escherichia coli*	*Bacillus subtilis*
Grow (+ or -)		
Contamination (+ or -)		
Growth pattern		

Question

1. What is the purpose of flaming the loop before and after use?
2. Why is aseptic technique important?
3. Summarize and compare characteristics of different inoculation methods.

第五节 微生物的纯种分离与培养技术

通常情况下，微生物以杂居状态存在。因此要对某一微生物进行研究，必须将其与其他微生物分离，进行纯培养，纯培养即由单一微生物生长得到的。纯种分离技术是微生物实验和科学研究中常用的技术。

混合样品的分离和纯化的过程基本包括两大步骤，使菌体浓度得到稀释，使微生物的细胞或孢子以单独的状态存在，在适宜的条件下形成菌落。如果将其接种到适当的培养基上就得到了纯种的微生物。

分离纯化的方法包括两类：一类是达到菌落纯水平，另一类是达到细胞纯水平。

平板划线法、平板涂布法和倾注分离法因方法简便、设备简单、分离效果良好，所以被一般实验室广泛采用；分离单细胞以达到菌株纯化的方法在微生物遗传学研究中十分重要，但通常设备要求较高。本章重点介绍菌落纯的纯种分离方法。

Section 5 Isolation and Cultivation of Pure Cultures

Microorganisms exist in nature as mixed populations. However, to study microorganisms in the laboratory we must have them in the form of a pure culture, that is, one in which all organisms are descendants of the same organism. The technology of the isolation of pure cultures is often used in the microbial experiment and scientific research.

Two major steps are involved in obtaining pure cultures from a mixed population: First, the mixture must be diluted until the various individual microorganisms become separated far enough apart on an agar surface that after incubation they form visible colonies isolated from the colonies of other microorganisms.

There are two types of isolation methods: One is the level of pure colonies, and the other is a pure cellular level.

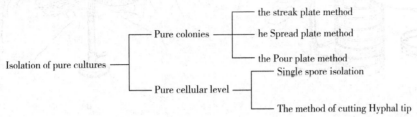

The streak plate method, the spread plate method and the pour plate method are widely used in the laboratory because of their simple methods, easy processing and good separation effects. The method of separating single cell is very important in microbial genetics research, but the equipment requirement is usually high. Therefore, this chapter mainly focuses on the level of pure colonies.

实验十七　微生物的平板划线分离法

【目的】
1. 掌握平板划线法的操作方法。
2. 了解平板划线法分离菌种的基本原理。

【基本原理】
常见的微生物分离方法是划线分离法。

当通过划线法分离微生物样品时，培养物在固体培养基表面通过划线被稀释。划线的目的是使细菌细胞间的空间足够大，这样单个细胞繁殖所产生的大量后代就会形成一个菌落。由此可见，菌落就是在固体培养基表面由单个微生物繁殖形成的肉眼可见的细胞集团，菌落也可以是由一群同类的微生物繁殖形成，用菌落形成单位表示。

挑取单菌落进行扩大培养，即成为纯培养。

【仪器与材料】
1. **菌种**　大肠埃希菌（*Escherichia coli*）和金黄色葡萄球菌（*Staphylococcus aureus*）混合液。
2. **培养基与试剂**　营养肉汤琼脂培养基。
3. **仪器**　超净工作台、恒温培养箱。
4. **其他**　接种环、无菌培养皿、接种钩、乙醇棉、记号笔等。

【方法与步骤】
1. **融化培养基**　将装有肉汤琼脂培养基的三角瓶加热至充分融化。

2. 倒无菌琼脂平板 将培养基冷却至50℃左右后,采用无菌操作法倒平板(每皿约倒15~20ml)(图17-1),水平静置待凝,即培养平板。

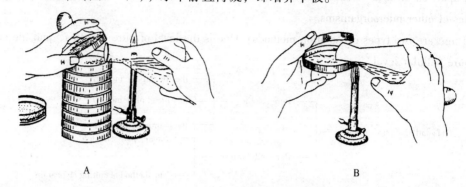

图 17-1 倒平板

Figure 17-1 Pour the plates

倒培养基时右手拿盛有培养基的三角烧瓶底部,左手持培养皿,小拇指与手掌之间夹住棉塞。在火焰旁拔取棉塞,瓶口通过火焰,此时稍打开皿盖(以容纳瓶口为限)。将瓶口放入皿内稍冷却后,将培养基倒入,盖好皿盖,水平静置、待凝。

3. 划线方法

A. 分区划线法

(1) 标记平皿 在平皿底部标记实验人姓名、组别、接种日期、所接种菌名等。

做分区标记 在皿底用记号笔划分成4个不同面积的区域,使D>C>B>A,且各区的夹角为120℃。

(2) 挑取混合样品 接种环灭菌、稍冷却,按无菌操作法沾取少量待分离的混菌液。

(3) 划线操作方式 划线操作可以手持平皿(图17-2A)或是在工作台上(图17-2B)操作。

①划A区:用左手微微打开平皿盖,右手持含菌的接种环(以握笔的方式),先在平板的A区轻柔地划4~6条平行线(图17-3),注意保持接种环与培养基平面平行,不要划破培养基,划线部分不要重合。

②划B区:烧掉接种环剩余的菌种,将灭菌的接种环在平板培养基边缘冷却,然后将平板转动一定角度,使B区被转至顶部位置,这样保证你每次都在"12:00"的位置操作。将接种环通过A区转至B区,在B区划6~7条致密的平行线。

③烧掉接种环剩余的菌种,将接种环在平板培养基边缘冷却,然后将平板转动一定角度,使C区被转至顶部位置,将接种环通过B区转C区,在C区划线。

④烧掉接种环剩余的菌种,将接种环通过C区转D区,在D区划线。注意划线时不要与前面区域的线条重合,烧去接种环的剩余菌液并将其放回原处。

⑤将划线平皿至于37℃培养箱,倒置培养24~48h,观察分离结果。

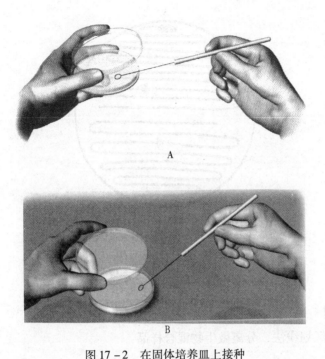

图 17-2　在固体培养皿上接种

Figure 17-2　Inoculation of a solid medium in a Petri plate

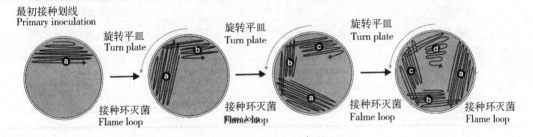

图 17-3　分区划线示意图

Figure 17-3　Streak plate technique for isolation of bacteria

B. 连续划线法

（1）标记平皿　在平皿底部标记实验人姓名、组别、接种日期、所接种菌名等。

（2）挑取混合菌样品　接种环灭菌、稍冷却，按无菌操作法沾取少量待分离的混菌液。

（3）划线　从平皿顶端开始，反复在平板上划线（不要重复），直划到平板中部，再将平板倒转方向，同样从平板另一端开始直划到中央为止（图 17-4）。

（4）将划线平皿至于 37℃ 培养箱，倒置培养 24~48h，观察分离结果。

4. 接种纯培养　从典型的单菌落中，挑取少量菌体至适当的斜面培养基上，经培养即为初步分离的纯种。

5. 培养　37℃，倒置培养 24~48h，获得纯培养。

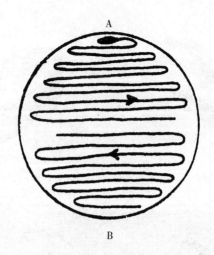

图 17-4 连续划线
Figure 17-4 Continuous streaking

【实验内容】

利用平板分区划线法，分离微生物混合样品。

【结果】

记录平板分区划线分离结果于下表，描述菌落特征。

菌落特征	大肠埃希菌	金黄色葡萄球菌
菌落大小		
菌落表面特征		
菌落形状		
菌落边缘特征		
菌落颜色		

【注意事项】

1. 在划线时，左手持培养皿底，不要过度打开平皿盖，以免空气中杂菌落入。

2. 划线前和划线后，一定要灭菌并冷却接种环。每完成一个区域的划线，一定要旋转平皿，保证每次都在时钟12点的位置划线操作，这样做能够保证接种环与培养基平面平行，而且不易划破培养基。

【思考题】

1. 在平板划线法中，怎样判断一个菌落是分离所得还是污染的菌落？

2. 如何防止平板被划破？

3. 采用分区划线法进行分离时应注意什么？为什么每次都需将接种环上的剩余物烧掉？

4. 了解下列各名词的含义：纯培养、无菌培养基、培养物、无菌技术、菌落。

Experiment 17 Isolation of Pure Culture-Streak Plate Technique

Objective
1. Understand the principle of streak plate technique.
2. Grasp the procedure of streak plate method.

Principle
The most common way of separating bacterial cells on the agar surface to obtain isolated colonies is the streak plate method.

When streaking for isolated colonies, the inoculum is diluted by streaking it across the surface of the agar plate. The purpose is to get the inoculum diluted to the point where there is only one bacteria cell deposited on the surface of the agar plate. When these lone bacteria cells divide and give rise to thousands of spring, an isolated colony is formed. Therefore, a colony is a population of cells that arises from a single bacterial cell. A colony may arise from a group of the same microbes attached to one another, which is therefore called a colony – forming unit (CFU).

Then, an isolated colony can be aseptically "picked off" the isolation plate and transferred to new sterile medium. After incubation, all organisms in the new culture will be descendants of the same organism, that is, a pure culture.

Apparatus and Materials
1. Specimens Mixed broth culture of *E. coli* and *S. aureus*.
2. Cultures Nutrient agar media.
3. Apparatus Clean bench, 37℃ and 28℃ incubator.
4. Others Inoculating loop, petri dishes, inoculating hook, alcohol cotton, marking pen and so on.

Methods and Procedures
1. Heat the flask to make the nutrient agar media melt.
2. Prepare nutrient agar plates Pour the plate after the media is cooled to about 50℃, pour it onto the plates with aseptic technique (15~20ml each plate) (Figure 17 – 1). Waiting to cool down, that is culture plate.

When pouring the medium, hold the bottom of culture flask with the right hand, hold a petri dish with the left hand, and catch tampon between little finger and the palm of hand. Drawing off tampon beside the flame, flame the mouth of the flask, and then open dish lid slightly. Put the flask into the plate, pour the medium after cooling slightly, cover dish lid, and let the plate stand horizontally.

3. Scraping line method

A. Streak plate method

(1) Label the bottom of one nutrient agar plate with your name and lab section, the date, and the source of the inoculum.

Label the bottom of the plate with four different areas, D > C > B > A, ensuring that the adjacent area is 120 degrees.

(2) Flame the inoculating loop to redness, allow it to cool, and aseptically obtain a loopful of the mixed broth culture.

(3) The streaking procedure may be done with the Petri plate on the workbench (Figure 17-2A) or held in your hand (Figure 17-2B).

a. To streak a plate (Figure 17-3), lift one edge of the Petri plate cover, and streak the first sector by making 4 to 6 lines without overlapping previous streaks. Keep the inoculating loop parallel with the agar; do not gouge the agar while streaking the plate. Hold the loop as you would hold a pencil or paintbrush, and gently touch the surface of the agar.

b. Flame your loop and let it cool. Turn the plate so the next sector is on the top, so you are always streaking the agar in the "12:00" position. Streak through one area of the first sector (area A), and then streak a few times (area B) away from the first sector.

c. Flame your loop and let it cool. Turn the plate again, and streak through one area of the second sector. Then streak the third sector (area C).

d. Flame your loop, streak through one area of the third sector, and then streak the remaining area of the agar surface (area D), being careful not to touch any of the areas already streaked. Flame your loop before setting it down.

e. Incubate the plate in an inverted position at 37℃ until discrete, isolated colonies develop (usually 24~48 hours).

B. Continuous scribing

(1) Label the bottom of one nutrient agar plate with your name and lab section, the date, and the source of the inoculum.

(2) Flame the inoculating loop to redness, allow it to cool, and aseptically obtain a loopful of the mixed broth culture.

(3) Streak over the surface of the medium, from the top section to the middle part (Figure 17-4). Do not overlap the previous streaks. Turn the plate in reverse direction, then scribe from the other end to the middle part in the same way.

(4) Incubate the plate in an inverted position at 37℃ until discrete, isolated colonies develop (usually 24~48 hours).

4. Pick off the isolation plate and transfer to new sterile nutrient slant.

5. Incubate the plate in an inverted position at 37℃ for 24~48 hours. After incubation, all organisms in the new culture will be descendants of the same organism, that is, a pure culture.

Experiment contents

Using the streak plate method of isolation, obtain isolated colonies from a mixture of microorganisms.

Results

Fill in the following table using colonies from the most isolated areas.

Colony characteristics	E. coli	S. aureus
Size		
Surface		
Colony form		
Margin		
Color		

Notes

1. When streaking, hold the bottom of the plate with left hand, lift the edge of the lid just enough to insert the loop.

2. Each time you flame and cool the loop between sectors, rotate the plate counterclockwise so you are always working in the "12:00 position" of the plate. This keeps the inoculating loop parallel with the agar surface and helps prevent the loop from digging into the agar.

Questions

1. How would you determine whether a colony was a contaminant on a streak plate?

2. How to prevent a plate cut?

3. What should be pay attention to when streaking on the plate? Why must the loop be sterilized after streaking?

4. Define the following terms: pure culture, sterile medium, inoculum, aseptic technique, and colony.

实验十八 涂布平板和倾注平板分离法

【目的】

1. 掌握用涂布平板法和倾注平板法分离微生物纯种的操作方法。
2. 了解用涂布平板法和倾注平板法分离微生物纯种的原理。

【基本原理】

涂布平板法和倾注平板法也是常用的菌种分离纯化方法，它们不仅可以用于微生物的分离纯化，还可用于微生物计数等。

涂布平板法　将经过适当稀释的一定体积的菌液加到已凝固的平板培养基表面，然后用无菌涂布棒迅速将其涂布均匀，如果涂布适宜，经培养后，可在平板表面得到单菌落，从而达到分离纯化的目的。涂布培养法还可用于微生物平板菌落计数。

倾注平板法　将待分离培养的菌液经过适当稀释后，取合适稀释度的少量菌液加到无菌平皿中，然后加入融化并冷却至50℃左右的培养基中，充分混匀，待凝固，在适宜的温度下培养一段时间，如果稀释得当，经过培养后可以从平板表面或内部长出单菌落，从而达到分离纯化的目的。倾注接种法也常用于微生物菌落计数，如水或牛奶等的活菌数测定。

在上面两种方法中，涂布平板法比较适合于好氧的细菌和有气生菌丝的放线菌的分离；而倾注平板法较适合兼性厌氧的细菌和酵母菌的分离。

【仪器与材料】

1. 菌种 大肠埃希菌（*Escherichia coli*）和金黄色葡萄球菌（*Staphylococcus aureus*）混合液。

2. 培养基与试剂 肉汤琼脂培养基。

3. 仪器 超净工作台、恒温培养箱。

4. 其他 生理盐水、接种环、乙醇棉、记号笔、无菌试管、无菌平皿、无菌涂布棒、无菌移液管等。

【方法与步骤】

（一）涂布平板法

1. 培养皿编号 取9只无菌培养皿，分别编号10^{-3}、10^{-4}、10^{-5} 3种稀释度，各3皿。

2. 制备无菌平板 将融化并冷却至50℃左右的肉汤琼脂培养基倒入无菌平皿中，每皿倒15~20ml，共计9个平皿，平置待凝。

3. 稀释菌液 取5支无菌试管，依次编号为10^{-1}~10^{-5}。以无菌操作法，在各管中分别加生理盐水4.5ml。用1ml无菌移液管在待稀释的原始菌液中，精确移取0.5ml至10^{-1}的试管中（注意移液管的尖端不要触及10^{-1}溶液的液面），混匀。另取1ml无菌移液管，以同样的方式，在10^{-1}试管中反复吹吸数次，并精确移取0.5ml待测菌液至10^{-2}的试管中，如此稀释至10^{-5}试管（图18-1）。

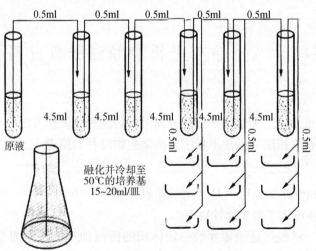

图18-1 涂布分离法

Figure 18-1 Spread plate technique

4. 加菌液并涂布均匀　分别用 1ml 无菌移液管精确吸取 10^{-3}、10^{-4}、10^{-5} 稀释液各 0.5ml，加至相应编号的无菌平板表面（图 18-2）。用无菌涂布棒在平板表面轻轻涂布，使菌液均匀铺满整个平皿表面（图 18-3）。

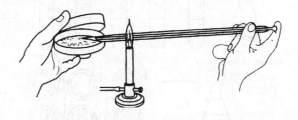

图 18-2　加菌液方法
Figure 18-2　Bacterium suspension adding method

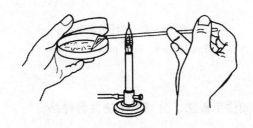

图 18-3　涂布方法
Figure 18-3　Spreading method

5. 培养　将培养皿倒置于 37℃ 恒温培养箱中培养 24h，观察结果。

6. 接种纯培养　从典型的单菌落中，挑取大肠埃希菌和金黄色葡萄球菌的单菌落至适当的肉汤斜面培养基上，经培养即为初步分离的纯种。

7. 培养　将培养皿倒置于 37℃ 恒温培养箱中培养 24h，观察结果。

8. 接种纯培养　挑取单菌落，用灭菌后的接种环分别挑取大肠埃希菌和金黄色葡萄球菌的单菌落至肉汤琼脂试管斜面，经培养后保存。

（二）倾注平板法

1. 培养皿编号　取 9 只无菌培养皿，分别编号 10^{-3}、10^{-4}、10^{-5} 等 3 种稀释度各 3 皿，并在平皿底部标记实验人姓名、组别、接种日期、所接种菌名等。

2. 稀释菌液　同涂布平板法。

3. 吸取菌液　分别用 1ml 无菌移液管精确吸取 10^{-3}、10^{-4}、10^{-5} 稀释液各 0.5ml，加至相应编号的无菌培养皿中。

4. 倾注培养基　向各培养皿中加入融化并冷却至 50℃ 左右的琼脂培养基约 15ml，轻轻摇转，静置，使凝成平板（图 18-4）。

5. 培养　将培养皿倒置于 37℃ 恒温培养箱中培养，24~48h，观察结果。

6. 接种纯培养　从典型的单菌落中，挑取大肠埃希菌和金黄色葡萄球菌的单菌落至适当的肉汤斜面培养基上，获得纯培养。

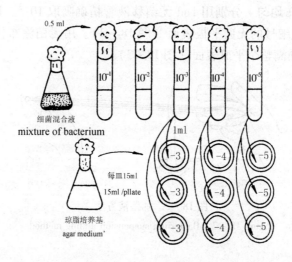

图 18-4 倾注分离法
Figure 18-4 Pour plate technique

【实验内容】

利用平板涂布法和倾注平板法，分离微生物混合样品。

【结果】

将平板涂布法和倾注平板法的结果记录在下表中。

菌落分离结果		涂布平板法			倾注平板法		
		10^{-3}	10^{-4}	10^{-5}	10^{-3}	10^{-4}	10^{-5}
数量/皿	金黄色葡萄球菌						
	大肠埃希菌						
形态特征	金黄色葡萄球菌						
	大肠埃希菌						

【注意事项】

在做涂布分离法实验时，倒平板时培养基温度不宜过高，否则易在平皿表面形成冷凝水，影响菌落分离效果。

【思考题】

1. 为从一个微生物混合材料中获得纯培养物，请设计两种实验程序。
2. 经分区划线可以分离到的微生物，采用倾注分离法却无法分离，为什么？
3. 在倾注平板法中，不同层次上的菌落特征有无区别？为什么？

Experiment 18 Spread Plate Technique and Pour Plate Technique

Objectives

1. Understand the principle of spread plate technique and the pour plate technique.

2. Grasp the procedures for spread plate technique and the pour plate technique.

Principles

Spread plate method and pour plate method are commonly used as isolating and purifying methods, which can be used for separation and purification and microbial counting.

Spread plate method (or spin plate method): Pure culture can also be obtained from spread plate. Small volumes of diluted microbial mixture are pipetted onto the surface of agar plates. A sterile, bent-glass rod is then used to spread the bacteria evenly over the entire agar surface. If the dilution is proper, the dispersed cells develop into isolated colonies. The Spread plate technique can also be used as microbial colony counting.

Pour plate method: The bacteria are mixed with melted agar until evenly distributed and separated throughout the liquid. After incubation, discrete bacterial colonies can then be found growing both on the agar and in the agar. You can use this technique as part of the plate count method of water or milk.

In the above two methods, the spread plate technique is more suitable for the separation of aerobic bacteria and actinomycetes. The pour plate technique is used for separating facultative anaerobic bacteria and yeast.

Apparatus and Materials

1. Specimens Mixed broth culture of *E. coli* and *S. aureus*.

2. Cultures Nutrient agar media.

3. Apparatus Clean bench, 37℃ and 28℃ incubator.

4. Others Saline (0.85%), inoculating loop, alcohol cotton, marking pen, sterile tube, sterile plate, sterile spreader, sterile pipette, petri dishes, inoculating hook, and so on.

Methods and Procedure

Ⅰ. Spread plate technique

1. Label the Petri dish Prepare 9 sterile Petri dishes, label three sets of sterile plate with the dilution (10^{-3}, 10^{-4} and 10^{-5})

2. Prepare the sterile plate Melt the nutrient agar media and cool to 50~55℃, add 15~20ml of it to each plate and curdle.

3. Prepare 5 sterile tube with 4.5ml of sterile saline solution, label them with the dilution (10^{-1}~10^{-5}).

Remove a pipette, attach a bulb, and aseptically transfer 0.5ml of the original broth inoculum to dilution tube number 1. Mix well.

Using a different pipette, transfer 0.5ml from dilution number 1 to dilution tube number 2. Mix well. Do the same way until to 10^{-5} (procedure showns in Figure 18-1).

4. With a 1ml pipette, transfer 0.5ml of diluted solution of 10^{-3}, 10^{-4} and 10^{-5} to three sets of sterile plates respectively (Figure 18-2) and spread (Figure 18-3).

5. Incubation Incubate all plates in an inverted position at 37℃ for 24h.

6. Pick off the isolated colonies of *E. coli* and *S. aureus* and transfer to new sterile nutrient agar slant. Incubate at 37℃ for 24~48 hours. After incubation, all organisms in the new culture will be descendants of the same organism, that is, a pure culture.

II. Pour plate technique

1. Label the Petri dish Prepare 9 sterile Petri dishes, label three sets of empty, sterile plate with the dilution (10^{-3}, 10^{-4} and 10^{-5}).

Label the bottom of one nutrient agar plate with your name and lab section, the date, and the source of the inoculum.

2. Serial dilutions: made by the same way as the spread plate technique.

3. With a 1ml pipette, transfer 0.5ml of diluted solution of 10^{-3}, 10^{-4} and 10^{-5} to three sets of sterile petri dish respectively.

4. Pour 15~20ml of agar medium which has been melt and then cooled to 50℃ onto the Petri dish with different concentration of serial dilutions. Mix well and let the agar harden in the plates (Figure 18-4).

5. Incubation Incubate all plates in an inverted position at 37℃ for 24h.

6. Pick off a single isolated colony of each of the two organisms in your original mixture and aseptically transfer them to new sterile nutrient slant. Incubate at 37℃ for 24~48h.

Experiment content

Separate the mixed sample using the spread plate and pour plate technique.

Results

Fill in the following table using colonies from the most isolated areas.

Separation results		The spread plate technique			The pour plate technique		
		10^{-3}	10^{-4}	10^{-5}	10^{-3}	10^{-4}	10^{-5}
Quantity/ plate	*E. coli*						
	Staphylococcus aureus						
Morphological characters	*E. coli*						
	Staphylococcus aureus						

Note

With the Spread plate method, high medium temperature is not suitable, as it is easy to form condensate on the surface of medium, which will bring bad isolation effects.

Question

1. Design two different practical procedures to obtain pure culture from mixed sample.

2. Could some bacteria grow on the streak plate and not be seen if the pour plate technique is used? Explain.

3. How do the colonies on the surface of the pour plate differ from those suspended in the agar?

实验十九　厌氧微生物的培养技术

【目的】

熟悉厌氧微生物培养的原理和方法。

【基本原理】

氧气分子的存在与否，对微生物的生长至关重要。根据微生物与氧的关系，可将微生物分为专性好氧菌、专性厌氧菌、耐氧菌、微好氧菌和兼性厌氧菌。

专性好氧微生物，在有氧气的条件下生长，最典型的代表是产碱杆菌。专性厌氧微生物需要在无氧气条件下生长，分子氧对其有剧毒（如，巴氏梭菌）；耐氧微生物不能利用氧气但是能耐受氧气；微好氧微生物，如弯曲菌属，其生长需要一定量的 CO_2 和少量的氧气（5% ~10%）；多数微生物在有氧气和无氧气的条件下均能生长，这些微生物被称为兼性厌氧菌，大肠埃希菌是典型的代表。

细菌在有氧代谢过程中，常产生具有强烈杀菌作用的超氧阴离子和过氧化氢等毒性代谢产物，多数微生物细胞具有分解毒性氧的酶。专性好氧和兼性厌氧微生物体内有超氧化物歧化酶和过氧化氢酶或触酶。前者能分解超氧阴离子为过氧化氢，后者可继续分解过氧化物。厌氧微生物是一类在其细胞呼吸过程中不能以氧气作为末端电子受体的微生物，专性厌氧菌缺少这些酶，故在有氧气时不能生长。

在实验室，我们一般可以通过去除培养环境中的游离氧气或使用还原培养基培养厌氧微生物。实际上，常常是两种方法结合培养厌氧菌。

厌氧微生物的常见培养方法见图 19 -1。

本实验主要介绍厌氧罐培养法和还原性培养基培养法。

图 19 -2 为 GasPak 厌氧罐的结构示意图和建立厌氧环境的原理。

【仪器与材料】

1. 菌种　培养 24h 的蜡状芽孢杆菌、大肠埃希菌营养肉汤培养物、培养 48h 的生孢梭菌营养肉汤培养物。

2. 培养基　营养琼脂培养基、硫代乙醇酸盐培养基。

3. 仪器　GasPak 厌氧培养罐、GasPak 孵育袋。

4. 其他　产气袋、厌氧指示剂、催化剂（钯粒）、接种环、试管架、乙醇棉、记号笔等。

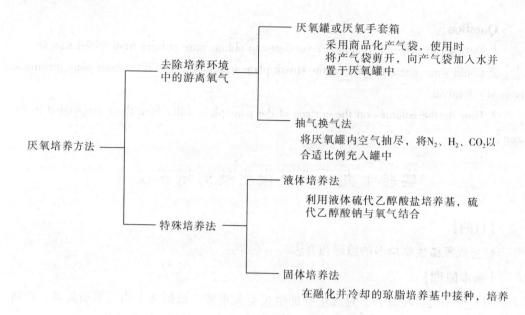

图 19-1 厌氧微生物的常见培养方法

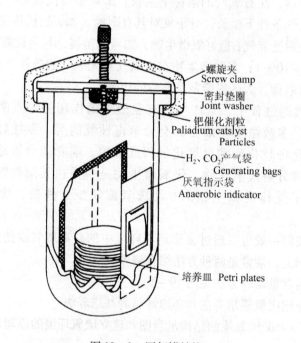

图 19-2 厌氧罐结构
Figure 19-2 Structure of anaerobic incubator

【方法与步骤】

（一）利用 GasPak 厌氧罐培养厌氧微生物

1. 分区和标记 取两个含有营养琼脂培养基的平皿，将每个平皿分成两部分，分别标记大肠埃希菌、生孢梭菌。

2. 接种 采用无菌操作技术分别在平板的每个区域划线接种对应的菌种。

3. 培养

（1）将接种的平板倒置放入厌氧培养罐中。

（2）在厌氧罐罐盖下面的催化剂室中加入催化剂（钯粒）。

（3）剪开气体发生袋的一角，将其放入厌氧罐内的金属架上，向袋中加入10ml水，再剪开指示剂袋，使指示剂暴露（还原态为无色，氧化态为蓝色），立即放入厌氧罐中，迅速盖好厌氧罐盖，旋紧螺旋，密闭厌氧罐。

（4）将厌氧罐至于37℃恒温培养箱中培养，注意观察厌氧指示条的颜色变化，从蓝色变为无色，表示罐内为厌氧化境。

（5）另取两个营养琼脂培养基平板，按照上述1~3的步骤进行分区、标记、接种。接种后将其置于37℃恒温培养箱，在有氧存在下进行培养。

（二）利用 GasPak 孵育袋培养厌氧微生物

GasPak 产气袋（图19-3）是一完整的产气系统，在加水后，可产生高浓度二氧化碳（10%）的厌氧环境。该系统不需要加热，产生的水蒸气少，透视度好，适合于菌落观察。每一产气袋可放3个培养皿，该产气袋适用于初步分离、培养厌氧菌或微好氧菌。

1. 将接种的平皿放在可重复密封的 GasPak 袋中。

2. 将 GasPak 特制的反应袋和指示剂放在 GasPak 袋中。

3. 重新密封 GasPak 袋，并置于特定温度培养。

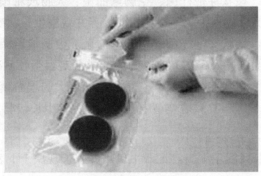

图19-3　GasPak 孵育袋
Figure 19-3　GasPak Pouch

（三）利用硫代乙醇酸盐流体培养基培养厌氧菌

接种厌氧菌前，观察试管培养基上方是否出现粉色，必须确保硫代乙醇酸盐培养基新鲜，无溶解氧。如果出现粉色说明培养基不新鲜，含有溶解氧。处理方法：

将装有培养基试管的试管帽旋松，然后将其加入沸水浴中加热10min，以除去培养基中溶解氧，冷却至50℃后接种。

1. 标记　取4支装有硫乙醇酸盐流体培养基的试管，在试管壁上用记号笔分别标记准备接种的菌名（粪产碱杆菌、梭状芽孢杆菌、空肠弯曲菌、大肠埃希菌等）。

2. 培养　接种后的试管置于37℃过夜培养。

3. 标记平皿　取两个营养肉汤琼脂平板，将每一个平皿的底部分成四个区，一个平皿标记"好氧（培养）"，另一个平皿标记"厌氧（培养）"。

4. 接种　在每一个平皿的对应区域，分别接种产碱杆菌、梭状芽孢杆菌、空肠弯曲菌、大肠埃希菌等（图19-4）。

图19-4　平板接种
A. 粪产碱杆菌　B. 梭状芽孢杆菌　C. 空肠弯曲菌　D. 大肠埃希菌
Figure 19-4　Petri plate inoculation
A. *Alcaligenes faecalis*　B. *Clostridium sporogenes*　C. *Campylobacter jejuni*　D. *Escherichia coli*

5. 培养　将标记"好氧培养"的平皿，放在37℃培养箱，倒置培养24～48h。将标记"厌氧培养"的平皿，放置在GasPak厌氧罐中（前面已降解使用方法），37℃，倒置培养24～48h。

【实验内容】

1. 将粪产碱杆菌、梭状芽孢杆菌、空肠弯曲菌、大肠埃希菌等分别接种于营养肉汤琼脂培养基表面，倒置放入厌氧罐中，37℃培养并观察生长状况。

2. 将粪产碱杆菌、梭状芽孢杆菌、空肠弯曲菌、大肠埃希菌等分别接种于营养肉汤琼脂培养基表面，倒置放入37℃培养箱中，培养并观察生长状况。

3. 将粪产碱杆菌、梭状芽孢杆菌、空肠弯曲菌、大肠埃希菌等分别接种于硫代乙醇酸盐流体培养基中，置于37℃培养箱中，培养并观察生长状况。

【结果】

记录试管培养生长状况于下表。记录平皿培养生长现象于下表。

菌种	硫乙醇酸盐流体培养基	营养琼脂培养基		结论
		有氧气条件	厌氧条件	对氧气的需求
粪产碱杆菌				
梭状芽孢杆菌				
空肠弯曲菌				
大肠埃希菌				

【思考题】

1. 为什么厌氧菌可以在硫乙醇酸盐培养基中生长？

2. GasPak厌氧培养罐中放入美兰指示剂的目的是什么？如何判断密封罐内有氧还是无氧？

Experiment 19 Cultivation for Anaerobic Bacteria

Objective

Be familiar with the principle and method of cultivation for anaerobic microorganisms.

Principle

The presence or absence of molecular oxygen (O_2) can be very important to the growth of bacteria. According to the relationship between microbes and oxygen, bacteria can be divided into obligate aerobe, anaerobe, aerotolerant anaerobes, microaerophiles and facultative anaerobe.

Some bacteria, called obligate aerobic bacteria, require oxygen. *Alcaligenes* is an example. Obligate anaerobes cannot tolerate the presence of oxygen that is lethal (e.g., *Clostridium pasteurianum*). Aerotolerant anaerobes cannot use oxygen but tolerate it fairly well. Some bacteria, such as *Campylobacter*, the microaerophiles, grow best in an atmosphere with increased carbon dioxide and lower concentrations of oxygen (5% ~ 10%). The majority of bacteria are capable of living with or without oxygen. These bacteria are called facultative anaerobes. *Escherichia coli* is a facultative anaerobe.

Some toxic intermediate reduction products, such as superoxide radical, hydrogen peroxide, are always produced during the metabolism in the presence of oxygen. Many microorganisms possess enzymes that afford protection against toxic O_2 products. Obligate aerobes and facultative aerobes usually contain the superoxide dismutase (SOD) and catalase, which catalyze the destruction of superoxide radical (O_2^-) and hydrogen peroxide (H_2O_2), respectively. Peroxidase can also be used to destroy hydrogen peroxide. The anaerobic microorganism is a kind of microorganisms which cannot make oxygen as the terminal electron acceptor. All obligate anaerobes lack both enzymes or have them in very low concentrations and therefore cannot tolerate O_2.

In the laboratory, we can culture anaerobes either by excluding free oxygen from the environment or by using reducing media. Many anaerobic culture methods involve both processes.

Common methods for culture of anaerobic microorganisms are as Figure 19 – 1.

This experiment mainly introduces the anaerobic incubator culture method and reducing media method.

The structure of GasPak anaerobic incubator and its operating principle are shown as Figure 19 – 2.

Apparatus and Materials

1. Specimens   *Bacillus cereus* nutrient broth culture and *Escherichia coli* nutrient broth culture of 24h, *Clostridium sporogenes* nutrient broth culture of 48h.

2. Cultures   Nutrient agar medium, thiolglycollate medium.

3. Apparatus   GasPak anaerobic incubator and Pouch.

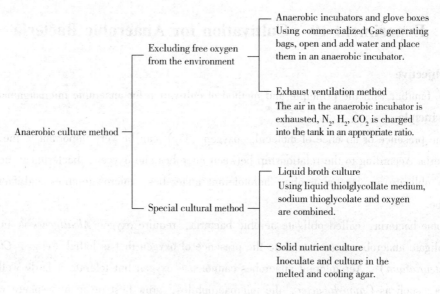

Figure 19-1 Methods for culfure of anaerobic microorganisms

4. Others Gas generating bag, anaerobic indicator, catalyst (palladium particle), inoculating loop, test tube rack, alcohol cotton, mark pen, etc.

Methods and Procedures

Ⅰ. Culture anaerobes by GasPak anaerobic incubator

1. Label the plates Take two nutrient agar plates and label the bottom of the plate with two different sections. Label the two sections with *Escherichia coli* and *Clostridium sporogenes* respectively.

2. Inoculation Aseptically inoculate the corresponding strains to each section of the plate respectively.

3. Incubation

(1) Put the incubated plates in an inverted position into the anaerobic incubator.

(2) Add the catalyst (palladium particle) in the catalyst chamber below anaerobic incubator cover.

(3) Cut a corner of the gas generating bag, place the bag in a metal frame of the anaerobic incubator and add 10ml of water to the bag. Cut the indicator bag to exposure indicator (colorless in reduced state, blue in oxidation state) and immediately place into the anaerobic incubator. Cover the anaerobic incubator, fasten the screw and make the anaerobic incubator closed.

(4) The anaerobic incubator is placed at 37℃. During the incubation, observe the color change of the anaerobic indicator from blue to colorless, which means the incubator, is anaerobic environment.

(5) Take another two nutrient agar plates and do the same way as in step 1 to 3. Incubate the plates at 37℃ in an inverted position.

II. Culture anaerobes by GasPak Pouch

The GasPak Pouch (Figure 19-3) is a self-contained system, which after the addition of water, will produce an anaerobic atmosphere containing approximately 10% CO_2. This heat-free technology minimizes condensate, which provides excellent bacterial colony visualization. Each pouch will hold up to 3 standard Petri dishes.

1. Place inoculated plates inside the resealable pouch.
2. Add one GasPak anaerobe sachet and anaerobic indicators into the GasPak Pouch.
3. Seal pouch by pressing zipper together and incubate for culture.

III. Cultivate anaerobic bacteria by thioglycollate medium

Before inoculation, ensure the fluid thioglycollate medium fresh and without dissolved oxygen, by observing the upper layer of the tube. Pink color indicates the medium is not fresh, which is containing dissolved oxygen. Then, the culture medium will be processed as follows: With the tube cap unscrewed, then put the tube containing thioglycollate medium in boiling water bath for 10min, the dissolved oxygen in the culture medium would be removed. Cool the medium to 50℃ and aseptically inoculate the corresponding strains.

1. Label four tubes of thioglycolate broth, and aseptically inoculate one with a loopful of *Alcaligenes faecalis*, one with *Clostridium sporogenes*, one with *Campylobacter jejuni*, and one with *Escherichia coli*.
2. Incubate the tubes at 37℃ overnight.
3. With a marker, divide two nutrient agar plates into four sectors on the bottom of the plates. Label one plate "Aerobic" and the other "Anaerobic".
4. Streak a single line of *Alcaligenes*, *Clostridium sporogenes*, *Campylobacter jejuni*, and *Escherichia coli* in the appropriate sector of each plate (Figure 19-4).
5. Incubate the "Aerobic" plates inverted, at 37℃ for 24~48h. Place the "Anaerobic" plates, at 37℃, inverted, in the GasPak anaerobic incubator (we have demonstrate how the incubator is rendered before).

Experiment contents

1. Inoculate *Alcaligenes*, *Clostridium*, *Campylobacter* and *Escherichia* to the surfaces of nutrient broth agar plates. Incubate at 37℃, inverted, in the GasPak anaerobic incubator.
2. Inoculate *Alcaligenes*, *Clostridium*, *Campylobacter* and *Escherichia* to the surfaces of nutrient broth agar plates. Incubate at 37℃, inverted.
3. Inoculate *Alcaligenes*, *Clostridium*, *Campylobacter* and *Escherichia* to thioglycolate broth tubes. Incubate at 37℃.

Results

Record the appearance of growth in each tube. Record the growth on each plate.

Specimens	Fluid Thioglycollate Medium	Nutrient agar		Conclusion
		Aerobic	Anaerobic	Demand for oxygen
Alcaligenes faecalis				
Clostridium sporogenes				
Campylobacter jejuni				
Escherichia coli				

Question

1. Why will obligate anaerobes grow in thioglycolate?

2. What's the purpose of the methylene blue indicator in GasPak anaerobic incubator? How to determine the aerobic or anaerobic state in a sealed GasPak Pouch?

实验二十 噬菌体的分离与纯化

【目的】

1. 掌握从自然环境中分离纯化大肠埃希菌噬菌体的原理。
2. 熟悉用双层琼脂平板法分离纯化噬菌体的方法。

【基本原理】

噬菌体是感染细菌、放线菌和真菌的病毒。噬菌体广泛存在于自然界，凡是有细菌的场所，就有其特异的噬菌体存在。例如河水、阴沟污水和粪便是各种肠道细菌尤其是大肠埃希菌的栖息地，故可分离到大肠埃希菌的噬菌体。

噬菌体分离纯化的基本原理是：①噬菌体对宿主具有高度特异性，可以利用其宿主作为敏感菌株在液体或固体培养基中培养它们；②在固体培养基中，可利用噬菌斑进行噬菌体分离纯化。

在融化的培养基中，将噬菌体和敏感细菌混合，将其倾注到底层肉汤固体培养基上，噬菌体侵入宿主细菌细胞后进行复制而导致细胞裂解，释放出噬菌体，进而感染更多的敏感菌，释放更多的新的噬菌体。在有宿主菌生长的琼脂平板上，噬菌体可裂解宿主菌而形成透明的肉眼可见的空斑，即噬菌斑（plaque）。而宿主菌未被裂解之处呈现浑浊现象。

噬菌斑形状、大小、边缘以及透明度等特征均随噬菌体的种类而异，故不仅可用于噬菌体的检出和定量，还可用于噬菌体的分离和鉴定。

【仪器与材料】

1. 菌种 大肠埃希菌（*Escherichia coli*）、37℃培养 18～24h 营养琼脂斜面培养物。

2. 噬菌体样品 阴沟污水或池塘污水。

3. 培养基 三倍浓缩的肉汤培养基（见附录一，4）、肉汤液体培养基、半固体培养基试管（上层培养基，见附录，3ml/试管）、肉汤琼脂培养基（底层培养基）。

4. 仪器 离心机、细菌过滤器、真空泵、抽滤装置、水浴锅。

5. 其他 无菌涂布棒、无菌吸管、无菌培养皿、三角瓶等。

【方法与步骤】

(一) 噬菌体的分离

1. 培养敏感菌 制备菌悬液 取经活化的大肠埃希菌斜面1支，从中挑取1环菌体接种至肉汤液体培养基中，培养至对数期，使菌悬液在600nm的OD值约为0.6。

2. 增殖噬菌体样品

（1）取1ml上述敏感菌培养液于装有50ml三倍浓缩肉汤琼脂培养基的三角瓶中，37℃震荡培养4~6h。

（2）培养后向其中加入100ml污水，37℃继续培养12~14h。

3. 获得噬菌体悬液（或裂解液）

（1）离心 将上述增殖的混合培养液于2500~3000 r/min离心20min，得上清液。

（2）过滤 取上清液用细菌过滤器过滤，装置如图20-1，收集滤液。

（3）滤液无菌检查 取少量滤液接种到10ml营养肉汤培养基中，37℃培养过夜，同时以另一瓶未接种的营养肉汤培养基为阴性对照。如果培养液经培养未变浑浊，表明滤液已无菌。

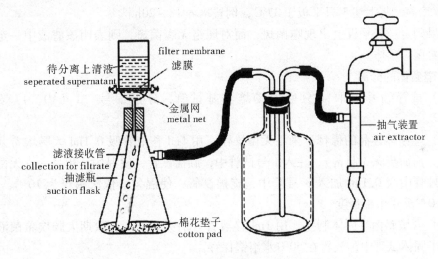

图20-1 负压抽滤装置

Figure 20-1 Vacuum filtration device

4. 噬菌体的检出

（1）试管法 于幼龄的细菌液体培养物（4~6h）内加上述过滤液一滴，37℃培养30~60min，如液体由浑浊变澄清，证明有噬菌体存在。

（2）固体平板法（图20-2）

1）倒培养基 取3只无菌培养皿，分别倒入融化并冷却到50℃左右的肉汤琼脂培养基（15~20 ml 养皿），待凝固。

2）涂布接种 在上述底层培养基表面，分别加入数滴培养至对数期的大肠埃希菌液，涂匀。

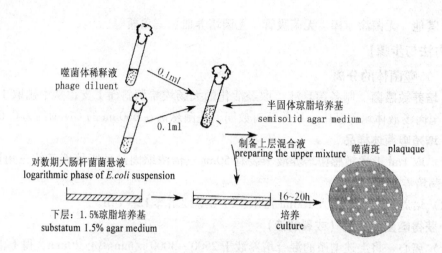

图 20 – 2　双层琼脂平板法分离噬菌体

Figure 20 – 2　Procedure for isolating bacteriophage

3）取 2 只涂布接种的平板，在平板上分别分散滴加 5~8 滴噬菌体裂解液。第 3 只平板滴加无菌生理盐水作为对照。

4）培养　将上述 3 只平板于 37℃，倒置培养 18~20h。

结果判定：若平板上出现噬菌斑，而对照组无噬菌斑，则表明该滤液中一定含有大肠埃希菌噬菌体。

5. 噬菌体的分离

（1）试管编号　取 4 只盛肉汤培养基试管，分别编号"1（10^{-1}）"到"4（10^{-4}）"。

（2）噬菌体样品的稀释　采用无菌操作，用 1ml 移液管吸 0.1ml 大肠埃希菌噬菌体悬液（前面步骤 3 制备），注入 1 号试管中，旋摇试管，使混匀。用另一支无菌移液管从 2 号管中吸 0.1ml 加入 3 号管中，旋摇试管，使混匀，依次稀释为 10^{-1}、10^{-2}、10^{-3}、10^{-4} 等 4 个稀释度。

（3）含敏感菌半固体制备　用 1ml 移液管分别吸 0.1ml 对数期大肠埃希菌液，加入 4 支半固体试管中，放置在 50℃ 水浴锅保存。

（4）倒底层琼脂平板　取无菌平皿 4 只，每皿倒入 10ml 肉汤琼脂培养基（底层培养基），并在皿底依次标明 10^{-1}、10^{-2}、10^{-3}、10^{-4} 稀释度。

（5）制备噬菌体和敏感菌的混合液　操作从标记 4（10^{-4}）的肉汤试管开始。采用无菌操作，吸取 0.1ml 10^{-4} 的噬菌体稀释液，加到含有敏感菌的半固体试管中，立即搓试管充分混匀，并倒在标注对应稀释度（10^{-4}）的底层琼脂平板表面，平置待凝。然后，用同一个移液管吸取 0.1ml 标记 3（10^{-3}）的噬菌体稀释液，加到含有敏感菌的半固体试管中，立即搓试管充分混匀，并倒在标注对应稀释度（10^{-3}）的底层琼脂平板表面，平置待凝。其他平皿的操作方法同上。

（6）培养　将上述平皿于 37℃ 倒置培养至噬菌斑出现。

（二）噬菌体的纯化

初步分离的噬菌体往往不纯，噬菌斑的大小、形态常不一致，所以还需要进行噬菌体的纯化。

用接种针挑取典型的噬菌斑，接种至含有大肠埃希菌的肉汤培养基中，37℃培养18～24h，以增殖噬菌体。然后重复上述分离步骤，直至平板上出现的噬菌斑形态、大小一致，则表明已获得纯的大肠埃希菌噬菌体。

【实验内容】

1. 采集污水水样。
2. 利用双层琼脂平板法分离大肠埃希菌噬菌体。
3. 纯化噬菌体样品。

【结果】

1. 将双层平板法分离噬菌体的结果填入表 20 – 1。

表 20 – 1

	阴性对照	平板1	平板2
噬菌斑是否形成			
噬菌斑形态特征			

2. 将噬菌体的分离与纯化结果记录在表 20 – 2。

表 20 – 2

噬菌体稀释度	10^{-1}	10^{-2}	10^{-3}	10^{-4}
计数噬菌斑				
噬菌斑特征				

【注意事项】

在采用双层平板法进行噬菌体分离纯化时，敏感菌与噬菌体样品的混匀吸附时间不能过长。

【思考题】

1. 为什么要在污水中接种大肠埃希菌？
2. 为什么在同一敏感菌的平板上会出现形态、大小不同的噬菌斑？
3. 能否用芽孢杆菌分离特异性的大肠埃希菌噬菌体？

（徐　威）

Experiment 20　Isolation and Purification of Bacteriophages

Objective

1. Grasp the principle of isolating and purifying *Escherichia coli* phage from natural environment.
2. Be familiar with the double agar plate method of isolating and purifying bacteriophage.

Principle

Bacteriophage is viruses which infects with bacteria, actinomyces and fungi. It widely exists in nature. All the places where bacteria exist, there is specific bacteriophage. Some particular environments, such as water, sewer wastewater, feces are the habitats where various intestinal bacteria exist especially *Escherichia coli*. Phages can be separated from them.

The principle: ①Bacteriophage is highly specific for the host: Bacteriophage can be grown in liquid or solid cultures of specific bacteria. ②Isolating and purifying bacteriophage by plaque: When solid media are used, the plaque forming method allows the bacteriophage to be isolated.

Host bacteria and bacteriophages are mixed together in melted agar, which is then poured into a Petri plate containing hardened nutrient agar. Each bacteriophage that infects a bacterium multiplies, releasing several hundred new phages. The new phages infect other bacteria, and more new phages are produced. All these bacteria in the area surrounding the original bacteriophage are destroyed, leaving a clear area, or plaque, against a confluent "lawn" of bacteria. The lawn of bacteria is produced by the growth of uninfected bacteria.

Different type of plagues has different characteristics such as the shape, size, edge, transparency and so on. Plaques can be used not only for the detection and quantification, but also the separation and identification of the bacteriophage.

Apparatus and Materials

1. Specimens *Escherichia coli*, cultured 18~24h in nutrient agar slant medium at 37℃.

2. Bacteriophage samples Sewage as source.

3. Cultures 3×nutrient broth medium (Appendix I, 4), nutrient broth, semi-solid agar tube (or upper medium, see Appendix, 3ml/tube), nutrient broth agar medium (substratum or bottom medium).

4. Apparatus Centrifuges, bacterial filters, vacuum pumps, sterile membrane filter assemblies, water bath.

5. Others Sterile glass spreader, sterile pipette, sterile petri dishes, Erlenmeyer flasks, etc.

Methods and Procedures

Ⅰ. Isolation of bacteriophage

1. Sensitive bacteria incubation Take an activated *Escherichia coli* slant, pick one ring of *E. coli* to nutrient broth and incubate to logarithmic phase, making the OD_{600nm} value about 0.6.

2. Proliferation of phage samples

(1) Take 1ml of sensitive bacteria inoculum to Erlenmeyer flask with 50ml of 3× nutrient broth medium, incubate at 37℃ for 4~6h with shaking.

(2) After incubation, add 100ml sewage to the above proliferated solution, incubate for another 12~14h.

3. Obtaining bacteriophage suspension (or bacteriophage lysate)

(1) Centrifugation The mixed proliferation cultures are centrifuged at 2500 – 3000 r/min for 20min to remove most bacteria and solid materials, with the supernatant remain.

(2) Filtration Filter the supernatant through a membrane filter (Figure 20 – 1). Decant the clear liquid into a screw capped tube.

(3) Sterility test Add a small amount of filtrates to 10ml of nutrient broth, incubate at 37℃ overnight. While take another bottle of nutrient broth as the negative control. If the cultured solutions is not clear, indicating that the filtrate is sterile.

4. Detection of Bacteriophage

(1) Broth – Clearing Assay Add a drop of the filtrate into the nutrient broth inoculum (4~6h), incubate at 37℃ for 30~60min. It would prove the presence of phage if the turbid liquid became clear.

(2) Plaque – Forming Assay (Figure 20 – 2)

1) Pouring plate: Take 3 sterile Petri dishes, pour nutrient broth medium (about 50℃) respectively onto the plates with aseptic technique (15~20ml each plate), waiting to cool down.

2) Spreading plate: Add to a few drops of the sensitive bacteria incubation onto the surface of the plate (made by the last step) and spread it.

3) Drop 5~8 drops of the bacteriophage lysate on the surface of 2 plates above. The third plate as blank control is added with the sterile saline.

4) Incubation Incubate all plates in an inverted position at 37℃ for 8~20h.

Interpretation of the results: If the plaque appeared on the plates, while no plaque on the blank control, it indicates that the filtrate containing the *E. coli* phage.

5. Isolation of bacteriophage

(1) Label the broth tubes "1 (10^{-1})" to "4 (10^{-4})".

(2) Serial dilutions of bacteriophage: Aseptically add 1ml of phage suspension (from step 3) to tube 1. Mix by carefully aspirating up and down three times with the pipette. Using a different pipette, transfer 1ml to the second tube, mix well. Continue until the fourth tube.

(3) Soft agar tube with *E. coli*: With a pipette, add 0.1ml of *E. coli* to the soft agar tubes and place them back in the water bath.

(4) Pouring the substratum agar plate: Prepare 4 sterile Petri dishes, label them with the dilution $10^{-1} \sim 10^{-4}$. Melt the nutrient agar media (as the substratum) and cool to 50~55 ℃, add 10ml of it to each plate and curdle.

(5) With a pipette, start with broth tube 4 and aseptically transfer 0.1ml from tube 4 to a soft agar tube, Mix by swirling, and quickly pour the inoculated soft agar evenly over the surface of Petri plate 4. Then, using the same pipette, transfer 0.1ml from tube 3 to a soft agar tube, mix and pour over plate 3. Continue until you have completed tube 1.

(6) Incubation : Incubate all plates in an inverted position at 37℃ until plaques develop.

II. Purification of Bacteriophage

The initial isolated phage is always impure, such as the size and the uneven shape of the plaque, etc. So it is necessary to purify the bacteriophage.

Pick the typical plaque with a inoculating needle and inoculate to *E. coli* nutrient broth, then incubate at 37℃ for 18~24h. Repeat the isolation process until the sizes and morphology of the isolated plaques consensus.

Experiment contents

1. Collecting the sewage water.
2. Isolate the *E. coli* phage using the double-deck agar plate technique.
3. Purify the phage samples.

Results

1. Fill in Table 20-1 with the isolated bacteriophage by means of double-deck plate method.

Table 20-1

	Negative control	Plate 1	Plate 2
Plaque formation			
Plaque morphological			

2. Fill in Table 20-2 with the isolated and purified bacteriophages.

Table 20-2

Bacteriophage dilution	10^{-1}	10^{-2}	10^{-3}	10^{-4}
Count the number of plaques				
Plaque morphology				

Note

The mixing time is not too long, when isolating *E. coli* bacteriophage using the double-deck plate method.

Questions

1. Why did you add *Escherichia coli* to sewage?
2. There are different sizes or morphological plaques isolated in the plate? Why?
3. Can you isolate *E. coli* bacteriophage from a species of *Bacillus*?

（徐　威）

第六节　微生物菌种保藏技术

微生物菌种为发酵法生产抗生素、氨基酸、酶制剂、酿造产品及现代生物技术药物等产品做出了巨大的贡献，成为人类宝贵的财富。然而，微生物菌种在传代繁殖过程中也会不断受到环境条件的影响,多数情况下会出现退化现象。因此,需要采用适宜的方法保持其重要的优良性状,以利于生产和科研的应用。可以说菌种保藏是一切微生物工作的基础。

由于各种微生物菌种生理生化特性不同,对环境条件适宜能力各异,保藏方法也不一样。通常分为两大类:一类是保藏时间较短的,如营养斜面法、液体石蜡法等,这些方法使菌种在保藏期间不能完全停止代谢活动,只能使代谢活动降至较低水平;另一类是保藏时间较长的,如甘油管法、沙土法、冷冻干燥法等,这些保藏法使菌种完全处于休眠状态,代谢活动停止,但生理生化的潜在能力并没有改变,一旦移接到适宜培养基中,其就会表现出典型特征。

Section 6　Technology of Preservation of Pure Cultures

Microorganisms are valuable wealth for human beings, because they made a great contribution to the production of antibiotics, amino acid, zymin, brewing products, modern biotech drugs and so on. However, pure cultures of those microorganisms (strains) may be influenced constantly during their breeding process and frequently emerge the degradation phenomena. For this reason, it is important to preserve their important merits by appropriate methods. Therefore, it is a fundamental process for microorganism preservation.

The methods of preservation of pure cultures diversify for their various physiological – biochemical characteristics and adaptive capacities to various conditions. These preservation methods could be classified to 2 categeries: one is for short period of culture preservation such as the methods of subculturing and covered with liquid paraffin. The other is for long period of culture preservation such as the methods of preserved in glycerin, mixed with sand and soil and preserved by freeze – drying. Strains preserved by these methods may entirely be in dormant state, but keep their physiological – biochemical potential abilities unchanged. Their typical characteristics may exhibit again once they were inoculated to the suitable medium.

实验二十一　菌种保藏

【目的】
1. 掌握菌种保藏的目的、基本原理。
2. 熟悉几种简易菌种保藏的常规方法。

【基本原理】
微生物具有容易变异的特性,因此,在菌种保藏过程中,必须使微生物的代谢处于最不活跃或相对静止的状态,才能在一定的时间内使其不发生变异而又保持生活能力。

菌种保藏的目的是为了把从自然界分离到的野生型,或者经过人工选育得到的变异型纯种,采用多种方法使菌种存活,不丢失,不污染杂菌,不发生或少发生变异,保持菌种原有的各种优良培养特征和生理活性，有利于生产、科研的正常进行，是一项重要的微生物学基础工作。

微生物菌种保藏的基本原理是使微生物的生命活动处于半永久性的休眠状态，也就是使微生物的新陈代谢作用限制在最低的范围内。干燥、低温和隔绝空气是使微生物代谢能力降低的重要因素，是保证获得这种状态的主要措施。有针对性地创造干燥、低温和隔绝空气的外界条件，是微生物菌种保藏的基本技术。尽管菌种保藏方法很多，但基本都是根据这三种主要措施设计的。

菌种保藏方法大致可分为以下几种。

1. 传代培养保藏法 这是最简单的保藏方法。对需氧菌可用斜面培养，对厌氧菌可进行穿刺培养或疱肉培养基培养等，培养后于4℃冰箱内保存。此法为实验室和工厂菌种室常用的保藏法，优点是操作简单，使用方便，不需特殊设备；缺点是传代次数多，容易变异、污染杂菌。若菌种经常使用，可应用此法。

保藏时间依微生物的种类而不同。霉菌、放线菌及有芽孢的细菌可保存2~4个月，移种一次；酵母菌可保存2个月；细菌最好每月移种一次。

2. 液体石蜡保藏法 它是在斜面培养物和穿刺培养物上面覆盖1cm灭菌的液体石蜡后，置于4℃直立保存的方法。这样可防止固体培养基的水分蒸发而引起的菌种死亡，另一方面液体石蜡可阻止氧气进入，使好氧菌不能继续生长，从而延长了菌种保藏的时间，此法实用效果好。霉菌、放线菌、芽孢菌可保藏2年以上不死亡，酵母菌可保藏1~2年；一般无芽孢细菌也可保存一年左右；甚至用一般方法很难保藏的脑膜炎球菌，在37℃恒温箱内，亦可保藏3个月之久。此法的优点是操作简单，不需特殊设备；缺点是在保存和运输过程中，菌种必须直立放置，所占空间较大。

3. 甘油管保藏法 甘油管保藏法是利用甘油作为保护剂，甘油渗透入细胞后，能强烈降低细胞的脱水作用，且在-70℃条件下，细胞代谢水平极低却仍保持生命存活状态，从而达到长时间保藏菌种的目的。在基因工程实验中，常用于保藏含质粒载体的大肠埃希菌，一般可保存0.5~1年，甚至更长时间。一般甘油的终浓度可采用15%~40%。

4. 沙土管保藏法 去除水分可以降低微生物的代谢速度，因而这一方法对那些形成孢子的微生物比如放线菌和真菌更加有效。因此在抗生素生产中应用最广、效果最好。操作是将微生物吸附在适当的载体，如土壤、沙子上，而后进行干燥。可保存2~10年，但应用于营养细胞效果不佳。

5. 超低温冷冻保藏法 可分低温冰箱（-20~-30℃，-50~-80℃）、干冰乙醇快速冻结（约-70℃）和液氮（-196℃）保藏法等。尤其液氮冷冻保藏法，除适用于一般微生物的保藏外，对支原体、衣原体、氢细菌、难以形成孢子的霉菌、噬菌体及动物细胞等一些用冷冻干燥法难以保存的微生物，均可用此法长期保藏。其优点是菌株性状不易变异，但需特殊设备。

6. 冷冻干燥保藏法 先使微生物在低温度（-70℃左右）下快速冷冻，然后在减压下利用升华现象除去水分（真空干燥），利用有利于菌种保藏的一切因素，使微生物始终处于低温、干燥、缺氧的条件下，因而它是迄今为止最有效的菌种保藏法之一。

用冷冻干燥保藏的菌种，其保藏期可达数年至数十年，其菌种要特别注意纯度，不能污染杂菌，这样再次使用该菌种时才不会出差错。细菌和酵母菌菌种要求培养到稳定期，一般细菌培养 24~48h，酵母菌培养 3d，放线菌与霉菌一般需培养 7~10d，若用对数生长期菌种进行保藏，其存活率反而会降低。此方法保存孢子比其营养体效果更好。

在冷冻过程中进行，为了避免恶劣条件对微生物的损害，需使用保护剂来制备细胞悬液，可以防止因冷冻或水分不断升华对细胞的损害。常用作保护剂的有脱脂牛奶、血清、糖类、甘油和二甲基亚砜等。

对于目前尚不能在人工培养基上生长的微生物，如病毒、立克次体、螺旋体等，可用寄主保藏法，即将它们在活的动物、昆虫、鸡胚内感染并传代，此法相当于一般微生物的传代培养保藏法。

【仪器与材料】

1. 菌种　细菌、酵母菌、放线菌和霉菌，或基因工程用含质粒载体的大肠埃希菌。

2. 培养基及试剂　固体斜面培养基、甘油、灭菌水、液体石蜡、河沙和黄土、灭菌脱脂牛乳、冰块、食盐、干冰、95%乙醇、10%盐酸、无水氯化钙等。

3. 仪器　干燥器、真空泵、低温冰箱（-30℃）、超低温冰箱（-70℃）、液氮冷冻保藏器。

4. 其他　灭菌吸管、灭菌培养皿、安瓿管、40目与100目筛子、喷灯、油纸等。

【方法与步骤】

1. 传代培养保藏法

（1）斜面低温保藏法　将菌种接种于固体斜面培养基上，置适宜温度培养，待菌充分生长后，移至4℃的冰箱中保藏。

（2）半固体穿刺法　将半固体培养基注入小试管（0.8cm×10cm），使培养基距离试管口约2~3cm深。用灭过菌的接菌针采用无菌操作挑取菌体，在半固体培养基顶部的中央直线穿刺到半固体培养基的1/3深处，在适宜条件下培养后，熔封试管或是塞上橡皮塞，移至4~6℃的冰箱中保藏。

2. 液体石蜡保藏法

（1）将液体石蜡高温灭菌（121.3℃，30min）。

（2）将菌种接种至斜面培养基中培养，使其充分生长。

（3）用无菌吸管吸取液体石蜡注入斜面，其用量以高出斜面顶端1cm为准。

（4）保持试管直立，置4℃保存。

从液体石蜡下面取培养物移种后，接种环在火焰烧灼时，培养物容易与残留的液体石蜡一起飞溅，应特别注意。

3. 甘油管保藏法

（1）将甘油进行121.3℃高压蒸汽灭菌20min，备用。

（2）将携带质粒载体的大肠埃希菌菌种接种至含氨苄青霉素（100μg/ml）的营养液体培养基试管中培养，使其充分生长。

（3）用无菌吸管吸取0.85ml大肠埃希菌培养液，置于1.5ml无菌Eppendorf管中，再加入0.15ml无菌甘油（即甘油终浓度为15%），使其与菌液充分混合均匀，然后将此甘油管置于液氮中速冻。

（4）将已冷冻的含菌甘油管置于-70℃冰箱中保存。

（5）转接：无菌操作用接种环刮取冻结的甘油管培养物表面，划线接种于含氨苄青霉素（100μg/ml）的营养琼脂培养基平板上，37℃培养过夜使之活化。

4. 沙土管保藏法

（1）取河沙加入10%稀盐酸浸泡，去除有机杂质。

（2）倒去酸水，用自来水冲洗至中性，烘干。用40目筛子过筛，去掉粗颗粒，备用。

（3）取非耕作层的瘦黄土，磨细，100目筛子过筛。

（4）按一份土、三份沙掺合均匀，装入小试管中，每管装1cm左右，灭菌。

（5）抽样进行无菌检查，每10支沙土管抽一支，将沙土倒入肉汤培养基中，37℃培养48h，若仍有杂菌，则需重新灭菌。

（6）接种方法分干法接种和湿法接种

干法接种（多用于保藏放线菌和部分真菌）：将斜面上已生长好的培养物，用接菌环或接菌铲刮取孢子，接种于无菌的沙土管中，搅拌均匀，注意勿接入培养基。

湿法接种：选择培养成熟的（一般指孢子层生长丰满的）优良菌种，将已生长好的菌种斜面培养物用3~5ml无菌水洗下，制成均匀的菌悬液，再用无菌吸管吸取菌悬液加到沙土管中，每管加10滴约0.5ml。接种后的沙土管放于真空泵中抽干以除去沙土管中的水分，抽干时间越短越好。

（7）每10支抽取一支，用接种环取出少数沙粒，接种于斜面上，培养后观察生长情况和有无杂菌。

（8）若经检查没有问题，用火焰熔封管口，放冰箱或室内干燥处保存。每半年检查一次活力和杂菌情况。

（9）菌种复活培养时，取沙土少许移入液体培养基内培养。

5. 液氮超低温冷冻保藏法

（1）准备安瓿管　用于液氮保藏的安瓿管，需要采用硼硅酸盐玻璃制成的，能耐受温度突然变化而不致破裂。安瓿管的大小通常为7.5mm×100mm，能容1.2ml液体。将空安瓿管塞上棉塞，高压灭菌。

（2）准备保护剂与灭菌　保护剂如10%的甘油或10%二甲基亚砜。含甘油溶液需经高压灭菌，含二甲基亚砜溶液则采用过滤除菌。保存细菌、酵母菌或霉菌孢子等容易分散的细胞时，则将空安瓿管塞上棉塞；若保存霉菌菌丝体，则需在安瓿管内预先加入保护剂如10%二甲基亚砜蒸馏水溶液或10%的甘油蒸馏水溶液，加入量以能浸没以后加入的菌块为限。0.1MPa（121.3℃）高压蒸汽灭菌20min。

（3）接入菌种　将菌种用10%的甘油蒸馏水溶液制成菌悬液，装入无菌的安瓿管；霉菌菌丝体可用无菌打孔器，从平板内切取菌落圆块，放入装有保护剂的安瓿管内，然后用火焰熔封，浸入水中检查有无漏洞。

（4）冻结　将已封口的安瓿管以每分钟下降1℃的慢速冻结至-30℃（避免细胞急剧冷冻在细胞内会形成冰的结晶，降低存活率）。

（5）保藏　将冻结至-30℃的安瓿管立即放入液氮冷冻保藏器的小圆筒内，再将小圆筒放入液氮保藏器内。液氮保藏器内的气相为-150℃，液态氮内为-196℃。

（6）恢复培养　需要用保藏的菌种时，将安瓿管取出，立即放入38~40℃水浴急速解冻，直到全部溶化为止。再采用无菌操作技术，打开安瓿管，将内容物移入适宜的培养基上培养。

6. 真空冷冻干燥保藏法

（1）准备安瓿管　安瓿管可用长颈球形底的，材料采用中性硬质玻璃为宜。安瓿管先用2% HCl浸泡8~10h，再用自来水冲洗多次，最后用蒸馏水洗2~3次后烘干。将印有菌名和接种日期的标签放入安瓿管内（有字的一面应朝向管壁），管口塞上棉花，0.1MPa（121.3℃）高压蒸汽灭菌30min。

（2）准备保护剂-脱脂牛奶　先将鲜牛奶煮沸，除去上面的一层脂肪，然后用脱脂棉过滤，并在3000 r/min离心15min。如果一次不行，再离心一次，直至除尽脂肪为止。牛奶脱脂以后，在55.2 kPa（113℃）条件下高压蒸汽灭菌30min，并做无菌检验。

（3）准备保藏菌种　所用菌种要特别注意其纯度，不能有杂菌污染。用无菌吸管吸取2~3ml脱脂牛奶加到待保藏的菌种斜面内，用接种环将菌苔刮下，轻轻搅动，使其均匀地悬浮在牛奶内成菌悬液（但应注意，切勿将琼脂刮到牛奶中），每支安瓿管分装0.2ml（一般装入量为安瓿管球部体积的1/3为宜）。

（4）预冻　将分装好的安瓿管放于低温冰箱或干冰乙醇液中（-25~-40℃之间）冷冻，使菌悬液结冰。预冻的目的是使菌悬液在低温条件下结冰，使水分在冻结状态下直接升华，避免在真空干燥时因菌悬液沸腾而造成气泡外溢。

（5）真空干燥　将装有已冻结菌悬液的安瓿管置于真空干燥箱中真空干燥，一般抽到真空度0.0267~0.0133 MPa维持6~8h，样品可被干燥。

（6）封管　取出安瓿管，接在封口用的抽气装置上抽气，当真空度达到0.0267MPa时，继续抽气数分钟，用火焰在细颈处烧熔封口（图21-1）。做好的安瓿管应放置在低温避光处保藏。

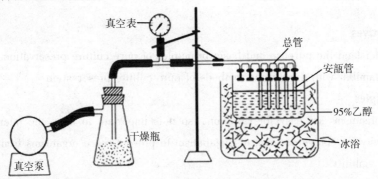

图21-1　真空冷冻干燥保藏法

Figure 21-1　Freeze-drying

(7) 恢复培养 如果要从中取出菌种恢复培养,可先用75%乙醇将管的外壁消毒,然后将安瓿管上部在火焰上灼烧,滴几滴无菌水,使管子破裂。再用接种针直接挑取松散的干燥样品,在斜面接种。也可先将无菌液体培养基加入安瓿管中,使样品溶解。然后再用无菌吸管取出菌液至合适的培养基中进行培养。

此法为菌种保藏最有效的方法之一,对一般生命力强的微生物及其孢子以及无芽孢细菌都适用,即使对一些很难保存的致病菌,如脑膜炎球菌与淋病球菌等亦能保存。

【实验内容】

1. 采用沙土保藏法保藏产生抗生素的放线菌菌种。
2. 选用真空冷冻干燥保藏法保藏研究用菌种。

【结果】

比较各种菌种保藏法并填写下表。

保藏法	适合保藏微生物的类型	保藏温度	保藏时间
斜面低温保藏法			
半固体穿刺法			
液体石蜡保藏法			
沙土管保藏法			
甘油管保藏法			

【思考题】

1. 经常使用的细菌种,用哪一种保藏方法既好又简便?
2. 细菌用什么方法保藏的时间长而又不易变异?
3. 产孢子的微生物常用哪一种方法保藏?
4. 在真空冷冻干燥保藏法中,为什么必须先将菌悬液预冻后才能进行真空干燥?

(苏 昕)

Experiment 21   Preservation of Pure Cultures

Objectives

1. Understand the purpose and basic principles of pure culture preservation.
2. Be familiar with the routine methods of pure culture preservation.

Principles

Microorganisms are easy for variation, so it is important to keep the microorganisms in their inactivity or relative dormant of metabolism to prevent those organisms from variation but retain their viability.

The purpose of culture preservation is to maintain the viability of wild cultures isolated from the nature or variable pure cultures obtained by artificial selection. Also it is essential to maintain the original features and physiological activities of those cultures, which could be best used in production and scientific research.

The basic principle of culture preservation is to keep the microorganisms in a semi-perpetual dormancy stage, which can minimize their metabolism. The most important factors of decreasing the metabolism are desiccation, low temperature and isolation from air. Therefore, the majority of preservation methods are designed according to the above three techniques.

The methods for culture preservation can be classified as follows.

1. Subcultur Preservation　　It is the simplest method of preservation. For aerobes, they are cultivated on agar slants. While for anaerobes, they are cultivated by stab culture or cooked meat medium. Cultures could be stored at 4℃ for several weeks. It is the common method in the laboratory and factory. The advantages are easy to operate, convenient to use and need no special equipment, while the disadvantages are easy strain variation and potential contamination after many passages.

The preservation period differs from microorganism species. Molds, Actinomycetes and Zygotomere could be preserved for 2~4 months. Yeasts could be preserved for 2 months. For bacteria, it is better to subculture once a month.

2. Liquid paraffin preservation　　Sterile liquid paraffin was loaded into the slant or stab cultures with 1cm higher than the agar and keep the tubes erectly at 4℃. This could prevent the culture death from water evaporation. For another thing, liquid paraffin can keep the oxygen out, so as to prolong the preservation period. This method is very practical and also has good effect. We can preserve molds, actinomycetes and spores for more than 2 years; yeasts for 1~2 years; and bacteria (without spore) for about 1 year. Even the *Meningococcus* which is difficult to preserve by ordinary method, can be maintained for 3 months in the incubator at 37℃. The advantage of this method is also easy to operate and need no special equipment, the disadvantage is that the tubes must be kept erect.

3. Glycerol Preservation　　As a protectant, glycerol could be infiltrate into the bacteria cell and strongly reduce cell dehydration effect. Under the condition of -70℃, the bacteria cells could keep their survival status at low metabolism, so as to be preserved for long time. In genetic engineering experiments, *E. coli* containing plasmid vectors could be preserved using this method. Those bacteria could generally be preserved for 6 months to one year, even longer. The finae concentration of glycerol generally can be used between 15%~40%.

4. Sand Preservation method　　Removal of water can reduce the microbial metabolism. This method is applicable to the spores of molds, actinomycetes and bacteria. So it is widely used in the antibiotic industry. Microorganisms are adsorbed to suitable carriers, such as soil or sand, and then dessicate. Samples could be preserved about 2~10 years. But the method is not applicable to the vegetative cells.

5. Liguid nitrogen Preservation　　Cultures can be frozen in low temperature refrigerator (-20~-30℃, -50~-80℃), dry-ice and alcohol (about -70℃) and liquid nitrogen (-196℃). Preserved in liquid nitrogen could be used for common microbes, also for

those difficult to preserve by freeze-drying, such as mycoplasma, chlamydia, hydrogen bacteria, molds difficult to form spores, bacteriophages or some zooblasts. The advantage is with little strain variation but special equipments required.

6. Freeze-drying method Freeze the microorganisms rapidly at low temperature (about $-70°C$), then dessicate the water under reduced pressure with sublimation (vacuum desiccation). This is so far one of the most effective methods to preserve cultures, because it could keep the microorganisms in the condition of low temperature, dry and free from oxygen.

Cultures preserved by freeze-drying can be maintained for several decades, with caution to their purity to prevent contamination for later use. Bacteria and yeasts are required to be cultivated to stationary phase, generally, 24~48 hours for bacteria, 3 days for yeasts, 7~10 days for actinomycetes and molds. If cultures in logarithmic (log) phase are used, the survival rate will decrease. It is better to maintain spores instead of vegetative cells.

In the process of the latter two methods mentioned above, protecting agents are needed to prepare the cell suspension to keep the cells from damaging. The commonly used protecting agents are skim milk, serum, saccharide, glycerol, dimethylsulfoxide (DMSO) and so on.

Microorganisms such as virus, rickettsiaes and spirochetes etc, which cannot be cultivated in artificial media, could be preserved by host preservation. Specimens are subcultured into live animals, insects or chicken embryos. This method is equivalent to subculturing.

Apparatus and Materials

1. Specimens Bacteria, yeasts, actinomycetes, molds, or *Escherichia coli* containing plasmids.

2. Cultures and Reagents Agar slant culture, glycerol, sterile water, liquid paraffin, sandy and soil, skim milk, ice, salt, dry-ice, 95% alcohol, 10% HCl, anhydrous $CaCl_2$, etc.

3. Apparatus Desiccator, vacuum pump, lower temperature refrigerator ($-30°C$), ultra-cold freezer ($-70°C$), liquid nitrogen frozen incubator.

4. Miscellaneous Sterile pipette, sterile Petri dish, 1.5ml sterile Eppendorf tube, ampoules, sieve (40 mesh and 100 mesh), blowtorch, oil paper, etc.

Methods and Procedures

1. Subculturing

(1) Slants at low temperature The cultures are inoculated to the agar slant and cultivated at optimal temperature. Keep the cultures in the refrigerator at 4°C after it is well grown.

(2) Stab cultivation Transfer the semi-solid media to a small tube (0.8cm × 10cm) with 2~3cm away from the top of the tube. Using aseptic inoculating needle to transfer bacteria from agar slants and stab to 1/3 depth of the solid media without touching the walls. After cultivating in optimal conditions, seal or plug the tube and store it at 4~6°C.

2. Covered with liquid paraffin

(1) Place liquid paraffin in the autoclave for sterilization for 30 minutes (121°C).

(2) Inoculate proper cultures to the slant and keep it well grown.

(3) Pour liquid paraffin into the agar slant, with 1cm higher than the upper edge of the slant.

(4) Keep the tube erectly at 4℃.

3. Preserved in glycerin

(1) Place glycerol in the autoclave for sterilization for 20 minutes (0.1MPa, 121℃).

(2) Inoculate the plasmid containing *E. coli* to broth culture media containing ampicillin (100μg/ml) and keep it well grown.

(3) Pipette 0.85ml of *E. coli* culture into 1.5ml sterile Eppendorf tube and add 0.15ml of sterile glycerol (the end concentration of glycerol is 15%). Vortex the mixture and put it into liquid nitrogen to freeze.

(4) Preserve the frozen tube in -70℃ refrigerator.

(5) Subculture: Scrap the cultures in glycerol by sterilize inoculating loop and inoculate it in agar medium plate containing ampicillin (100μg/ml) and place it in a 37℃ incubator overnight.

4. Mixed with sand and soil

(1) Using 10% HCl to wash out organic impurity in river sand.

(2) Remove the acid solution, wash the sand to neutral, dry and sieve with 40 mesh.

(3) Mill and grind thin loess, sieve with 100 mesh.

(4) Mix loess with sand in a ratio of 1 to 3. Put them into small test tubes about 1cm high each and then sterilize.

(5) Perform a sterility test by sampling one out of the ten tubes randomly. Mix the content with broth culture media and place them in a 37℃ incubator for 48 hours. Resterilize if there are any bacteria.

(6) For inoculating, there are two methods as follows:

Dry method (for the preservation of actinomycetes and some molds): Transfer a loop of strong mature strains and inoculate them to the sandy soil tube, mix throughout. Do not inoculate the culture media into the tube.

Wet method: Select strong mature strains and wash them with 3~5ml of sterile water in order to get spore solutions. Add 10 drops (0.5ml) of these solutions to each sandy soil tube and mix them. Put these tubes into a vacuum desiccator to try to get rid of the water as fast as possible.

(7) Take one tube out of ten tubes, inoculate some sands onto a slant, and observe their growth.

(8) If there is no problem, seal the tubes and preserve them in the refrigerator or some dry places. Check their activity and purity once half a year.

(9) When it is claimed to use them, take some sands into the liquid culture.

5. Ultra – cold frozen in liquid nitrogen

(1) Prepare the ampoules Ampoules should be made in silicon borate glass and crackless due to sudden change of the temperature. Tampon the ampoules and sterilize them.

(2) Prepare the protective agents and sterilize Prepare 10% glycerol or 10% DMSO as protective agents. Sterilize glycerol using autoclaving while DMSO using filtration. It is necessary to add the protecting agents such as 10% glycerol in ampoules and sterilize for 20 min at 0.1MPa for preservation of mold mycelium.

(3) Inoculation Prepare spore solution using 10% glycerol mixed with bacteria strains. Put them into sterile ampoules. For mold mycelium, cut a piece of colony block into the ampoule with protective agent. Seal the tubes on flame.

(4) Freezing Freeze the ampoules slowly with 1℃ drop per minute till −30℃. (The crystal may be formed in cells to lower the livability of them if freeze them rapidly.)

(5) Preservation Place the ampoules into liquid nitrogen incubator (gas nitrogen −150℃ and liquid nitrogen −196℃).

(6) Return to culture Take out the ampoules and thaw in 38 ~ 40℃ water bath. Open up the ampoules and transfer the content into the proper culture media.

6. Freeze – drying

(1) Prepare the ampoules Choose the ampoules made in hard glass with long neck and round bottom. Immerse them in 2% HCl for 8 ~ 10 hours, then wash with tap water for several times. Desiccate the ampoules after wash them with distilled water for 2 ~ 3 times. Label the ampoules with strain name and inoculation date. Tampon the ampoules and sterilize them at 0.1MPa for 30min.

(2) Prepare the protective agents – skim milk Boil the fresh milk, skim, filter, and centrifugate at 3000r/min for 15min. Repeat until the fat is removed completely. Sterilize it at 55.2kPa for 30min and then test the sterilization results.

(3) Prepare the strain Pay attention to the purity of the strain. For bacterial slant, add 2 ~ 3ml defatted milk into a slant to get concentrated solution 0.2ml per ampoule (about 1/3 of the ampoule volume).

(4) Pre – freeze Place the ampoules in low temperature refrigerator or dry ice – alcohol (−25 ~ −40℃) to make the solution frozen.

(5) Vacuum drying Desiccate the ampoules while in the frozen state using a high vacuum to sublime the water, maintaining 6 ~ 8 hours in 0.0267 ~ 0.0133 MPa.

(6) Seal the tubes Take out the ampoules, connect with special equipment to produce vacuum in them, then fuse the neck seal the ampoule in fire (Figure 21 – 1). Preserve them in refrigerator or in the dark at room temperature.

(7) Return to culture Firstly, sterilize the outer of the ampoules with 75% alcohol. Then, burn the upper of the ampoule and add drops of distilled water to make the tube crack. Inoculate the loose powder on the slant. Alternatively, pour the sterile liquid culture

media into ampoule to dissolve the sample and then inoculate the bacteria solution to proper culture media.

This method is one of the most effective methods for preservation. It is used for common microbes, even for what is difficult to preserve, such as *Meningococcus* and *Gonococcus*.

Experiment contents

1. Preserve the strains of antibiotics – producing actinomycetes by the method of mixed with sand and soil.
2. Preserve the strains using in research by the method of freeze – drying.

Results

Compare the above methods of preservation of pure cultures and fill in the following table.

methods	strains	temperature	time
Slants at low temperature			
Stab cultivation			
Covered with liquid paraffin			
Mixed with sand and soil			
Preserved in glycerin			

Questions

1. Which method can be used to preserve the bacteria simply?
2. Which method is used to preserve the bacteria without mutation for a long time?
3. Which method is used for spore – forming bacteria?
4. In the method of freeze – drying, why must to pre – freeze the concentrated solution containing strains first, then to vacuum drying?

（苏　昕）

第七节　微生物的生长繁殖技术

Section 7　Microbial Growth Techniques

实验二十二　微生物显微镜直接计数法
　　　　　　——血球计数板计数法

【目的】

1. 掌握应用血球计数板测定青霉菌孢子浓度的方法。
2. 了解血球计数板的构造、原理和使用方法。

【基本原理】

显微直接计数法，即利用血球计数板在显微镜下直接计数，是一种常用的微生物计数方法,各种单细胞菌体纯培养悬浮液、单孢子悬浮液以及各种微生物细胞的原生质体等均可采用血球计数板计数。将孢子悬液(或菌悬液)放在血球计数板载玻片与盖玻片之间的计数室内,由于载玻片上的计数室盖上盖玻片后的容积($0.1mm^3$)是一定的,所以可以根据在显微镜下观察到的微生物数目换算为单位体积内的微生物数目。

血球计数板是一块特制的厚载玻片,载玻片上有由4条槽构成的3个平台。中间的平台较宽,其中间又被一短横槽分隔成两部分,每个半边平台上面各有一个计数室(图22-1)。计数室的刻度有两种:一种是计数室分为16个中方格(图22-2),每个中方格又分成25个小方格(图22-3);另一种是一个计数室分成25个中方格,每个中方格又分成16个小方格。不管是哪一种计数室,计数室都由400个小方格组成。

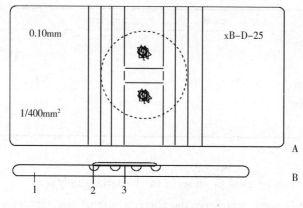

图22-1 血球计数板的构造
A. 正面图　B. 侧面图
1. 血球计数板　2. 盖玻片　3. 计数室

Figure 22-1　Structure of a blood counting chamber
A. Front view　B. Side view
1. Blood counting chamber　2. Cover slip　3. Counting room

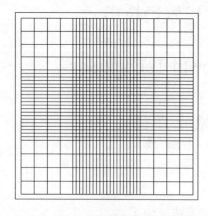

图22-2 血球计数板放大后的构造

Figure 22-2　Magnification of a counting chamber

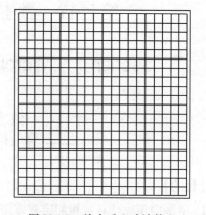

图22-3 放大后血球计数室

Figure 22-3　Magnification of the counting room

中央为计数室，每一个计数室大方格的面积为 1mm²，盖上盖玻片后，载玻片与盖玻片之间的距离为 0.1mm，所以每个计数室的体积为 0.1mm³。在计数的时候，若计数室由 16 个中方格组成，一般计数左上、左下、右上、右下的 4 个中方格（共计 100 个小格）的细胞数或孢子数；如果计数室由 25 个中方格组成，除计数上述 4 个中方格，还需计数中央一个中方格的菌数或孢子数（共计 80 个小格）。在计数的过程中要不断地调节细准焦螺旋，以便能看到计数室内不同深度的细胞或孢子。凡是落在中格左方和上方双线上的孢子或细胞都计算在内，而落在下方或右方双线上的孢子或细胞均不计算在内。最后可求出每小格的平均细胞数或孢子数，按照下列公式算出原孢子悬液的孢子浓度或原细胞悬液的细胞浓度。

样品中菌（或孢子）数（个/ml）= 每小格的平均数 × 400 × 稀释倍数 × 10000

在这一公式中，10000 代表 1ml 的容积（即 1000mm³）是一个计数室容积（0.1mm³）的 10000 倍。

【仪器与材料】

1. 菌种 青霉菌（*Penicillium sp.*）72h 沙氏培养基斜面培养物。

2. 仪器 普通光学显微镜、恒温震荡培养箱。

3. 其他 血球计数板、滴管、擦镜纸、移液管、接种环、盛有玻璃珠的三角瓶、含有脱脂棉的无菌注射器、试管等。

【方法与步骤】

1. 青霉菌孢子悬液的制备

（1）取一支孢子成熟的青霉菌斜面培养物，用无菌移液管吸取 5ml 无菌水加入菌管中。

（2）采用无菌操作技术，用接种环将菌管内斜面上的孢子轻轻刮下，注入盛有玻璃珠的三角瓶中，同样方法将另 15ml 无菌水分 3 次再加入菌管中，将剩余的孢子全部刮干净，将孢子液全部合并于三角瓶中，水平旋转振荡 30min，然后倒入含有脱脂棉的无菌注射器中过滤，得单孢子悬浮液。

2. 显微镜下孢子浓度测定

（1）稀释 定量取出孢子悬液，加到无菌干燥的试管中，按照一定倍数将其稀释，稀释的程度一般以血球计数板每小格内含有 5~10 个菌或孢子最为合适。

（2）加样 取洁净干燥的血球计数板盖上盖玻片，用无菌滴管从盖玻片的边缘滴一滴孢子稀释液，则孢子稀释液自行渗入，注意不要产生气泡。

（3）计数 静止 5min，使孢子沉降不再流动，然后即可在显微镜下观察计数。先在低倍镜下找到计数室的位置，然后换成高倍镜计数。如发现孢子悬液太浓，应重新稀释并计数。计数一个样品要从两个计数室中的计算结果取平均值，以减少实验误差。

（4）清洗血球计数板 计数完毕，将血球计数板取下，在水龙头上用水柱冲洗，切忌用硬物洗刷，然后自然吹干，镜检观察是否有孢子或其他沉积物，直至洗刷干净。

【实验内容】

以青霉菌为材料，采用血球计数板计数法测定青霉菌的孢子浓度。

【结果】

将实验结果填入下表，计算出每毫升青霉菌孢子悬液中含有孢子的数目。

计数次数	每个中方格菌数（或孢子数）（个）						每个小方格菌数（或孢子数）平均值	稀释倍数	个/ml
	1	2	3	4	5	平均值			
第一次									
第二次									
平均值									

【注意事项】

此方法计数的是活菌体和死菌体的总和，因此常常称其为总数测定法。

【思考题】

1. 血球计数板直接计数法的原理和方法是什么？
2. 血球计数板的实验误差分析有哪些？

（蔡苏兰）

Experiment 22　Direct Microscopic Count——Blood Counting Chamber Method

Objectives

1. Grasp the measuring method to determine the concentration of spores of *Penicillium sp.* using blood counting chamber.

2. Be familiar with the principal and manipulation of blood counting chamber method.

Principles

The direct microscopic count is a common method used to determine the total number of microorganism cells under the microscope. During the procedure, a blood counting chamber (hemocytometer) is applied. Uni – cellular microorganisms, spores and protoplasts can be counted by this method. Because the testing volume in the blood counting chamber is constant ($0.1mm^3$), the total microbial number can be calculated according to the cell number on the blood counting chamber.

The blood counting chamber is a special piece of thick glass slide which contains three platforms separated by four grooves. The middle platform is divided into 2 parts with a counting chamber laid on each (Figure 22 – 1). There are two kinds of counting chamber: one is divided to 16 squares each of which contains 25 small squares (Figure 22 – 2); the other is divided into 25 squares with 16 small squares on each. No matter what kind of blood counting chamber it is, the number of small squares is 400.

The volume of the counting chamber is $0.1mm^3$ because the area of each chamber is $1mm^2$, and the height between the slide and the cover slip is 0.1mm. When using a 16 –

square counting chamber, 4 squares located in upper left, lower left, upper right and lower right (in total 100 smaller squares) need to be counted. While using the 25 - square counting chamber, another square in the middle also needs to be counted besides the 4 squares mentioned above (in total 80 smaller squares). When counting, the fine adjustment knob needs to be constantly adjusted in order to focus on the spores located in different depth. For the cells located on double lines of the chamber, following rules are usually applied: The cells which located on the top lines or left lines will be counted while the ones on the bottom lines or the right lines are omitted. Finally the average concentration of the original cell suspension can be calculated according to the following formula.

The number of cells (or spores) per milliliter = average number of cells (or spores) in each smaller square × 400 × dilution ratio × 10000. The 10000 means that 1ml (1000mm^3) is 10000 times to the counting chamber volume (0.1mm^3).

Apparatus and Materials

1. Specimens   72h Sabouraud's agar slant cultures of *Penicillium sp.*

2. Apparatus   Incubator and bright - field light microscope.

3. Others   Blood counting chamber, dropper, lens paper, pipette, inoculating loop, flask contained glass beads, sterile syringes contained absorbent cotton and tubes, etc.

Methods and Procedures

1. Preparing spores suspension of *Penicillium sp.*

(1) Using a sterile pipette, transfer 5ml sterile water to a slant culture of *Penicillium sp.*

(2) With aseptic technique, scrape the spores from the agar surface gently with an inoculation loop to make spore suspension. Then pour the suspension into a flask contained glass beads. Do it for another 3 times and 20ml in total of spore suspension will be obtained. After shaking for 30 minutes, the single spore suspension could be obtained by filtering through a sterile funnel containing absorbent cotton.

2. Counting of the spores concentration

(1) Diluting   Transfer quantitative spore's suspension to sterile dry test tubes and dilute it in series until only 5 ~ 10 cells lied in each smaller square.

(2) Loading   Siphon a drop of dilution onto a clean, dry blood count chamber at the edge of the cover slip covered on the chamber. Be careful not to produce bubbles.

(3) Counting   Count the spore number of *Penicillium sp.* after stilling for 5 min. Focus the counting chamber at low power first, then move to high power. If the spore concentration was too high, it should be diluted and counted again. The average of the two counting chamber is needed to reduce the experimental error.

(4) Cleaning the blood counting chamber   After counting, rinse the blood counting chamber with water and insure that not any spores or other sediments are left in the squares. Notice not to use any tough implement to clean the blood counting chamber surface.

Experiment content

Determine the average concentration of the spores of *Penicillium sp.* with blood counting chamber.

Results

Record the results in the following table and calculate the average concentration of the original spore suspension.

	Spore number in every square						Average number of spores in a smaller square	Dilution ratio	Spores concentration (per milliliter)
	1	2	3	4	5	Average			
First counting									
Second counting									
Average									

Notes

In this method, living or dead cells cannot be distinguished under the microscope. So it is used to count the total cells.

Questions

1. What is the principles and manipulations of the direct microscopic count?
2. What would cause the error when using a blood counting chamber?

（马晓楠）

实验二十三 微生物间接计数法——平板菌落计数法

【目的】

1. 掌握平板菌落计数方法基本原理。
2. 熟悉平板菌落计数法的常见方法和操作步骤。

【基本原理】

稀释平板菌落计数法是一种最常用的活菌计数方法。其依据是：微生物在高度稀释条件下，在固体培养基上形成的一个菌落是由一个单细胞繁殖形成的，因此一个菌落形成单位代表一个细胞。在计数的时候，首先将待测样品做系列稀释，使待测样品中的微生物细胞成单个细胞存在，再取一定量的稀释菌液接种到固体培养基平板中（平板接种培养有两种方法：涂布平板法和倾注平板法），使其均匀地分布于培养基表面或内部，经适宜条件培养后，单个细胞生长繁殖形成菌落，计数菌落数目，即可换算出样品中的含菌数。

平板菌落计数法常用于生物制品的检验、土壤含菌量的测定以及食品、水源的污染程度的检验等。

【仪器与材料】

1. 菌种 大肠埃希菌（*Escherichia coli*）18h 肉汤悬液。

2. 培养基与试剂 肉汤琼脂培养基。

3. 仪器 恒温培养箱。

4. 其他 无菌培养皿（90mm）、无菌移液管、无菌玻璃三角爬、试管等。

【方法与步骤】

（一）倾注平板计数法

1. 平板编号 分别取9支盛有4.5ml无菌水的试管，依次标记10^{-1}、10^{-2}、10^{-3}、10^{-4}、10^{-5}、10^{-6}、10^{-7}、10^{-8}、10^{-9}，另取无菌平皿9套，分别编号10^{-7}、10^{-8}、10^{-9}各3皿。

2. 待测样品稀释液的制备 用一支1ml无菌移液管精确吸取0.5ml大肠埃希菌悬液加到10^{-1}试管中，并用此移液管将管内悬液反复吸吹3次，使菌悬液混合均匀。另取一支1ml无菌移液管精确吸取0.5ml加到10^{-2}试管中，反复吸吹3次，其余各管依此类推。整个稀释过程见图23-1。若待检测样品为固体时，一般准确称取待测样品10g，放入装有90ml无菌水的250ml三角瓶中，充分振荡20min，使微生物细胞分散，静置30s，即是10^{-1}稀释液，其余稀释操作法与上述相同。

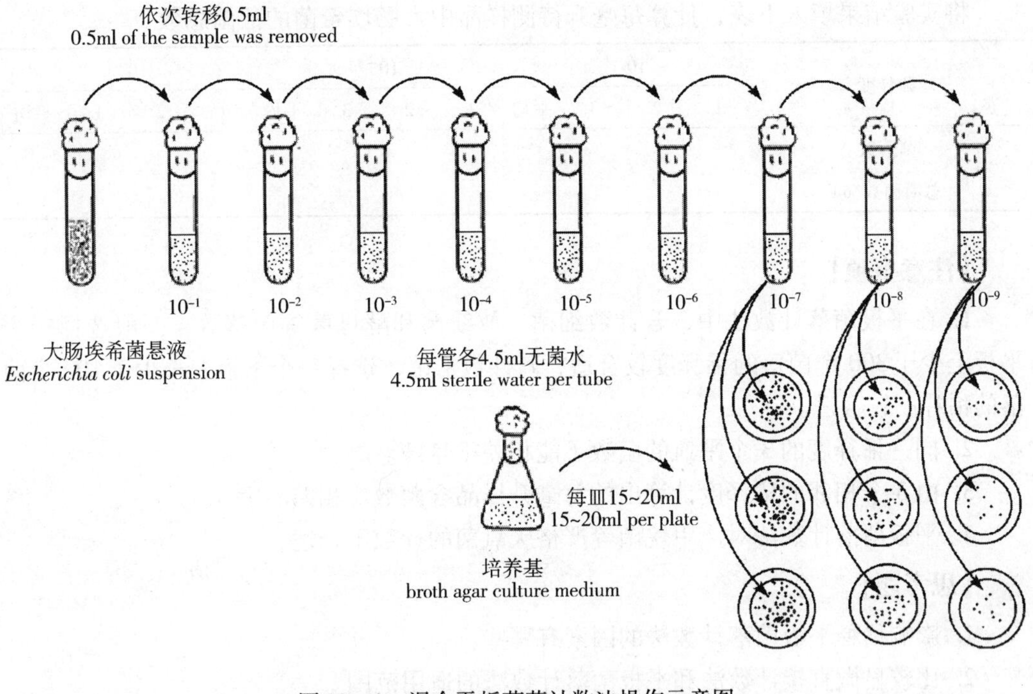

图23-1 混合平板菌落计数法操作示意图

Figure 23-1 Operation schematic diagram of pour plate count

3. 取样和制作混菌平板 用1ml无菌移液管分别精确吸取10^{-7}、10^{-8}、10^{-9}的稀释液0.5ml，对应加入已编号的无菌培养皿中（由低浓度向高浓度取稀释液时，可不必更换移液管）。在每皿中倒入融化并冷却至45~50℃的肉汤琼脂培养基15~20ml，迅速轻轻转动混匀平板，待培养基凝固后倒置，于37℃恒温培养箱中培养48h后计数。

4. 计数菌落数 菌落长出后取出培养皿，计数同一稀释度的 3 个平皿的菌落数，并计算其平均值，按下面公式换算每毫升样品的总活菌数。

总活菌数/ml = 同一稀释度的 3 个平皿的菌落数的平均值 × 稀释倍数 × 2

（二）涂布平板计数法

涂布平板计数法和倾注平板计数法基本相同，所不同的是：涂布平板计数法先将培养基融化后倒入无菌平板中，待凝固后编号，然后用无菌移液管吸取 0.5ml 菌液稀释液对号加到不同稀释度编号的对应培养皿上；然后采用无菌操作技术，用无菌三角爬将平板上的菌液涂布均匀（每个稀释度更换一个三角爬）；将涂布过的平板置于桌面放置 20min，使菌液渗入培养基中，最后将平板倒置于 37℃ 培养箱中培养 48h 后计数。

【实验内容】

以大肠埃希菌悬液为材料，采用平板菌落计数法测定每毫升样品中大肠埃希菌的活菌数量。

【结果】

将实验结果填入下表，计算每毫升待测样品中大肠埃希菌的活菌总数。

稀释度	10^{-7}				10^{-8}				10^{-9}			
	1	2	3	平均	1	2	3	平均	1	2	3	平均
菌落数												
总活菌数/ml												

【注意事项】

1. 在平板菌落计数法中，若计数细菌、放线菌和酵母菌等菌落数，一般选择每个平板上少于 300 个菌落的稀释度较合适；若计数霉菌，选择每个平板上长少于 100 个菌落的稀释度较合适。

2. 同一稀释度的 3 个平皿的菌数不能相差很悬殊。

3. 由 3 个相邻的稀释度计算出的每毫升样品含菌数应相差不大。

4. 平板菌落计数法对产甲烷菌等严格厌氧菌的计数不合适。

【思考题】

1. 影响稀释平板菌落计数法的因素有哪些？
2. 比较显微直接计数法和平板菌落计数法的适用范围。

（蔡苏兰）

Experiment 23   Indirect Microscopic Count——Plate Count

Objectives

1. Grasp the principle of plate count.

2. Understand the procedures of plate count.

Principles

Viable plate count, a common method in microorganism counting, is based on that every bacterial colony arises from an individual cell that has undergone cell division on solid medium. In this method serial dilutions of a sample containing viable microorganisms are inoculated on agar plates. The microbial suspension is either spread onto the agar surfaces after solidification or mixed with the medium first and then poured into plates to solidify. After incubating at suitable conditions, colonies can be counted, and the number of bacteria in the original sample can be calculated.

Viable plate counting is widely used in biological products, determining the number of microorganism in soil and the degree of contamination in food and water.

Apparatus and Materials

1. Specimens   18h broth culture of *Escherichia coli*.
2. Cultures   Nutrient broth agar.
3. Apparatus   Incubator.
4. Others   Petri plate (90mm), sterile pipette, sterile glass swab and tubes, etc.

Methods and Procedures

1. Pour plate count

(1) Labeling plates and tubes   Label 9 tubes containing 4.5ml sterile water with 10^{-1} to 10^{-9} and label 9 plates with 10^{-7}, 10^{-8}, 10^{-9} respectively (3 plates for every dilution).

(2) Series dilution   Dilutions are achieved by adding quantitive volume of *Escherichia coli* suspension to the 4.5ml-water-containing tubes. 0.5ml of *Escherichia coli* suspension was firstly added to the tube labled with 10^{-1} to make the 10^{-1} dilution. After mixing adequately, 0.5ml of the suspension from the 10^{-1} tube was transferred to the 10^{-2} tube and the dilutability was 10^{-2}. All the dilution procedure is described in Figure 23-1. If the sample was solid, 10g sample could be removed to a 250ml flask contained 90ml sterile water and shaking for 30mins. The dilutability was also 10^{-1}. The other dilutability could be achieved by the method mentioned above.

(3) Preparation of plates contained *Escherichia coli*   Using a sterile pipette, transfer 0.5ml of diluted suspension of *Escherichia coli* to a sterile plate. Then pour about 15~20ml of medium at 45~50℃ and blend immediately. After solidification, incubate the agar plate upside down at 37℃ for 48h.

(4) Counting   After incubation, count the number of colonies of the same dilutability and calculate the average of bacteria colonies. Then calculate the concentration of microorganisms in the original suspension according to the following formula.

The total number of living bacterium per milliliter = the average of bacteria colonies of the same dilutability × the dilution factor of the plate counted × 2

2. Spread plate count   The spread plate count is basically the same principle of pour plate count. The procedure is described as follows: first to prepare a sterile nutrient agar plate. Then

add 0.5ml of bacteria suspension onto the agar surface and use sterile swab to spread uniformly. (One swab is only for one dilution) When laid aside for 20 minutes to allow the suspension to infiltrate the medium, incubate the plates at 37℃ up side down.

Experiment content

To test the average concentrations of the original *E. Coli* suspension by plate count.

Results

Record the results in the following table and calculate the average concentrations of the original bacteria suspension.

Dilutability	10^{-7}				10^{-8}				10^{-9}			
	1	2	3	Average	1	2	3	Average	1	2	3	Average
Colony number												
Total viable cell number per milliliter												

Notes

1. In this test, no more than 300 colony forming units are fit to count the bacteria, actinomycetes or yeast, while no more than 100 colony forming units are fit to count the mold.

2. The number of bacteria colony in the three plates of the same dilutability cannot differ extremely.

3. The original concentration of bacteria calculated from the three adjacent dilutabilities cannot differ extremely.

4. This method cannot be used to count the colonies of obligatory anaerobic bacteria such as *Methanogens sp.*

Questions

1. What are the influencing factors when using the plate count?

2. Compare the applicability of the microscopic direct count with that of the plate count.

（马晓楠）

实验二十四　细菌生长曲线的测定——比浊法

【目的】

1. 掌握光电比浊计数法测定大肠埃希菌生长曲线的方法。

2. 了解光电比浊计数法的原理和细菌生长曲线的特点。

【基本原理】

细菌的生长一般指群体的生长，常常具有一定的规律性。描述细菌在液体培养基中的生长规律的曲线叫生长曲线（growth curve），即将一定量的细菌接种到一定容积的液体培养基中，在适宜的条件下进行培养，以培养时间为横坐标，以菌数的对数为纵坐标进行作图得到的曲线。典型的生长曲线可分为延迟期、对数生长期、稳定期和衰亡期四个时期。

细菌在液体培养基生长过程中，由于原生质含量的增加，会引起培养物混浊度的增高。细菌悬液的混浊度和透光度成反比，与光密度成正比，透光度或光密度可借助光电比浊计精确测出，因此，可用光电比浊计测定细胞悬液的光密度（OD值），表示该菌在特定实验条件下细菌的相对数目，进而反映出其相对生长量。

【仪器与材料】

1. 菌种　大肠埃希菌（*Escherichia coli*）16~18h 肉汤琼脂培养基斜面培养物。

2. 培养基与试剂　肉汤液体培养基。

3. 仪器　721型分光光度计、摇床。

4. 其他　试管、1ml 无菌移液管等。

【方法与步骤】

1. 菌种准备　以无菌操作技术将大肠埃希菌接入装有100ml肉汤液体培养基的三角瓶中，37℃，150r/min，振荡培养18h。

2. 接种、培养　取预先编号的17支分别装有10ml肉汤液体培养基的大试管，采用无菌操作技术，用移液管向每个试管准确加入大肠埃希菌悬液0.2ml，轻轻振荡混匀，另设一空白对照组（不加大肠埃希菌悬液）。将接种后的17支试管置于摇床上，37℃，150r/min，振荡培养。

3. 比浊测定　每隔一定时间间隔取样摇匀，将大肠埃希菌培养液进行适当稀释，使光密度在0.1~0.65之间，以没有接种的空白对照组液体培养基调零点，在550nm波长，1cm比色杯中依次进行OD值测定。

4. 绘制生长曲线　以光密度（OD值）为纵坐标，培养时间为横坐标，绘制大肠埃希菌的生长曲线。

【实验内容】

以大肠埃希菌为实验材料，采用光电比浊计数法测定大肠埃希菌的生长曲线。

【结果】

将测定的OD值填入下面表格中，绘制大肠埃希菌的生长曲线。

培养时间（h）	0	1	1.5	2	2.5	3	4	5	6	8	10	12	14	16	18	20
OD_{550}																

【注意事项】

1. 在测定过程中，尽量控制光密度值在0.1~0.65之间，以得到较高的精确度。

2. 颜色过深的样品或在样品中还含有其他物质的悬液，不能用比浊法测定细胞生长量。

3. 如果我们将本实验中不同培养时间的大肠埃希菌悬液采用稀释平板菌落计数法进行测数，可测定不同培养时间的活菌总数，以大肠埃希菌悬液光密度值为横坐标，以活菌数目为纵坐标，可绘制一条标准曲线。这样在测定任意培养时间菌悬液的光密度值后，可以在标准曲线上查出活菌总数。该方法较稀释平板菌落计数法节省时间，已经广泛应用在工业生产上。

【思考题】

微生物光电比浊计数法的原理是什么？

（蔡苏兰）

Experiment 24 Bacterial Growth Curve

Objectives

1. Grasp the procedures of turbidity measurement of *Escherichia coli* growth.

2. Understand the principle of turbidity measurement method and the features of bactemial growth carve.

Principles

When inoculating quantitative bacteria seed into quantitative liquid medium, bacteria growth curve can be gained by plotting incubation time versus logarithm of viable cell count. A typical growth curve can be described as four phases which are lag phase, log phase, stationary phase and death phase respectively.

In liquid medium, the turbidity of culture would increase with the protoplasm's increasing. The concentration of cells in bacteria suspension is in proportion to the optical density, but in inverse proportion to the transmittance in given range. So optical density (OD) of the bacteria suspension can be used to determine the relative number of bacteria at specific experimental conditions to reflect the relative growth.

Apparatus and Materials

1. Specimens 16~18h nutrient agar slant cultures of *Escherichia coli*.

2. Cultures Nutrient broth medium.

3. Apparatus Spectrophotometer and table concentrator.

4. Others Tubes and 1ml sterile pipette, etc.

Procedures

1. Preparation of *Escherichia coli* suspension Inoculate *Escherichia coli* into 100ml of broth culture medium in a flask and incubate at 150r/min, 37℃ for 18h.

2. Inoculation and incubation Prepare 17 pre-labeled tubes containing 10ml of nutrient broth culture medium. Add 0.2ml of *Escherichia coli* suspension to each tube respectively, and mix gently. Prepare another tube containing 10ml of nutrient broth medium only as a blank. Incubate the 17 tubes at 37℃ with 150r/min.

3. Measuring OD_{550} value Take out the tubes and shake gently at regular time, then measure OD_{550} value of each tube using the sterile broth medium as a blank. The inoculum can be diluted with the medium when necessary to keep the OD_{550} value in the range of 0.10~0.65.

4. Plot the growth curve Plot the growth curve of *Escherichia coli* with OD values as ordinate and incubation time as abscissa.

Experiment content
Test the growth of *Escherichia coli* by turbidity measurement.

Results
Records the OD values in the following table and plot the growth curve of *Escherichia coli*.

Incubation time (h)	0	1	1.5	2	2.5	3	4	5	6	8	10	12	13	14	16	18	20
OD_{550}																	

Notes
1. The OD_{550} value should be controlled in the range of 0.1 ~ 0.65 to insure the accuracy.

2. This method is not suitable to measure the bacteria growth if the medium was dark or contained other insoluble substances.

3. If the number of live cells was tested by viable plate count at different time point in this test, another curve can be obtained with the OD_{550} of *Escherichia coli* suspension as abscissa and the live cell number as ordinate. This curve can help to get the number of live cells according the OD_{550} value conveniently and has been widely used in industrial production to save time.

Questions
What is the principle of turbidity measurement method?

（马晓楠）

实验二十五　霉菌生长的测定——重量法

【目的】
1. 掌握重量法测定霉菌生长的方法。
2. 了解重量法测定霉菌生长的原理。

【基本原理】
霉菌（molds）是丝状真菌的统称，它们的生长主要表现为菌丝的伸长和分枝，它们互相缠绕，很难分清个体与个体之间的界限。甚至在液体培养基中进行通气搅拌或震荡培养时还会产生菌丝球，所以传统的个体计数法或光电比浊法无法真实地反映出其生长状况。霉菌的生长量测定通常采用重量法，即以菌丝的重量增长来衡量生长状况。重量法包括湿重法和干重法。若测定通过过滤或离心收集霉菌的菌丝后，将其称重，即为霉菌的湿重；若将收集到的菌丝进一步经80℃烘干后称重，则为霉菌的干重。此法适用于不易于形成均匀悬液的微生物的测定，如放线菌、霉菌等。

【仪器与材料】
1. 菌种　青霉菌（*Penicillium sp.*）72h沙氏培养基斜面培养物。
2. 培养基与试剂　马铃薯葡萄糖液体培养基。

3. 仪器 分析天平、恒温摇床、电热干燥箱。

4. 其他 定量滤纸、接种环、无菌移液管、试管、盛有玻璃珠的三角瓶、含有脱脂棉的无菌注射器等。

【方法与步骤】

1. 制备孢子悬液 取一支孢子成熟的青霉菌斜面培养物，用无菌移液管吸取 5ml 无菌水加入菌管中。采用无菌操作技术，用接种环将菌管内斜面上的孢子轻轻刮下，注入到盛有玻璃珠的三角瓶中，同样方法将另 15ml 无菌水分 3 次再加入菌管中，将剩余的孢子全部刮干净，将孢子液全部合并于三角瓶中，水平旋转振荡 30min，然后倒入含有脱脂棉的无菌注射器中过滤，得单孢子悬浮液。

2. 接种 将孢子悬液按 2%～4% 接种量接入 50ml 摇瓶马铃薯葡萄糖液体培养基中，28℃震荡培养 5～8d。

3. 测定 取品质和大小相同的定量滤纸两张，分别在分析天平上称重（A_1 和 A_2）。取其中一张定量滤纸（A_1），将培养一定时间的青霉菌培养物进行过滤，收集菌体，称重（B），再置 80℃ 干燥箱中，烘干至恒重（C）。另取一张定量滤纸（A_2），用滤液润湿后称重（D），然后也置 80℃ 干燥箱中，烘干至恒重（E）。

菌体的湿重 =（B – A_1）–（D – A_2）

菌体的干重 = C – E

【实验内容】

以青霉菌为材料，采用湿重法和干重法测定霉菌的生长。

【结果】

记录 A_1、A_2、B、C、D，并计算培养液中青霉菌的湿重和干重。

【注意事项】

1. 如果霉菌的孢子悬液浓度低或接种量少，在摇瓶中培养往往仅形成几个菌丝球，营养物质消耗慢。

2. 若制备的孢子悬液浓度较低，可静置一段时间，然后去掉上层部分液体以提高孢子悬液的浓度。

【思考题】

测定过程中要注意哪些操作步骤？

【拓展】

如果设计好 N 个时间点，使用 $3N$ 个摇瓶同时震荡培养，在每个时间点取出 3 个摇瓶按上述方法测定菌丝的湿重或干重，然后以培养时间为横坐标，以菌丝的湿重或干重为纵坐标进行作图则可绘制出霉菌的生长曲线。

（蔡苏兰）

Experiment 25   Mold Growth Measurement – Weighing Method

Objectives

1. Grasp the procedures of weighing method to determine mold growth.
2. Be familiar with the principle of weighing method to determine mold growth.

Principles

Molds are described as filament – like or filamentous fungi, because they form long filament – like, thread – like or strands of cells called hyphae. Molds grow through hypha extending. The mycelia twine around each other and make it difficult to distinguish the boundaries. When aeriferously cultured in liquid medium, the mycelium pellets may form. Traditional counting methods (e.g., plate count or turbidity measurement) cannot truly reflect the growth of molds. The weighing method was usually used to test molds growth. When filtering or centrifuging the fermentation broth of molds, the wet weight can be gained by weighing the mycelium. If the above gathered mycelium continued to be heated dry at 80℃, the dry weight can be gained. This weighing method is suited to measure the growth of actinomycetes or molds which cannot form uniform suspension.

Apparatus and Materials

1. Specimens   72h Sabouraud's agar slant cultures of *Penicillium sp.*
2. Cultures   Potatodextrose broth medium (PDB)
3. Apparatus   Incubator, analytical balance and electric oven.
4. Others   Quantitative filter paper, inoculating loop, pipette, tubes, conical flask containing glass beads and sterile syringe contained absorbent cotton, etc.

Methods and Procedures

1. Preparation of spores suspension   Using a sterile pipette, transfer 5ml of sterile water to a mature slant culture (with spores) of *Penicillium sp.* With aseptic technique, scrap the spores from the agar surface gently with an inoculation loop to make spore suspension. Then pour the suspension into a flask contained glass beads. Do it for another 3 times and 20ml in total of spore suspension will be obtained. After shaking for 30 minutes, the single spore suspension could be obtained by filtering through a sterile funnel containing absorbent cotton.

2. Inoculation   Inoculate quantitive (about 2% ~ 4%) spore suspension into 50ml of PDB, then incubate at 28℃ for 5 ~ 8 days.

3. Measurment   Weigh two pieces of quantitative filter paper by analytical balance respectively (A_1 and A_2). Filter the fermentation broth with the A_1 paper and then weigh the paper including the mycelium on it (B). Put them in a drying oven at 80℃ and until the weight becomes constant (C). The other paper (A_2), after wetting by filtrate, also need to be weighed (D) and then dried at 80℃ to a constant weight (E).

Wet weight = (B − A$_1$) − (D − A$_2$)

Dry weight = C − E

Experiment content

Test the wet weight and dry weight of fermentation broth of *Penicillium sp.*

Notes

1. If the spore concentration was low, mycelium pellets usually form and nutrients are consumed slowly.

2. This can be improved by stilling the suspension for a while and then removing the upper layer of liquid to concentrate.

Questions

What are the precautions in this test?

Extending

Choose N time points during incubation, and inoculate quantitative mold seed into $3N$ quantitative liquid medium respectively. For each time point, measure and calculate the average of wet weight or dry weight of three conical flasks, mold growth curve can be gained by plotting incubation time versus logarithm of wet weight or dry weight.

（马晓楠）

第八节　环境微生物的分布规律

Section 8　Distribution of Enviromental Microorganisms

实验二十六　空气中微生物检查

【目的】

1. 了解空气中微生物的分布规律和检查法。

2. 进一步加深理解"消毒"、"灭菌"和"无菌"的概念。

【基本原理】

空气中缺乏微生物可直接利用的营养物质和足够的水分，又受阳光直接照射，所以不是微生物生长繁殖的适宜场所。空气中微生物数量较少，主要来自土壤尘埃、人和动物的呼吸道及口腔排出物。只有对干燥和日光抵抗力较强的菌类和细菌的芽孢才能暂时存留。空气中微生物的分布亦有很大的地区差异。

空气中的微生物种类主要是细菌和霉菌，细菌为一些球菌、杆菌，尤其是一些产芽孢的细菌如枯草芽孢杆菌和产色素的细菌等。此外，也可能有一些病原性细菌，如结核分枝杆菌、溶血性链球菌、脑膜炎球菌、百日咳杆菌等，可以造成传染病的传播与流行。甲型链球菌常作为空气污染的指标。

空气中的微生物是微生物实验、生物制品及发酵工业污染的重要来源，也是食物变质的原因之一。当空气中的微生物借助重力落在适合于它们生长的固体培养基表面时，在适宜的条件下培养，每一个单个的菌体或孢子就会形成一个肉眼可见的孤立的菌群——菌落。通过观察菌落的特征（大小、形态、颜色、边缘等）便可以大致鉴别空气中存在的微生物种类和数量。

【仪器与材料】

1. 培养基 肉汤琼脂培养基、沙氏培养基。

2. 仪器与试剂 培养箱。

3. 其他 无菌培养皿、三角瓶等。

【方法与步骤】

1. 制备无菌平板（肉汤琼脂培养基平板和沙氏培养基平板）。

2. 将平板拿到预测地点，打开平皿盖子，暴露于空气中 15~20min，然后盖好平皿盖，注明实验地点，班组。

3. 将上述肉汤琼脂培养基平板倒置于 37℃ 培养箱中，培养 24h。将沙氏培养基平板倒置于 28~30℃ 培养箱中，培养 48h，观察结果。计数平板上生长的菌落数。

【实验内容】

采用重力法检测微生物在特定固体培养基表面的细菌菌落数和真菌菌落数。

【结果】

将实验结果填入下表，并描述菌落特征。

	菌落平均数	菌落特征					
		大小	形状	表面	颜色	边缘	质地
细菌培养基							
真菌培养基							

【注意事项】

平板菌落数的多少与暴露于空气中的时间和空气的污染程度有关。

【思考题】

分析空气中微生物的种类和数量。

（蔡苏兰）

Experiment 26　Distribution of Microbes in the Air

Objectives

1. Understand the distribution characteristics and detecting method of the microorganisms

in the air.

2. Know the essentiality of disinfection and sterilization.

Principles

There are many microbes in the air because of the movement of human being and natural phenomena. But air is not a favorable environment for microbial colonization. Several factors are responsible for this hostile microenvironment. First, the air is subject to periodic drying. Lack of moisture drives many resident microbes into a dormant state. Second, the ultraviolet in the sunlight have vital effects on the microorganisms. Therefore, only some spores of fungi and bacteria, which can resist aridity and sunlight, can survive in the air temporally. The distribution of microorganisms in the air also differed in different region.

Airborne microbes were mainly bacteria and molds. Bacteria were mainly coccus and bacillus, particular the bacteria which can form spores or pigments. In addition, there may be some pathogenic bacteria, such as *mycobacterium tuberculosis*, *hemolytic streptococcus*, *meningococcus* and *bordetella pertussis*, which can cause the spread of infectious diseases. Group A streptococcus is often used as a index of air pollution.

The airborne microbes usually result in the contamination in the fermentation industry and microbial experiments. It is also a cause to result in food deterioration. When falling on the surface of the solid medium, colonies could form after cultivating at suitable conditions. Then the amount and the species of airborne microbes could be identified by observing the characteristics of colonies.

Apparatus and Materials

1. Cultures   Nutrient agar broth and Sabouraud's agar.

2. Apparatus   Incubator.

3. Others   Petri dishes and flask, etc.

Methods and Procedures

1. Make sterile agar plate (nutrient agar broth and Sabouraud's Agar media).

2. Expose the agar plates to the air for 15min to 20 min and recover the plate. Label with the student ID No. and the place clearly.

3. For the nutrient agar broth plates, incubate at 37℃ for 24h then count the colony number. For the Sabouraud's Agar plates, incubate at 28℃ for 48h then count the colony number.

Experiment content

Test the numbers of bacteria and fungi in the air by gravitational measurement.

Results

Record the results in the following table and describe the colony characteristics.

Average number of colonies	Colony characteristics					
	Size	Shape	Surface	Pigment	Margin	Texture
Bacteria medium						
Fungi medium						

Notes

By plate count, the amount of colonies is related to the exposure time and the air pollution conditions.

Question

Please analyze the amount and the species of airborne microbes.

（马晓楠）

实验二十七　人体表面及口腔中的微生物检查

【目的】

了解人体正常菌群的含义和检查法。

【基本原理】

正常人体的体表及与外界相通的腔道，如口腔、鼻咽腔、眼结膜、肠道、泌尿生殖道等部位存在不同种类和数量的微生物，这些微生物通常对人体无害，成为人体的正常微生物群，称之为正常微生物群或正常菌群。一般情况下，正常菌群与宿主、正常菌群中各种微生物之间互相制约又互相依存，构成了一种微生态平衡。微生态平衡对人类健康、疾病与保健都具有十分重要的意义。在一定条件下，正常菌群和宿主间的生态平衡可被打破而造成微生态失调，使原来不致病的正常细菌成为条件致病菌而引起疾病，这种由条件致病菌所导致的感染也通称为二重感染（superinfection）。

【仪器与材料】

1. **培养基与试剂**　肉汤琼脂培养基、75%乙醇。
2. **仪器**　培养箱。
3. **其他**　无菌培养皿、三角瓶、乙醇棉球、镊子等。

【方法与步骤】

1. 手指消毒前后的细菌检查

（1）制备肉汤琼脂培养基无菌平板，待凝固后使用。标记清楚"消毒前"及"消毒后"。

（2）将未消毒的手指，在肉汤琼脂培养基平板上涂抹（约占平板面积的1/2）。

（3）将手指用乙醇棉球擦拭，待手指干后，立即在平板的另一半涂抹。

（4）将上述平板置于37℃培养箱中，倒置培养24h，观察结果。

2. 飞沫中的细菌检查

（1）制备肉汤琼脂培养基无菌平板，待凝固后使用。

（2）在火焰旁无菌区将皿盖打开，将平皿置于口前，对准口腔位置咳嗽2次后，将平皿盖子盖好。

（3）将上述平板置于37℃培养箱中，倒置培养24h，观察结果。

【实验内容】

利用肉汤琼脂平板检测手指消毒前后的微生物数量，及口腔中的微生物数量。

【结果】

1. 将观察到的手指消毒前后的结果填入下表。

	菌落平均数	大小
手指消毒前		
手指消毒后		

2. 将飞沫中细菌检查结果填入下表。

	菌落平均数	大小	形状	表面	颜色	边缘	质地
口腔							

【思考题】

通过本次实验,在防止培养物污染方面,你学到些什么?

(蔡苏兰)

Experiment 27　Distribution of Microbes on Human Body

Objectives

Understand the concept of normal flora and the method to detect the microorganisms on human bodies.

Principals

In a healthy human, the surface tissues (e. g., skin and mucous membranes) are constantly in contact with environmental microorganisms and become readily colonized by certain microbial species. The mixture of microorganisms regularly found at any anatomical site is referred to as the normal microbiota, the indigenous microbial population, or the normal flora. They have a commensal relationship with human body. Usually, there was an interdependent relationship between normal flora with the host, the environment outside and other normal flora microorganisms, which form a kind of micro ecological balance (eubiosis). Eubiosis was in great significance to human health, pathema and health care. Under certain conditions, the eubiosis between normal flora and hosts could be break, and the non-pathogenic bacteria became conditional pathogenic ones. The infection caused by conditional pathogenic bacteria is also known as super-infection.

Apparatus and Materials

1. Cultures and Reagents　Nutrient agar broth and 75% alcohol.

2. Apparatus　Incubator.

3. Others　Petri dishes, flask, alcohol cotton and forceps.

Methods and Procedures

1. Normal flora on fingers before and after disinfection label the plate with "Before/After" disinfection.

(1) Make sterile agar plate.

(2) Turn on the burner. In the aseptic area, smear half of the agar surface with your finger before disinfection.

(3) Smear the other half of agar surface with the same finger after disinfection with alcohol cotton.

(4) Incubate at 37℃ for 24h then count the colony number.

2. Normal flora in oral cavity

(1) Make sterile agar plate.

(2) Turn on the burner. In the aseptic area, open the plate lid and cough twice to the agar surface then recover the plate.

(3) Incubate at 37℃ for 24h then count the colony number.

Experiment content

Detect the number of the colonies on the finger before and after disinfection and in the oral cavity on nutrient agar plate.

Results

1. Record the results of examination of normal flora on the finger before and after disinfection in the following table.

	Number of colonies	Size
Normal flora on the finger before sterilization		
Normal flora on the finger after sterilization		

2. Record the results of normal flora in oral cavity.

	Number	Size	Shape	Surface	Pigment	Margin	Texture
Colonies on agar plate							

Question

From this practical, What would you learn to prevent the culture medium from contamination?

(马晓楠)

第九节　细菌的生理生化反应

Section 9　Biochemical Tests of Bacteria

实验二十八　糖类发酵试验

【目的】

1. 掌握糖发酵的原理及其在肠道细菌鉴定中的重要作用。

2. 了解不同细菌对单糖的发酵能力。

【基本原理】

各种细菌的酶系统不同，发酵各种单糖的能力各异，其产生的分解产物也不同，即有的只产酸，有的既产酸又产气，有的菌对某种单糖不能利用，借此可协助鉴别菌种，尤其是在肠道细菌的鉴别中经常使用。本试验主要检测大肠埃希菌、产气肠杆菌、伤寒沙门菌。虽然大肠埃希菌和产气肠杆菌分解乳糖和葡萄糖均产酸和产气，但其代谢途径不同。大肠埃希菌分解乳糖或葡萄糖产生丙酮酸，丙酮酸裂解生成乙酰 CoA 与甲酸，甲酸在酸性条件下，经甲酸解氢酶可进一步裂解生成 H_2 和 CO_2；而产气肠杆菌分解乳糖和葡萄糖产生丙酮酸后，可将两分子丙酮酸脱羧生成一分子乙酰甲基甲醇。而伤寒沙门菌不能分解乳糖，能分解葡萄糖，另因伤寒沙门菌不含甲酸解氢酶，故不能将甲酸进一步分解为气体，所以发酵葡萄糖的结果是只产酸不产气。

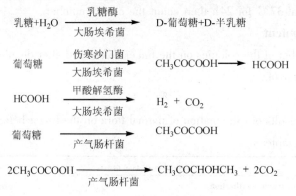

【仪器与材料】

1. 菌种 大肠埃希菌（*Escherichia coli*）、产气肠杆菌（*Enterobacter aerogenes*）、伤寒沙门菌（*Salmonella typhosa*）肉汤琼脂 18～20h 斜面培养物。

2. 培养基与试剂 单糖发酵培养基（见附录一，10），配制两种，一种加葡萄糖，一种加乳糖，分别称葡萄糖发酵培养基、乳糖发酵培养基。其中指示剂为溴麝香草酚蓝（BTB）（见附录四，3），变色范围 pH6.0～7.6（黄→蓝）。

3. 仪器 恒温培养箱。

4. 其他 试管（带小导管）、接种环、酒精灯等。

【方法与步骤】

1. 实验准备 配制单糖发酵培养基，分装至试管（带小导管，装量以没过小导管为宜），灭菌后备用。

2. 接菌 以无菌操作技术将大肠埃希菌、产气肠杆菌、伤寒沙门菌斜面培养物分别接入两种单糖发酵培养基中（注：接菌前注意观察未接菌的培养基的状况：液体澄清、呈中性的绿色、小导管无气泡。

3. 培养 将已接菌的各试管和未接菌的对照管置 37℃ 恒温箱培养 24h。

4. 观察 保留一支未接菌的培养基试管作同步对照实验。并观察结果。

5. 结果判断

（1）对照管 液体澄清、呈中性的绿色、小导管无气泡。

（2）试验管　若葡萄糖发酵培养基或乳糖发酵培养基试管中液体混浊，颜色变为黄色，且小导管内有气泡，证明此菌能发酵葡萄糖产酸又产气；若葡萄糖发酵培养基乳糖发酵培养基试管中液体混浊，颜色变为黄绿色，且小导管内无气泡，证明此菌能发酵葡萄糖或乳糖产酸但不产气。

【实验内容】

检测大肠埃希菌、产气肠杆菌和伤寒沙门菌对葡萄糖与乳糖的发酵能力。

【结果】

观察记录实验结果，列入下表。

细菌单糖发酵实验结果

菌名	葡萄糖发酵	乳糖发酵
大肠埃希菌		
产气肠杆菌		
伤寒沙门菌		

【思考题】

结合实验现象及原理说明产气肠杆菌葡萄糖发酵结果，虽然也是产酸又产气，但试管中液体颜色与大肠埃希菌的颜色不同，为什么？

Experiment 28　Utilization of Carbohydrates

Objectives

1. Grasp the principals of mono-saccharide fermentation test and the roles of the test in identification of intestinal bacteria.

2. Be familiar with the different abilities of various bacterial species to use mono-saccharide

Principle

Bacterial organisms differ in their abilities to use mono-saccharide for owning various enzyme systems. That means when provided with the same substrate (mono-saccharide), different organisms will produce various products. For example, some bacteria only produce acids, some produce acids and gases and some can even not use the mono-saccharide at all, which can be used in identification of mycobacterium especially in gut bacteria identification. This test is mainly used for the detection of *Escherichia* coli, *Enterobacter aerogenes* and *Salmonella typhi*. Though lactose and glucose can be decomposed by *Escherichia coli* and *Enterobacter aerogenes* to produce acids and gases, the metabolic pathways are different. In *Escherichia coli* or *Enterobacter aerogenes*, after the decomposition of glucose to pyruvic acid, *Escherichia coli* further degrades it into Acetyl CoA and formic acid. With the existence of formic hydrogenlyase, formic acid was transformed to CO_2 and H_2 which are finally released. While

Enterobacter aerogenes utilizes the butylene glycol pathway and produces acetoin and CO_2. *Salmonella typhi* can decompose glucose but not lactose. For lacking formic hydrogenlyase, the formic acid cannot be further degraded, so no CO_2 and H_2 will be produced.

$$lactose + H_2O \xrightarrow[\textit{Escherichia coli}]{lactase} D\text{-}glucose + D\text{-}galactose$$

$$glucose \xrightarrow[\textit{Escherichia coli}]{\textit{Salmonella typhi}} CH_3COCOOH \longrightarrow HCOOH$$

$$HCOOH \xrightarrow[\textit{Escherichia coli}]{hydrogenlyase} H_2 + CO_2$$

$$glucose \xrightarrow[\textit{Enterobacter aerogenes}]{} CH_3COCOOH$$

$$2CH_3COCOOH \xrightarrow[\textit{Enterobacter aerogenes}]{} CH_3COCHOHCH_3 + 2CO_2$$

Apparatus and Materials

1. Specimens   18~20h nutrient agar slant cultures of *Escherichia coli*, *Enterobacter aerogenes* and *Salmonella typhi*.

2. Cultures and Reagents   Monosaccharide fermentation broth (Appendix I, 10) (Lactose fermentation broth and glucose fermentation broth), 0.4% BTB (Appendix IV, 3), etc.

3. Apparatus   Incubator.

4. Others   Tubes, pipette, inoculating loop, burner, etc.

Methods and Procedures

1. Preparation   Prepare monosaccharide fermentation broth medium and transfer into tubes containing an inverted Durham tubes (the media should immerse the Durham tube). Sterilize at 55.21 kPa for 15~20 min.

2. Inoculation   Inoculatethe sterilized media with *Escherichia coli*, *Enterobacter aerogenes* and *Salmonella typhosa* respectively.

3. Incubation   Incubate all tubes at 37℃ for 24h.

4. Observation   Observe the tubes directly and compare them with a blank one.

5. Experimental phenomena

Blank tube: Clear medium, green color and no gas in the Durham tube.

Test tubes: Opacitas medium, yellow or green – yellow color and gas in the Durham tube proved that the strain can ferment the mono – saccharide to produce acid and gases. Opacitas meium yellow color but no gas in the Durham tube proved that the strain can ferment the mono – saccharide to produce acid but can not produce gases. While clear fermentation broth medium, green color and no gas in the Durham tube proved that the strain cannot ferment the mono – saccharide.

Experiment content

Test the ability of using mono-saccharide of *E. Coli*, *Enterobacter aerogenes* and *Salmonella typhosa*.

Results

Records the phenomena in the following table.

Results of mono-saccharide fermentation

Bacteria	Glucose fermentation test	Lactose fermentation test
Escherichia coli		
Enterobacter aerogenes		
Salmonella typhosa		

Question

Both *Escherichia coli* and *Enterobacter aerogenes* can ferment glucose to produce acid and gases; however the colors of two fermentation broths were different, why?

实验二十九　吲哚实验

【目的】

1. 掌握吲哚试验的基本原理及其检测方法。
2. 了解不同细菌利用氮源的能力。

【基本原理】

不同细菌所含酶系统不同，某些细菌因含有色氨酸酶，能够分解培养基内蛋白胨中的色氨酸，产生吲哚，吲哚与柯氏试剂中对-二甲基氨基苯甲醛结合，形成玫瑰红色化合物，即玫瑰吲哚，其具体反应如下：

色氨酸 $\xrightarrow[H_2O]{色氨酸酶}$ 吲哚 + NH_3 + $CH_3COCOOH$

2 吲哚 + 对二甲基氨基苯甲醛 → 玫瑰吲哚 + H_2O

【仪器与材料】

1. 菌种 大肠埃希菌（*Escherichia coli*）、产气肠杆菌（*Enterobacter aerogenes*）肉汤琼脂 18~20h 斜面培养物。

2. 培养基与试剂 蛋白胨水培养基（见附录一，12）、柯氏试剂（见附录四，4）等。

3. 仪器 恒温培养箱。

4. 其他 试管、滴管、接种环、酒精灯等。

【方法与步骤】

1. 实验准备 配制蛋白胨水培养基，分装至小试管中（约 1/4 试管高度），灭菌后备用。

2. 接菌 以无菌操作技术将大肠埃希菌、产气肠杆菌斜面培养物分别接入蛋白胨水培养基中。（注：接菌前注意观察未接菌的培养基的状况：液体澄清、呈黄色。并同时保留一支未接菌的培养基试管作同步对照实验）

3. 培养 将已接菌的各试管和未接菌的对照管置 37℃ 恒温箱培养 24h 后观察结果。

4. 加指示剂（柯氏试剂），观察结果 取上步培养后的对照管和试验管，分别加入 5~10 滴柯氏试剂，充分振荡后静置观察颜色变化。

5. 结果判断

（1）对照管 液体澄清、呈黄色。

（2）试验管 试管中液体混浊，颜色为黄色，证明均有菌生长。

阳性结果：若加指示剂（柯氏试剂）后试管中液体上层即柯氏试剂层出现玫瑰红色，证明其为吲哚试验阳性。

阴性结果：若加入柯氏试剂后试管中液体未出现红色，证明其为吲哚试验阴性。

【实验内容】

检测大肠埃希菌和产气肠杆菌代谢色氨酸产生吲哚的能力。

【结果】

观察记录实验结果，列入下表。

细菌吲哚试验结果

菌名	吲哚试验现象与结果
大肠埃希菌	
产气肠杆菌	

【注意事项】

玫瑰吲哚与水不互溶，而溶于有机溶剂，可被萃取在所加入的柯氏试剂层中。所以，观察结果时，振荡后应静置观察，只在液面上层出现玫瑰红色。

【思考题】

在观察结果时，为什么吲哚试验阳性试验结果只在液体上层即柯氏试剂层出现玫瑰红色？

Experiment 29 Indole Test

Objectives

1. Grasp the principals and the detection method of the Methyl Red (MR) test.

2. Be familiar with the ability of different bacterial species to produce Indole using tryptophan.

Principle

Organisms those posses the enzyme tryptophanase can break down the amino acid tryptophan to indole. When indole reacts with para – dimethyl – aminobenzaldehye (Kovac's reagent) a pink – colored complex is produced.

$$\text{Tryptophan} \xrightarrow[H_2O]{\text{tryptophanase}} \text{Indole} + NH_3 + CH_3COCOOH$$

$$2\,\text{Indole} + \text{Para-dinmethyl-amin obenzaldehye} \longrightarrow \text{Rosy Zndol} + H_2O$$

Apparatus and Materials

1. Specimens 18 ~ 20h nutrient agar slant cultures of *Escherichiu coli* and *Enterobacter aerogenes*.

2. Cultures and Reagents Peptone broth medium (Appendix I, 12), Kovac's reagent (Appendix IV, 4), etc.

3. Apparatus Incubator.

4. Others Tubes, pipette, inoculating loop, burner, etc.

Methods and Procedures

1. Preparation Prepare sterile peptone broth medium in small test tubes with 1/4 height.

2. Inoculatation Inoculate the media with either *E. coli* or *Enterobacter aerogenes*.

3. Incubatation Incubate all tubes at 37℃ for 24h.

4. Observation Add 5 drops of Kovac's reagent. Mix the tube and Record the results.

5. Experimental phenomena

Blank tube: Clear medium, yellow color.

Test tubes: Opacitas medium and yellow color means they can grow in the medium. After adding the Kovac's reagent, red color appeared in the upper organic layer means positive result. While no red color shown means negative result.

Experiment contents

Test the ability of using tryptophan of *E. coli* and *Enterobacter aerogenes* to produce Indole.

Results

Records the phenomena in the following table.

Result of the Indole test

Bacteria	Phenomenon and result of the Indole test
Escherichia coli	
Enterobacter aerogenes	

Notes

The red color compound only appears in the upper organic layer, so it should be stilled for a while after adding the Kovac's reagent.

Question

Why does the red color compound appearing in the upper organic layer?

实验三十　甲基红实验

【目的】

1. 掌握甲基红试验的基本原理及其检测方法。
2. 了解不同细菌分解葡萄糖产酸的能力。

【基本原理】

大肠埃希菌和产气肠杆菌均属 G^- 短杆菌，并且都能分解葡萄糖、乳糖产酸产气，二者不易区别。但两者所产生的酸类和总酸量不同：大肠埃希菌分解葡萄糖可产生甲酸、乙酸、乳酸、琥珀酸等多种酸，而产气肠杆菌只产生乙酰甲基甲醇。从而大肠埃希菌产酸能力强，培养液酸性强，pH 值在 4.5 以下，加入甲基红指示剂呈红色，为甲基红试验阳性；产气肠杆菌将分解葡萄糖产生的两分子丙酮酸转变成 1 分子中性的乙酰甲基甲醇，故生成的酸类少，培养液最终 pH 值在 5.4 以上，加入甲基红指示剂呈桔黄色，甲基红试验阴性。

【仪器与材料】

1. 菌种　大肠埃希菌（*Escherichia coli*）、产气肠杆菌（*Enterobacter aerogenes*）肉汤琼脂 18~20h 斜面培养物。

2. 培养基与试剂　葡萄糖蛋白胨水培养基（见附录一，13）、甲基红指示剂（变色范围 pH4.5 以下呈红色，4.5~5.4 呈桔黄色，5.4 以上呈黄色）（见附录四，5）等。

3. **仪器** 恒温培养箱。
4. **其他** 试管、滴管、接种环、酒精灯等。

【方法与步骤】

1. 实验准备 配制葡萄糖蛋白胨水培养基,分装至小试管(约1/4试管高度),灭菌后备用。

2. 接菌 以无菌操作技术将大肠埃希菌、产气肠杆菌分别接入葡萄糖蛋白胨水培养基中(注:接菌前注意观察未接菌的培养基的状况:液体澄清、呈黄色。并同时保留一支未接菌的培养基试管作同步对照实验)。

3. 培养 将已接菌的各试管和未接菌的对照管置37℃恒温箱培养24h后观察结果。

4. 加指示剂(甲基红试剂)观察结果 取上步培养后的对照管和试验管,分别加入3~5滴甲基红试剂,充分振荡后静置观察颜色变化。

5. 结果判断

(1) 对照管 液体澄清,呈黄色。

(2) 试验管 试管中液体混浊,颜色仍为黄色,证明均有菌生长。

阳性结果:若加入甲基红试剂后试管中液体变成红色,证明其为甲基红试验阳性。

阴性结果:若加入甲基红试剂后试管中液体未变为红色,而呈桔黄色,证明其为甲基红试验阴性。

【实验内容】

检测大肠埃希菌和产气肠杆菌发酵葡萄糖的产酸能力。

【结果】

观察记录实验结果,列入下表。

细菌甲基红试验结果

菌名	甲基红试验现象与结果
大肠埃希菌	
产气肠杆菌	

【注意事项】

在配制培养基时,pH值要调节适当,不能过高,pH值过高,会影响对最终产酸量的判断,而出现假阴性,即pH值不够低未使甲基红变红。

【思考题】

此试验为什么用甲基红指示剂,用一般酸碱指示剂如溴麝香草酚蓝可以吗?为什么?

Experiment 30　MR Test

Objectives

1. Grasp the principals and the detection method of the Methyl Red (MR) test.

2. Be familiar with the ability of different bacterial species to produce acid using glucose.

Principles

Escherichia coli and *Enterobacter aerogenes* were both G^-, and both can decompose glucose or lactose to produce acids and gases. But the metabolic pathways were different. *Escherichia coli*, for example, use the mixed acid pathway, which produces acidic end products such as lactic, acetic, and formic acid. These acidic end products are stable and will make the media remain acidic with the pH at about 4.5. When the pH indicator methyl red is present in the media, red color will be seen. While *Enterobacter aerogenes* use different pathways to utilize glucose which will then produce unstable acidic products and quickly convert to neutral compounds, for example acetoin. The pH now is above 5.4. When adding methyl red, the culture will stay yellow.

Apparatus and Materials

1. Specimens   18~20h nutrient agar slant cultures of *Escherichiu coli* and *Enterobacter aerogenes*.

2. Cultures and Reagents   MR - VP broth medium (Appendix I, 13), Methyl red indicator (Appendix IV, 5), etc.

3. Apparatus   Incubator.

4. Others   Tubes, pipette, inoculating loop, burner, etc.

Methods and Procedures

1. Preparation   Prepare sterile MR - VP broth medium in small test tubes with 1/4 height.

2. Inoculation   Inoculate the media with either *E. coli* or *Enterobacter aerogenes*.

3. Incubation   Incubate all tubes at 37℃ for 24h.

4. Observation

After incubation: Add 3~5 drops of methyl red reagent. Mix the tube well. Observe the color change and record the results.

5. Experimental phenomena

Blank tube: Clear medium, yellow color.

Test tubes: Opacitas medium and yellow color means they can grow in the medium. After adding the methyl red reagent, the medium color turning to red means positive result. While the medium color remaining orange means negative result.

Experiment contents

Test the ability of using glucose of different bacterial species to produce acids.

Results

Records the phenomena in the following table.

Result of the V. P test

Bacteria	Phenomenon and result of the Methyl red tests
Escherichia coli	
Enterobacter aerogenes	

Notices

If the pH of MR broth medium was too high, the false result may appear.

Questions

Can other indicators, for example Bromothymol blue, be used in this test? Why?

实验三十一 乙酰甲基甲醇实验

【目的】

1. 掌握 V. P 试验的基本原理及其检测方法。
2. 了解不同细菌分解葡萄糖产生乙酰甲基甲醇的能力。

【基本原理】

本试验又称伏 – 普二氏试验（Voges – Proskauer），检测不同细菌分解葡萄糖产生乙酰甲基甲醇的能力。某些细菌在糖代谢过程中，经糖酵解途径产生丙酮酸，因不同细菌所含酶不同，进一步对丙酮酸的代谢也不同。产气肠杆菌分解葡萄糖产生丙酮酸后，可将两分子丙酮酸脱羧生成一分子乙酰甲基甲醇，乙酰甲基甲醇在碱性溶液中被空气中的氧气氧化，生成二乙酰，二乙酰和培养液中含胍基的化合物反应，生成红色化合物，称为 V. P 试验阳性。大肠埃希菌分解葡萄糖不生成乙酰甲基甲醇，故 V. P 试验阴性。

$$2CH_3COCOOH \longrightarrow CH_3COCHOHCH_3 + 2CO_2$$
乙酰甲基甲醇

$$CH_3COCHOHCH_3 \xrightarrow[+KOH]{-2H} CH_3COCOCH_3$$
二乙酰

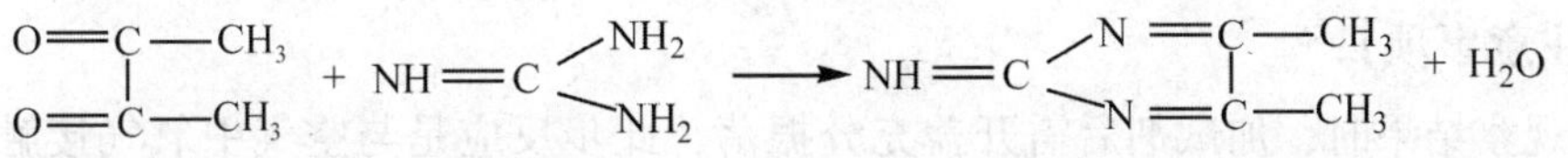

二乙酰 + 胍基 → 红色化合物 + H_2O

【仪器与材料】

1. 菌种 大肠埃希菌（*Escherichia coli*）、产气肠杆菌（*Enterobacter aerogenes*）肉汤琼脂 18~20h 斜面培养物。

2. 培养基与试剂 葡萄糖蛋白胨水培养基（与甲基红试验培养基相同）（见附录一，13）、40% KOH（或 NaOH）、6% α – 萘酚溶液等。

3. 仪器 恒温培养箱。

4. 其他 试管、移液管、接种环、酒精灯等。

【方法与步骤】

1. 实验准备 配制葡萄糖蛋白胨水培养基，定量分装至试管（2ml/管），灭菌后备用。

2. 接菌 以无菌操作技术将大肠埃希菌、产气肠杆菌斜面培养物分别接入葡萄糖蛋白胨水培养基中（注：接菌前注意观察未接菌的试管中培养基的状况：液体澄清、呈黄色。并同时保留一支未接菌的培养基试管作同步对照实验）。

3. 培养 将已接菌的各试管外加未接菌的对照管置37℃恒温箱培养24h后观察结果。

4. 加指示剂观察结果 取上步培养后的对照管和试验管，分别定量加入0.4ml 40%KOH（或NaOH），开盖充分振荡，与氧气接触氧化；再加入1ml 6%α-萘酚溶液充分振荡观察颜色变化（此步反应速度较慢，加热可加速反应。若培养基中所含胍基太少，可加入少量肌酸等含胍基的化合物）。

5. 结果判断

（1）对照管 液体澄清，呈黄色。

（2）试验管 试管中液体混浊，为黄色，证明均有菌生长。

阴性结果：若加指示剂后试管中液体未变成红色，证明其为V.P试验阴性。

阳性结果：若试管中液体变为深红色，证明其为V.P试验阳性。

【实验内容】

检测大肠埃希菌和产气肠杆菌代谢葡萄糖产生乙酰甲基甲醇的能力。

【结果】

观察记录实验结果，列入下表。

细菌 V.P 试验结果

菌名	V.P试验现象与结果
大肠埃希菌	
产气肠杆菌	

【注意事项】

在观察结果时，加试剂后需开盖充分振荡，此步反应是与空气中氧气接触氧化的过程。另外，此反应速度较慢，需5~10min才能出现结果。

【思考题】

甲基红试验和V.P试验的最终产物有何异同点？为什么会出现不同的最终代谢产物？

Experiment 31　V. P Test

Purpose

1. Grasp the principals and the detection method of the V. P test.

2. Be familiar with the ability of different bacterial species to produce acetoin using glucose.

Principles

The pyruvate produced by different bacteria could be metabolized to produce different products by various metabolic pathways. In some of these pathways unstable acidic products are produced which then quickly convert to neutral compounds, for example acetoin in *Enterobacter aerogenes*. The VP test detects organisms that utilize the butylene glycol pathway to produce acetoin. The acetoin end product is oxidized in the presence of potassium hydroxide (KOH) to diacetyl. Diacetyl then reacts with the amino acid containing Guanidyl to produce a red color compound. Therefore, being red is a positive result. Other organisms for example *Escherichia coli* cannot produce acetoin. This means after the reagents have been added, red color will not be present and the result is negative.

$$2CH_3COCOOH \longrightarrow CH_3COCHOHCH_3 + 2CO_2$$
$$\text{Acetyl methyl carbinol}$$

$$CH_3COCHOHCH_3 \xrightarrow[+KOH]{-2H} CH_3COCOCH_3$$
$$\text{Diacetyl}$$

$$\underset{\text{Diacetyl}}{\begin{matrix}O=C-CH_3\\O=C-CH_3\end{matrix}} + \underset{\text{Guanidyl}}{NH=C\begin{matrix}NH_2\\NH_2\end{matrix}} \longrightarrow \underset{\text{A red colour compound}}{NH=C\begin{matrix}N=C-CH_3\\N=C-CH_3\end{matrix}} + H_2O$$

Apparatus and Materials

1. Specimens   18 ~ 20h nutrient agar slant cultures of *Escherichia coli* and *Enterobacter aerogenes*.

2. Cultures and Reagents   MR – VP broth medium (Appendix I, 13), 40% KOH, 6% alpha – naphthol, etc.

3. Apparatus   Incubator.

4. Others   Tubes, pipette, inoculating loop, burner, etc.

Methods and Procedures

1. Preparation   Preparesterile MR – VP broth medium with 2ml in each tube.

2. Inoculation   Inoculate the media with either *E. coli* or *Enterobacter aerogenes*.

3. Incubation   Incubate all tubes at 37℃ for 24h.

4. Observation   After incubation: Add 0.4ml 40% KOH solution and mix it well with the cap open. Add 1ml of 6% alpha – naphthol. Mix well. Observe the phenomenon and record. Heating or adding some compound contained guanidino group for example the creatine in this step may accelerate the reaction to make the phenomenon more obvious.

5. Experimental phenomena

Blank tube: Clear medium, yellow color.

Test tubes: Opacitas medium and yellow color means they can grow in the medium. After adding reagents, the medium color turning to dark red means positive result. While no dark red color shown means negative result.

Experiment content
Test the ability of using glucose to produce acetoin in *Escherichia coli* and *Enterobacter aerogenes*.

Results
Records the phenomena in the following table.

Result of the V. P test

Bacteria	Phenomenon and result of the V. P test
Escherichia coli	
Enterobacter aerogenes	

Notices
After adding the KOH and alpha-naphthol reagents, the cap should be opened to ensure the oxidation of products which will make the proper phenomenon appear in 5~10min.

Questions
What are the similarities and differences between MR test and V. P test? Why?

（蔡苏兰）

实验三十二　枸橼酸盐利用试验

【目的】
1. 掌握柠檬酸盐利用试验的基本原理，检测方法与结果。
2. 了解不同细菌利用枸橼酸盐（柠檬酸盐）的能力。

【基本原理】
肠杆菌科各种细菌利用枸橼酸盐的能力不同，有的菌可利用枸橼酸钠作为碳源，有的则不能。某些菌分解枸橼酸盐形成 CO_2，培养基中钠离子而形成碳酸钠，使培养基碱性增加，根据培养基中的指示剂变色情况来判断试验结果。产气肠杆菌等能利用枸橼酸盐作为唯一碳源，并形成碱性，在含单一枸橼酸盐的培养基上能够生长。产气肠杆菌利用枸橼酸盐产生下列产物：枸橼酸盐→乙酸盐+甲酸盐+琥珀盐+ CO_2，利用有机酸和它们的盐作为碳源，进一步分解产生碱性碳酸盐和碳酸氢盐，使培养基由中性变为碱性，故使指示剂溴麝香草酚蓝（BTB）由淡绿色变为深蓝色，为枸橼酸盐利用试验阳性。大肠埃希菌不能利用枸橼酸盐，故不能生长，培养基仍为绿色，为枸橼酸盐利用试验阴性。指示剂可用1%的溴麝香草酚蓝乙醇溶液，变色范围pH6.0~7.6（黄→蓝）。

【仪器与材料】
1. **菌种**　大肠埃希菌（*Escherichia coli*）、产气肠杆菌（*Enterobacter aerogenes*）。
2. **试剂与培养基**　溴麝香草酚蓝溶液、枸橼酸盐琼脂培养基（附录一，14）。

3. 仪器 高压蒸汽灭菌锅。
4. 其他 试管、接种环、酒精灯。

【方法与步骤】

1. 试验准备 配制枸橼酸盐琼脂培养基，分装至试管，灭菌后摆成斜面备用。

2. 接种 以无菌操作技术将大肠埃希菌、产气肠杆菌分别接入枸橼酸盐斜面培养基中（注：接菌前注意观察未接菌的培养基的状况：斜面无菌生长时，颜色为绿色，pH 呈中性。并同时保留一支未接菌的培养基斜面做同步对照实验）。

3. 培养 将已接菌的各试管和未接菌的对照管置 37℃ 恒温箱培养 24 h 后观察结果。

4. 结果观察 无需加任何试剂，直接观察培养基颜色变化及细菌生长状况。

5. 结果判断

（1）对照管 斜面上未长菌，颜色仍为绿色。

（2）试验管

阴性结果：试管中斜面上未有菌生长，培养基颜色仍为绿色，证明该菌不能利用枸橼酸盐。

阳性结果：试管中斜面上有菌生长，培养基颜色变为蓝色，证明该菌能利用枸橼酸盐，产生碱性的碳酸盐。

【结果】

观察并记录实验结果，列入下表。

表 32-1 细菌枸橼酸盐利用试验结果

菌名	柠檬酸盐利用试验现象与结果
大肠埃希菌	
产气肠杆菌	

【思考题】

在鉴定大肠埃希菌时都要伴随鉴定产气肠杆菌，为什么？如何鉴别大肠埃希菌和产气肠杆菌？

Experiment 32　Citrate Utilization Test

Objective

1. Be familiar with the principle, methods and results of citrate utilization test.

2. Test if different bacteria could use citrate as the unique carbon source.

Principles

Enterobacteriaceae have different abilities to use citrate. *Enterbacter aerogenes* could use citrate as the unique carbon source to generate CO_2, which could combine with the sodium ion in the media to generate sodium carbonate. The media pH value could be increased after these reactions. Bromothymol blue (BTB) is a kind of indicator which has an indicating range from pH 6.0 (yellow) to pH 7.6 (blue). So in this test, the media could change from light green to dark blue.

While the *Escherichia coli* could not use the citrate to generate any product, so that it could not grow in the media only contains citrate. Also the media would maintain the green color.

Apparatus and Materials

1. Specimens   *Escherichia coli*, *Enterobacter aerogenes*.

2. Reagents   Citrate agar medium (Appendix I, 14), Brom thymol blue (Autoclave BTB).

3. Apparatus   Lamps.

4. Others   Inoculation loop, tubes, lamps.

Methods and Procedures

1. Preparation   To make citrate agar media, dispense to tubes, sterilize media to make the slants.

2. Inoculation   Inoculate the *E. coli* and *E. aerogenes* into two different tubes, and keep one citrate agar tube as control.

3. Culture   All three tubes were cultured at 37℃ for 24h.

4. Observation   To observe the color change and bacteria growth condition in those tubes, without adding any reagents more.

5. Experimental phenomena

(1) Control tubes   No growth of bacteria with green color.

(2) Test tubes

For negative results: There is none growth observed on the agar slant, with a green color unchanged, which indicated the citrate could be not used by this strain.

For positive results: The color of the slant changed from green to blue and growth lawn could be seen on the agar surface, which indicated that the citrate could be used to generate alkaline.

Results

Records the phenomena in the following table.

Bacteria	Phenomenon and resulf of the IMVIC test
Escherichia coli	
Enterobacter aerogenes	

Questions

Why the *E. aerougene* would always be tested when identifying the *E. coli*? How to differentiate between the *E. aerougene* and *E. coli*?

实验三十三　淀粉水解试验

【目的】
1. 掌握淀粉水解试验的原理及其试验方法。
2. 熟悉微生物点种接种法。
3. 了解细菌是否具有产生淀粉酶并利用淀粉的能力。

【基本原理】
不同微生物所含的酶系统不完全相同，其对复杂有机大分子的水解能力也不同。有些细菌和多数放线菌具有产生淀粉酶的能力，并可以将淀粉酶分泌到细胞外即产生胞外淀粉酶，淀粉酶可以使淀粉水解形成麦芽糖和葡萄糖，其可被细胞直接利用。碘液遇淀粉呈蓝色，而淀粉水解后遇碘不再变蓝而呈现透明圈。从菌落的周围是否出现蓝色，可判断该菌株是否水解淀粉；另外，还可以根据透明圈大小判断测试菌株水解淀粉能力的强弱。

【仪器与材料】
1. 菌种　大肠埃希菌（*Escherichia coli*）、枯草芽孢杆菌（*Bacillus subtilis*）、链霉菌（*Streptomyces* 4.92）。

2. 试剂与培养基　卢戈碘液、细菌淀粉水解培养基（附录）、链霉菌淀粉水解培养基（附录一，18）。

3. 仪器　高压蒸气压菌锅。

3. 其他　平皿、接种针、酒精灯等。

【方法与步骤】
1. 准备　配制淀粉琼脂培养基，灭菌后，倒入无菌平皿中，待凝固后制成无菌平板备用。

2. 接种　以无菌操作技术将大肠埃希菌、枯草芽孢杆菌、链霉菌4.92分别接入淀粉培养基中（注：用点种法接种，可以每个平板接一种菌，也可以一个平板同时接两种不同的菌）。

3. 培养　将已接细菌的各平板置37℃恒温箱培养24h；链霉菌平板生长繁殖速度缓慢，置28℃恒温箱培养3~5d。

4. 结果观察　取出培养后的平皿，打开平皿盖，滴加少量碘液于平板上，轻轻转动使碘液均匀铺满整个平板，观察现象。

【结果】
1. 观察接菌平板生长状况，点种处有菌落长出。
2. 检测结果　加入碘液后，菌落周围出现无色透明圈，则说明此菌可产生淀粉酶，将其周围培养基中淀粉水解，即为淀粉水解试验阳性；若菌落周围无透明圈，遇碘液后呈现蓝色，说明淀粉未被水解，此菌不产生淀粉酶，即为淀粉水解试验阴性；另外，针对阳性试验结果中，可以根据透明圈大小判断测试菌株水解淀粉能力的强弱。

列表记录实验结果，"+"表示阳性，"-"表示阴性。

表33-1 细菌淀粉水解试验结果

菌名	淀粉水解试验	现象（透明圈大小 mm）
大肠埃希菌		
枯草芽孢杆菌		
链霉菌 4.92		

【注意事项】

点种法是微生物实验中常用的实验技术，除用于本实验外，还可用于观察菌落特征的实验接种等。在本实验中，每个平皿不宜点种太多位置，以免形成的透明水解圈相互重合，不易量取透明圈大小。

【思考题】

此方法一般用于检测细菌所产生的胞外酶，除淀粉酶外，还可用于检测果胶酶、蛋白酶、纤维素酶等，是否可用于检测胞内酶，为什么？

Experiment 33　Starch Hydrolysis Test

Objectives

1. Be familiar with the principle and methods of starch hydrolysis.

2. Be familiar with the dibbling inoculation method.

3. Examine if the bacteria could produce amylase and hydrolyze starches.

Principles

Some bacteria and most actinomyces could produce amylase themselves, and secreted the enzymes out of the cell to produce extracellular amylase. Amylase could hydrolyze the starch into glucose and maltose, which could be used directly by the bacteria. When starch combines with iodine, the color changed to blue. When the starch was hydrolyzed, it could not be blue but indicated a transparent circle. It could be determined from the color change to identify wether the bacteria could hydrolyze starches. Also, it could be analyzed from the diameter of transparent circle to test the ability of a strain to hydrolyze the starch.

Apparatus and Materials

1. Strains　*Escherichia coli*, *Bacillus subtilis* and *Streptomyces* 4.92.

2. Reagents and media　Lugol iodine, Starch hydrolysis medium for bacteria (Appendix), Starch hydrolysis medium for *Streptemycete* (Appendix I, 18).

3. Apparatus　Autoclave.

4. Others　Petri dishes, needle inoculation, lamps.

Methods and Procedures

1. Preparation　To make aseptic petri plates with starch hydrolysis media.

2. Inoculation　　Inoculate the *E. coli*, *B. subtilis* and *Streptomyces* 4.92 into three different plates by dibbling method. (Spot inoculated one/two strains in one plate.)

3. Culture　　All inoculated bacteria plates were cultured at 37℃ in the incubator for 24h. Streptomycetes were cultured at 28℃ for 3~5 days.

4. Observation　　Flood the plate with iodine solution and observe.

Results

1. Observe the plate to identify the growth on the spotting site.

2. Test results　　When the iodine solution was flooded, the area where starch has been hydrolyzed by amylase will appear clear/transparency. While the development of a blue color indicated the presence of residual starch, indicated that the bacteria have no amylase.

Notes

Spot inoculation is a common technique, which could also be used in the inoculation of colony morphology observation. In this experiment, the amount of spot inoculations should be limited, so that the transparent zones could be measured.

Questions

This experiment could be use to test the bacteria extracellular enzymes, including amylase, pectinase, protease and cellulose. Could this method be used to test the intracellular enzyme, why?

实验三十四　产硫化氢试验

【目的】

1. 掌握产硫化氢试验的实验原理。
2. 熟悉细菌分解利用氮源并释放出硫化氢的检测方法。
3. 了解不同细菌利用氮源的能力。

【基本原理】

有些细菌能分解培养基中的含硫氨基酸，如胱氨酸、半胱氨酸、甲硫氨酸等产生硫化氢，硫化氢遇重金属盐如醋酸铅或硫酸亚铁，能生成黑色的硫化铅或硫化亚铁沉淀，从而可确定硫化氢的产生，则为硫化氢试验阳性。在试验中，可在培养基中加入氯化亚铁，培养后观察是否有黑色沉淀产生；也可在液体培养基中接种细菌，在试管棉塞下吊一块浸有醋酸铅的滤纸进行检测，细菌分解含硫氨基酸释放出H_2S，逸出的H_2S与滤纸上的醋酸铅反应形成黑色化合物。

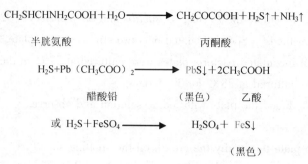

【仪器与材料】

1. 菌种　大肠埃希菌（*Escherichia coli*）、普通变形杆菌（*Proteus vulgaris*）。

2. 培养基　硫化氢试验培养基Ⅰ、硫化氢试验培养基Ⅱ（附录一，39，40）。

3. 仪器　高压蒸汽灭菌锅。

4. 其他　试管、接种环、接种针、酒精灯、醋酸铅试纸（附录一，38）。

【方法与步骤】

1. 方法一（穿刺接种法）

（1）接种　取硫化氢试验培养基Ⅱ 2支，分别穿刺接种大肠埃希菌和普通变形杆菌，保留一支未接菌的培养基试管做对照同步培养。

（2）培养　37℃培养24h。

（3）结果观察　观察培养后试管中菌种的生长情况及颜色变化，并与未接菌的对照管相比较。

2. 方法二（纸条法）

（1）接种　用新鲜斜面培养物接种硫化氢试验培养基Ⅰ。

（2）悬挂醋酸铅纸条　接种后，用无菌镊子夹取一条醋酸铅纸条用棉塞塞紧，使其悬挂于试管中，下端接近培养基表面，但不得接触液面，保留一支未接菌的培养基试管做对照同步培养。

（3）培养　37℃培养24h。

（4）结果观察　接种后3d、7d、14d分别观察培养物的生长情况及纸条颜色的变化。

【结果】

观察并记录实验结果，填于表34-1和表34-2。

1. 方法一　培养后的试管中如出现黑色沉淀线者为试验阳性。观察时注意穿刺接种细菌的线周围有无向外扩展的情况，如有，表示该菌有运动能力。

表34-1　产硫化氢试验结果（方法一）

试验	对照管	大肠埃希菌	普通变形杆菌
现象			
结果			

2. 方法二　观察培养后的试管中纸条颜色变黑者为阳性，不变者为阴性。

表34-2 产硫化氢试验结果（方法二）

试验	对照管	大肠埃希菌	普通变形杆菌
现象			
结果			

【思考题】

1. 方法一的阳性试验结果出现黑色沉淀，形成了什么产物？试说明反应原理。
2. 方法二的阳性试验结果滤纸条变黑，形成了什么产物？试说明反应原理。

Experiment 34　Hydrogen Sulfide Production Test

Objectives

1. Understand the abilities of the bacteria to use nitrogen resources.
2. Be familiar with the principle of hydrogen sulfide production.
3. Be familiar with the detection method to test H_2S.

Principles

Some bacteria could decompose the sulfur-containing amino acids, such as cystine, cysteine methionine to generate hydrogen sulfide. When hydrogen sulfide encountered heavy metal salts such as the lead acetate or ferrous sulfate, it could generate a black lead sulphide or iron sulfide precipitation, which could determine the generation of hydrogen sulfide. This is the positive result of hydrogen sulfide test. In this experiment, the culture medium could be supplemented with the ferrous chloride. The positive result could be obtained to see if a black precipitate was observed.

Also, it could be manipulated to inoculate bacteria into liquid media, and suspended lead acetate paper in a tube under a impregnated tampon to detect if the bacteria could decompose sulfur-containing amino acids to release H_2S. The escaping H_2S could react with the lead acetate on the paper to generate a black compound.

Apparatus and Materials

1. Strains　*Escherichia coli* and *Proteus vulgaris*.
2. Media　H_2S media Ⅰ, H_2S media Ⅱ (Appendix Ⅰ, 39, 40).
3. Apparafus　Autoclave.
4. Miscellaneous　Inoculation loop, invcwlation needle tubes, lamps, lead acetate paper (Appendix Ⅰ, 38).

Methods and Procedures

1. Stab inoculation method (Method 1)

(1) Inoculation　Stab inoculation *E. coli* and *P. vulgaris* into two H S tubes, keep a third tube for control.

(2) Culture　Incubate the tubes at 37℃ for 24h.

(3) Observation Observe the growth conditions and color in the tubes.

2. Lead acetate paper method (Method 2)

(1) Inoculation Inoculate the H_2S media I with fresh growth culture.

(2) Suspend the lead acetate paper Suspend the lead acetate paper in the inoculated tubes by aseptic forceps, which is close to the media. Keep a third tube for control.

(3) Culture Incubate the tubes at 37℃ for 24h.

(4) Observation Observe the media growth condition and lead acetate paper color change on 3d, 7d and 14d after inoculation.

Results

Fill in the blank in table 34 – 1 and 34 – 2, after the observation of the experiment results.

Questions

1. What black product had been generated in the stab region by using method 1? Please specify the principle.

2. What product had been generated on the black Lead acetate paper by using method 2? Please specify the principle.

实验三十五　明胶液化试验

【目的】

1. 掌握明胶液化试验的原理及检测方法。
2. 了解细菌对明胶的分解和利用的情况。

【基本原理】

不同微生物所含的酶系统不完全相同，其对复杂有机大分子的分解能力也不同。明胶是一种动物蛋白质，低于20℃呈固体，高于20℃自行液化呈液态。某些细菌及放线菌具有胶原酶可使胶原破坏后而失去凝固能力，虽在低于20℃的条件下亦不凝固。利用这一特点可进行菌种鉴别。变形杆菌、霍乱弧菌、铜绿假单胞菌、枯草芽孢杆菌等细菌和某些链霉菌能液化明胶；大肠埃希菌、沙门菌属则不能。

【仪器与材料】

1. **菌种**　大肠埃希菌（*Escherichia coli*）、产气肠杆菌（*Enterobacter aerogenes*）、枯草芽孢杆菌（*Bacillus subtilis*）、链霉菌（*Streptomyces* 4.393）。

2. **培养基**　明胶液化培养基（附录）。

3. **仪器**　高压蒸汽压菌锅。

4. **其他**　试管、接种环、酒精灯等。

【方法与步骤】

1. **准备**　配制明胶液化培养基，灭菌后，倒入无菌试管中，待凝固后备用。

2. **接种**　用穿刺接种法分别将大肠埃希菌、产气肠杆菌、枯草芽孢杆菌接种于明胶培养基试管中。而接种链霉菌时不用穿刺接种法，只接于表面（保留一支未接菌的对照管做对照试验）。

3. 培养 放置于20℃恒温箱中培养48h。若细菌在20℃下不生长，则应放在最适温度下（37℃）培养。链霉菌可培养于28℃，因其生长速度慢，需培养5d、10d和20d。

4. 结果观察 取出培养后的试管，观察培养基有无液化情况及液化后的形状，并与对照管相比较。

【结果】

因明胶在低于20℃时凝固，高于20℃时自行液化，若是在高于20℃下培养的细菌，观察时应放在冰浴中观察，若明胶被细菌液化，即使在低温下明胶也不会再凝固，此为阳性反应；若高于20℃时呈液化状，而在冰浴中则又呈凝固状，则为阴性反应。

明胶液化形状见图35-1。

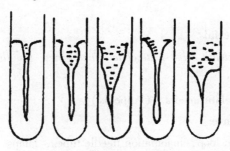

图35-1 明胶液化形态图

Figure 35-1 Morphology of gelatin hydrolysis

列表记录实验结果（表35-1）。"+"表示阳性，"-"表示阴性。

表35-1 细菌明胶液化试验结果

菌名	明胶液化实验现象与结果	
	20℃以上	冰浴中
大肠埃希菌		
产气肠杆菌		
枯草芽孢杆菌		
链霉菌4.393		

【注意事项】

1. 穿刺接种法是微生物实验中常用的接种方法，除用于本实验外，还可用于观察细菌是否具有鞭毛（即半固体培养基穿刺接种法）。在用接种针进行穿刺接种时，手要稳，只接一条细线为最佳，切勿搅动，以免使接种线不整齐而影响观察。

2. 因链霉菌生长速度慢，需延长培养时间，在第5d、10d和20d分别观察结果，最后判定其是否液化明胶。

【思考题】

此试验为什么要在冰浴中观察结果？

Experiment 35 Gelatin Hydrolysis Test

Objectives

1. Understand how the bacteria hydrolyze the gelatin.

2. Be familiar with the principle and test method for gelatin hydrolysis.

Principles

Different microorganisms have diversified enzyme systems, which have different hydrolyze abilities for the macro – organic complex. Gelatin is a kind of animal protein, which could solidify under 20℃, and liquefy above 20℃. Some gelatinase – containing bacteria could hydrolyze the gelatin to make the culture liquefy, such as *P. vulgaris*, *B. substilis*, while E. coli and Salmonella sp. could not.

Apparatus and Materials

1. Specimens *Escherichia coli*, *Enterobacter aerogenes*, *Bacillus subtilis* and *Streptomyces* 4.393.

2. Media Nutrient gelatin media. (Appedix)

3. Apparatus Autoclave.

4. Others Inoculation loop, inaculation needle tubes, lamps.

Methods and Procedures

1. Preparation Make the nutrient gelatin media, sterilization, and dispense to the tubes.

2. Inoculation Stab inoculation the *E. coli*, *E. areogenes*, *B. subtilis* into three different tube, inoculate the *streptomycetes* on the surface, and keep one more tube for control.

3. Culture Incubate the media at 20℃ for 48h. If the bacteria could not growth under 20℃, the tubes should incubate under 37℃. The streptomycetes could be incubated at 28℃ for 5d, 10d and 20d days.

4. Observation Observe the tubes to see if the media turned liquefying, compare the result with the control tube.

Results

Gelatin solidifies under 20℃, and liquefies above 20℃. If the bacteria were incubated above 20℃, it should be observed in the ice bath. If the gelatin has been liquefied, it could not resolidfy even in the low temperature. This is positive result for the presence of enzyme gelatinase. If the culture liquefied above 20℃ and resolidified in ice bath, it is an indication of negative result (Figure 35 – 1).

Notes

1. Stab inoculation could be also used to test if the bacteria having flagella by semi – solid

stab inoculation.

2. Streptomycete grows slowly, so the results should be observed on day 5, 10 and 20, to identify the gelatin liquefying.

Question

Why should the result been observed in the ice bath?

（徐慰倬）

第十节　免疫学实验技术

Section 10　Immunological Experimental Techniques

实验三十六　E 玫瑰花环试验

【目的】

1. 掌握 E 花环试验的原理和操作方法。
2. 熟悉淋巴细胞和外周血单核细胞的分离方法。
3. 了解显微镜下 E 玫瑰花环的形态。

【基本原理】

人类外周血 T 淋巴细胞表面有绵羊红细胞（sheep red blood cell，SRBC）的受体，即 CD2 分子（又称淋巴细胞功能相关抗原 2），是 T 淋巴细胞特异标志。在最适温度 4~5℃的体外条件下，T 淋巴细胞与绵羊红细胞的比例为 1∶50~1∶100 时，绵羊红细胞可粘附于 T 细胞周围，形成以单个 T 细胞为中心的玫瑰样的细胞团（吸附 3 个或 3 个以上绵羊红细胞），称为 E 玫瑰花环试验（erythrocyte rosette test）。形成此种花环的 T 细胞称为 E 花环形成细胞。

本试验的意义在于：可对 T 细胞进行鉴定与计数，以辅助测定机体的细胞免疫水平，为临床某些疾病的诊断和防治提供免疫方面的重要参考；观察受检者的细胞免疫功能及免疫增强剂的疗效；观察免疫抑制剂的效果；协助进行恶性肿瘤疗效的观察及预后判断等。

检测标本中 T 淋巴细胞的总数为总 E 花环实验（Et，t 为 total 的缩写）；检测对绵羊红细胞具有高亲和力的那部分 T 淋巴细胞称为活性 E 花环试验（Ea，a 为 active 的缩写）。对绵羊红细胞具有高亲和力是指这部分淋巴细胞在与较低比例绵羊红细胞混合，低速离心后不经低温放置，即能迅速形成玫瑰花样花环。在某些免疫缺陷病和各种恶性肿瘤的研究中，患者的 Et 玫瑰花环可以正常，但 Ea 玫瑰花环减少，因此 Ea 花环试验能敏感地反映机体的细胞免疫功能和动态变化。

正常人外周血绵羊细胞玫瑰花环（即 T 细胞）约为淋巴细胞总数的 68%±9.9%。

Ea 花环形成率正常值为 25%~40%。

Ea 花环形成率 = 形成花结的淋巴细胞数/淋巴细胞总数（形成花环 + 未形成花环）×100%

【仪器与材料】

1. 试剂　肝素抗凝血、Hank's 液、淋巴细胞分离液、绵羊红细胞悬液、灭活小牛血清。

2. 仪器　水平离心机、37℃培养箱、显微镜等。

【方法与步骤】

1. 取 2ml 人肝素抗凝血于试管中，加等量 Hank's 液稀释，手动轻轻摇匀。

2. 用吸管吸取稀释抗凝血，缓慢加入淋巴细胞分离液上，注意保持两种液体界面清晰。

3. 室温下 2000r/min，离心 20min，离心后分为 4 层。

4. 用吸管吸取血浆与分层液界面处单核细胞层，移入另一干净试管。

5. 加入 5 倍体积的 Hank's 液吹打混匀，1000r/min 离心 5min，弃上清，用少量回流液重悬沉淀细胞，制成淋巴细胞悬液。

6. 取 0.1ml 淋巴细胞悬液，加等量灭活小牛血清（0.1ml）加入等量 1% 绵羊红细胞悬液混匀，1000r/min 离心 5min，取出。

7. 缓慢旋转试管底部，轻轻摇匀，吸取少许滴于载玻片上，加盖玻片，镜下观察（Ea）。

【实验内容】

利用人肝素抗凝血与绵羊红细胞在适宜条件下制备 E 玫瑰花环并观察其形态。

【结果】

计数花环个数并计算 Ea 花环形成率。

【注意事项】

1. 实验血液新鲜分离，是为了保证淋巴细胞的活性，不影响 E 玫瑰花环形成率。

2. 绵羊红细胞要新鲜，并采用无菌脱纤维羊血。

Experiment 36　Erythrocyte Rosette Test

Objectives

1. Be familiar with the isolation methods for the lymphocytes and monocytes from peripheral blood.

2. Be familiar with the principle and manipulation protocol for the erythrocyte rosette test.

3. Be familiar with the morphology of erythrocyte rosette test under microscope.

Principles

The surface of human peripheral T lymphocytes contains the receptor CD2 (lymphocyte function associated antigen – 2), which may bind to the sheep erythrocytes (E receptor), and is the specific marker for T lymphocyte. When those cells are incubated at 4 ~ 5℃ in vitro, and the ratio of human T – lymphocyte versus sheep red blood cell varies from 1∶50 ~ 1∶100. The sheep red blood cells will then bind the human T – lymphocyte cells, to form a T – cell centered rose – like cells colony (which may absorb three or more sheep red cells). This is called the erythrocyte rosette test. The T – cells forming this rosette were called erythrocyte rosette forming cells.

The E – rosette test has been used extensively in T lymphocytes counting and identification, which may help to exam the cellular immunological condition in human body. This test could be used to set up immunological reference for the disease diagnostics, to monitor the patient's cellular immunological function and effects of immunopotentiators, to evaluate the effect of immunosuppressant, and to judge the curative effects of malignant neoplasm and the prognosis.

The total erythrocyte rosette test (Et) is performed to numerate the numbers of T lymphocyte, while the active erythrocyte rosette test (Ea) is performed to numerate the numbers of T lymphocytes with high affinities to the sheep red cells. The high affinities to sheep red cells is termed such lymphocytes could form the rosetteimmediately when mixed with low concentration of sheep red cell at low centrifugation speed without low temperature incubation. In some immunodeficiency diseases and malignant neoplasm researches, the patients may display a normal Et but a decreased Ea count, so that the Ea count could display the sensitive variance of human cell immunological function.

The normal human peripheral blood sheep red cell rosette (T lymphocyte) ratio is about 68% ±9.9% of the total lymphocytes. The average Ea rate is about 25% ~ 40%.

The forming ration of Ea = Lymphocytes formed the rosette/Total counts of lymphocytes (lymphocytes formed the rosette + lymphocytes not formed the rosette) ×100%

Apparatus and Materials

1. Reagents   Heparin treated anticoagulant blood, Hank's solution, lymphocytes isolation solution, sheep red blood cells suspension, inactive calf bovine serum.

2. Apparantus   Centrifugator, incubator, microscopes.

Methods and Procedures

1. Draw 2ml heparin – anticoagulant blood into tube, supplement equal volume of Hank's solutions, mix the sample gently.

2. Gently pipette the heparin – anticoagulant blood on the lymphocytes isolation solution.

3. Centrifugation at 2000 r/min for 20min.

4. Carefully remove the white mononuclear cells (containing the lymphocytes) layer into a new tube.

5. Wash the cells with 5 × volume of Hank's solution, centrifugation at 1000 r/min for 5min, discard the supernatant, and resuspend the lymphocytes.

6. Take 0.1ml of the lymphocyte suspension solution and inactive calf bovineserum, add equal volume of 1% sheep red blood cells, mix gently and centrifuge at 100 r/min for 5 min.

7. Gently rotate the bottom of the tube, take several drops onto the slides, lay thecoverslips and observe the samples under microscope.

Experiment content

Make the erythrocyte rosette under the proper condition with human heparin – anticoagulant blood and sheep red blood cells. Observe the test results under the microscopes.

Results

Count the rosette numbers and the rosette forming ratio.

Notes

1. The use of fresh blood is to guarantee the activity of lymphocytes.

2. The sheep blood should be fresh and filtered to move the fibres.

实验三十七　淋巴细胞增殖试验——体内法

【目的】

1. 掌握淋巴细胞转化的原理。
2. 熟悉体内法检测转化淋巴细胞的方法。

【基本原理】

T淋巴细胞和B淋巴细胞都具有丝裂原受体，在体内或体外遇到有丝分裂原刺激后，可转化为淋巴母细胞，称淋巴细胞转化实验。T淋巴细胞表面存在的相应丝裂原受体包括美州商陆素（PWM）、植物血凝素（PHA）受体和刀豆蛋白A（ConA）受体。B淋巴细胞表面常用的丝裂原受体为美洲商陆素（PWM）、脂多糖（LPS）和葡萄球菌A蛋白（SPA）等受体。基于这一原理，目前常用淋巴细胞转化试验以计算在PHA刺激下，T淋巴细胞转化成母细胞的数目，或测定被摄取含氚^{3}H标记的胸腺嘧啶核苷（^{3}H-TdR），来反映机体内T淋巴细胞的免疫功能。本实验可用于结核、布氏杆菌病、慢性肝炎等传染病细胞免疫功能的测定，以及判断肿瘤的疗效和愈后。

转化的淋巴细胞包括过渡型淋巴细胞和淋巴母细胞。过渡型淋巴细胞：比小淋巴细胞大，约10~20μm，核染色致密，但出现核仁，此为与成熟小淋巴细胞鉴别要点。淋巴母细胞：细胞体积增大，约20~30μm，比原来成熟的未转化小淋巴细胞大3~4倍或更大；核大呈蓝色，核质疏松，核内出现红色核仁1~3个；胞浆丰富，有时呈伪足突起。此类母细胞常多个聚在一起，也有散在的。淋巴细胞转化率的计算：计算过渡型和淋巴母细胞百分率。在正常情况下，PHA诱导的淋巴细胞转化率为60%~80%，如为50%~60%则偏低，50%以下则为降低。

利用淋巴母细胞的不同特点，目前有多种实验方法可用于淋巴细胞转化程度的检测。根据其形态学改变，可通过体内法和体外法检测；根据细胞内核酸和蛋白质合成增加的特点，可通过^{3}H-TdR掺入法检测；根据细胞代谢功能旺盛的特点，可通过

MTT 法进行检测。

本实验采用体内法。在体内,当外周血 T 淋巴细胞遇到 PHA 或 ConA 后可发生转化形成淋巴母细胞,通过采集外周血涂片染色,镜下计数 100~200 个淋巴细胞,计算其转化率。转化率高低可反映机体细胞免疫水平的高低,因此,常作为检测细胞免疫功能的指标之一。形态学方法简便易行,但结果受操作和主观因素影响较大。转化过程中,常见的细胞类型有:淋巴母细胞、过渡型淋巴母细胞、核分裂相细胞和成熟淋巴细胞等。计数时,过渡型淋巴母细胞和核分裂相细胞亦作为转化细胞。

转化率 = 转化的淋巴细胞数/(转化的淋巴细胞数 + 未转化的淋巴细胞数) × 100%

【仪器与材料】

1. 试剂 ConA 溶液:根据 ConA 的纯度配制成最适浓度,一般为 0.3~0.5mg/ml。姬姆萨染液(附录三,13)或瑞氏染液。

2. 仪器 离心机、显微镜、计数器等。

【方法与步骤】

1. 实验前 3d,每只小鼠腹腔注射 ConA 0.3~0.5mg。
2. 3d 后,通过摘除小鼠眼球采集外周血,加入预先加有肝素的试管中。
3. 涂片:取 1 小滴抗凝血滴在玻片中央,用推片将血液涂开,自然干燥。
4. 固定:取甲醇 1~2 滴滴在涂片上,自然干燥。
5. 染色:加姬姆萨或瑞氏染液 2 滴于涂片上,同时加 2 滴水,用吸管水平涂开,使染液均匀覆盖涂抹面,染色时间为 5~10min。
6. 自来水细水冲洗染液,然后用吸水纸轻轻吸干玻片上的液体。
7. 显微镜观察结果,计数淋巴转化细胞,计算转化率。

【实验内容】

利用刀豆蛋白 A(ConA)作为有丝分裂原刺激小鼠,体内法检测转化淋巴细胞。

【结果】

计数淋巴转化细胞,计算转化率。

Experiment 37　Lymphocyte Proliferation Assay

Objectives

1. Be familiar with the principle for the lymphocyte proliferation assay.
2. Be familiar with the detection method for lymphocyte proliferation.

Principles

Mitogen receptors exists on the surfaces of both T and B lymphocytes, which may transform to lymphoblast when encountered the mitogen in vivo or in vitro. Mitogen receptors expressed on the T lymphocytes are PWM、PHA and ConA receptors, while PWN, LPS and SPA receptors were usually expressed on B lymphocytes surfaces. Based on this principle, the present lymphocyte transformation experiment could be used to exhibit the T lymphocyte immunological function by numerating the cell amounts transformed to lymphoblast from T lymphocyte upon the stimulation of PHA, or detecting the ingested ^{3}H marked thymidine (^{3}H – TdR). This experiment could be used to detect the chronicle infections such as tuberculosis, Brucellosis, hepatitis, and evaluate the treatment and prognosis of tumor.

The transformed lymphocytes are consisted of transitional lymphocytes and lymphoblast. The transitional lymphocytes are bigger than the small lymphocytes, 10 ~ 20μm, with the dense coloring in nucleoplasm and nucleolus. While the lymphoblast is about 20 ~ 30μm, three to four times bigger than the mature but untransformed small lymphocytes, in which the nucleus is blue and the nucleoplasm is loose with 1 ~ 3 red nucleolus. The cytoplasm is abundant with expanded pseudopodia sometimes. These lymphoblast usually assemble together, but sometimes scattered. The transformation rate is calculated by the ratio of transitional lymphocytes and the lymphoblast. In the normal condition, the average lymphocytes transformation ratio induced by PHA varies from 60% ~ 80%, a lower range may vary from 50% ~ 60%. If the ratio is less than 50%, then the patient should go to the hospital.

There are many methods used to detect the levels of lymphocytes transformation. Morphologically, there are the tests performed in vivo and in vitro. ^{3}H – TdR incorporation test could be performed according to the increasing of nucleotides and protein synthesis in the lymphocytes, while MTT test could be performed according to the rushing metabolism in lymphocytes.

In this test, the in vivo method is used. The peripheral T lymphocytes may transformed into lymphoblast when encounter PHA or ConA. We could calculate this transformation ratio by performing the peripheral blood smear and count 100 ~ 200 lymphocytes by microscopy. The transformation ratio may indicate the immunological function, so that it could be used as a marker for cellular immunological function examination. The morphological method is easy to perform, but the results may vary according to the manipulation and other subjective factors. The common cells are consisted of lymphoblast, transitional lymphoblast, mitotic cells and mature lymphocytes. When the calculation is performed, the transitional and mitotic cells could be recognized as transformed cells.

Transformation ratio = transformed lymphocyte/ (transformed lymphocytes + untransformed lymphocytes) ×100%

Apparatus and Materials

1. Reagents   ConA solution, 0.3 ~ 0.5mg/ml, Giemsa solution (Appendix III, 13) or Wright's dye solution.

2. Apparatus   Centrifugator, microscope, counter, etc.

Methods and Procedures

1. 3 days before the experiment, intraperitonealy injected ConA 0.3 ~ 0.5mg into each mouse.

2. Orbital blood was collected into the heparin – anticoagulant tubes.

3. Smear   Take one drop of anticoagulant blood in the middle of slide, smear the blood and dry in the air.

4. Fixation   Take 1~2 drop of methanol on the slide and dry in the air.

5. Stain   Add 2 drops of Giemsa/Wright's solution on the slide, supplement 2 drops of water, spread the sample, dye for 5~10 min.

6. Wash the sample with tap water to remove excess dye. Use the water–absorption paper to make the smear dry.

7. Examine the stained slides under microscope. Count the transformed lymphocytes and transformation ratio.

Experiment content

Stimulate the mice by ConA as mitogen, detect the transformed lymphocytes in vivo.

Results

Count the transformed lymphocytes and calculate the transformation ratio.

实验三十八　凝集反应

【目的】

1. 熟悉玻片凝集与试管凝集的操作与凝集现象观察。
2. 熟悉免疫血清效价的测定。
3. 了解血清学反应的基本原则。

【基本原理】

凝集反应是经典的血清学反应，是在电解质存在的情况下，颗粒性抗原（如细菌或红细胞等）与相应的抗体结合生成肉眼可见的凝集块，又叫直接凝集反应。电解质（一般用生理盐水）的作用主要是消除抗原抗体结合物表面上的电荷，使其失去同电相斥的作用而转变为互相吸引，否则即使抗原与抗体产生结合亦不能聚合成明显的肉眼可见的凝集块。

直接凝集反应分为玻片凝集法与试管凝集法两种。玻片凝集法是一种定性方法，利用已知抗血清鉴定未知细菌常用玻片凝集法，其突出优点是极端快速，为诊断肠道传染病时鉴定病人标本中肠道细菌的重要手段。试管凝集法是一种定量法，可利用已知抗原测定人体内抗体的水平（效价），也是诊断肠道传染病的重要方法，例如诊断伤寒、副伤寒的肥达氏反应便是试管凝集反应。

试管凝集法是以一定浓度的抗原与一系列的抗体稀释液在试管中进行试验，主要用于抗体定量，说明抗体凝集抗原的能力，也称定量凝集反应。当抗体的某一稀释度能凝集抗原，而再进行稀释后的抗体则不能凝集抗原时，前者的稀释倍数为该免疫血清（抗体）的效价，如抗体的效价为6400×，即把抗体稀释到6400倍还有凝集抗原的能力，再稀释就无效。一般肉眼可观察凝集现象及上清液的透明度。

试管凝集反应常将试管放在一定温度的水浴箱中进行反应，因温度可促进抗原颗粒的分子运动，使细菌与细菌间互相碰撞的机会增多，从而使反应迅速产生，一般均放在37℃或56℃的水浴箱中进行反应，温度超过60℃则将引起抗体性质的损坏。

发生明显凝集反应（用"＋＋"表示）的最高稀释度即为该免疫血清的效价。

【仪器与材料】

1. 菌种和试剂　抗－A血清、抗－B血清、灭活的大肠埃希菌（7×10^8/ml）、待检测的动物血清（由大肠埃希菌抗原刺激的免疫血清）、生理盐水。

2. 其他　载玻片、小试管、试管架、吸管、水浴锅等。

【方法与步骤】

（一）玻片凝集法

1. 用记号笔把干净的载玻片分为两个区域，分别标记A和B。
2. 从标准血清试剂瓶中在A端挤一滴抗－A血清，B端挤一滴抗－B血清。
3. 将手指皮肤消毒，干燥后用无菌针迅速刺伤皮肤。
4. 将血清分别滴在载玻片两端，与抗－A和抗－B血清各自混合。
5. 挤压手指，用无菌棉签按住止血。
6. 反应数分钟后，在白色背景下观察结果。

（二）试管凝集法

1. 取6支试管编号，放在试管架上。
2. 加0.9ml生理盐水于第1试管内，其余试管各加0.5ml。加0.1ml大肠埃希菌免疫血清于第1试管内，与生理盐水混合均匀。从第1管吸0.5ml稀释免疫血清注入第2管，同样方法混匀，再自第2管吸取0.5ml注入第3管，以此类推至第5管。混匀后，自第5管吸出0.5ml弃去。第6管不加血清，作为阴性对照。最后，向每支试管中加入0.5ml大肠埃希菌菌悬液。表38－1是实验步骤。

表38－1　凝集反应步骤表

试管	1	2	3	4	5	6
生理盐水（ml）	0.9	0.5	0.5	0.5	0.5	0.5
大肠埃希菌稀释血清	0.1	0.5	0.5	0.5	0.5	—
细菌悬液	0.5	0.5	0.5	0.5	0.5	0.5
最终血清稀释度	1:20	1:40	1:80	1:160	1:320	0

3. 轻摇混匀，将试管架放在56℃水浴箱中1~2h。

【结果】

1. 记录　如果发生凝集反应，混合的流动血清会逐渐从浑浊变为澄清状态，出现红色的凝集块。如果保持浑浊，则未发生凝集反应。血型可以由此鉴定。（表38－2）

表38-2

血清＼血型	A	B	AB	O
抗A血清	+	-	+	-
抗B血清	-	+	+	-

2. 试管凝集反应结果

（1）对照组（6号管）：上清浑浊，抗原在管底凝集，轻轻摇动使细菌分散成均匀的浑浊液体。

（2）实验组：观察1～5号试管，管底形成边缘不整齐的凝集块为阳性结果，和对照管相同的为阴性结果。

（3）凝集反应程度分类

＋＋＋＋：液体澄清，凝集块完全沉于管底。轻轻摇动可清晰观察到大的凝集凝块。

＋＋＋：液体稍混浊，凝集块沉于管底，轻轻摇动可观察到较小的凝集凝块。

＋＋：液体明显浑浊，轻轻摇动可观察到更小的凝集块。

＋：液体浑浊，仔细观察可见微小凝集块。

－：对照组采用相同记录方法。

（4）凝集效价的测定：效价通常代表发生明显凝集反应（＋＋）的最高稀释度。

【注意事项】

1. 将载玻片分为A，B两区域，保持两端血清液体分离。
2. 取血前将手指消毒，不要在手指未干前刺破，以免乙醇损坏红细胞。
3. 及时观察实验结果。
4. 将血液污染物严格消毒，以避免传播疾病，再丢弃到指定地点。
5. 依次对试管编号，不同试剂的移液管不可混用。
6. 稀释抗血清时，确保混合均匀再加入下一号试管。
7. 精确加入试剂，避免气泡。

Experiment 38　　Coagulation Test

Objective

1. Be familiar with the manipulation protocol for slide and tube coagulation test.
2. Be familiar with the titrate detection of immunological sera.
3. Understand the reaction principles of serology.

Principle

Coagulation is a classical serological reaction, which is called direct coagulation test. In this test, the granular antigen (such as bacteria and red cells) may coagulate with the antibodies to form the visible clot by naked eyes. The electrolyte (usually saline) could be used to eliminate the electronic charges both on the surfaces of antigen and antibody, so as to change the electric repulsion to the absorption, to form the visible coagulation clot by the antigen – antibody reaction.

The direct coagulation reaction could be classified as slide coagulation and tube coagulation method. Slide coagulation is a qualitative test, which could identify the unknown bacteria by known antisera, to provide a quick diagnostic of the intestinal diseases. Tuber coagulation is a quantitative method to test the antibody titers by known antigen, which could be used to identify the intestinal diseases, such as the Widal tests for the Typhoid and Paratyphoid.

Tube coagulation method is performed by reacting a series of antibody dilution with the known concentration of antigen, to provide the quantitative measurement of the antibody. When some dilution of the antibody could coagulate with the antigen, and the subsequent dilution antibody could not coagulate with the same antigen, the former dilution could be decided as the titer of the antibody. If some antibody has a titer of 6400 ×, which means the antibody could be diluted into 6400 times and still could coagulate with the antigen. Usually the coagulation and the supernatant transparency could be visualized by the naked eyes.

Tuber coagulation test could be performed in the warm water bath, because the warm temperature may stimulate the molecular motion of antigen, so as to increase the collision opportunities between bacteria and bacteria, and the reaction velocity. Normally, the reaction could be performed in the water bath at 37℃ or 56℃. When temperature is above 60℃, the antibody would be denaturized.

The maximal dilution (shown as " + + ") which may coagulate with the antigen is termed as the titer for the antisera.

Apparatus and Materials

1. Strains and Reagents   Anti – A serum, anti – B serum, inactivated *Escherichia coli* (7×10^8/ml), animal sera stimulated by the *Escherichia coli*, saline.

2. Others   Slide, tubes, racks, pipette, and water bath.

Methods and Procedures

Ⅰ. **Slide coagulation methods**

1. Take a clean slide and divide it into two sections. Mark A and B on different sections.

2. Take one drop of standard anti – A serum on A section, and standard anti – B serum on B Section.

3. Stab the disinfected finger skin.

4. Drop bloods into two sections of the slide. Mix with anti – A and anti – B serum respectively.

5. Press finger to stop bleeding with aseptic tampon stick.

6. Keep the slide still for a few minutes. Observe the result on white background.

Ⅱ. **Tube coagulation methods**

1. Settle a series of 6 tubes on a rack.

2. Add 0.9ml of saline in Tube 1, and 0.5ml of saline in the rest of tubes. Add 0.1ml of *E. coli* immunized serum in Tube 1, mix throughoutly. Pipette 0.5ml of diluted serum from Tube 1 to Tube 2, serial diluted till Tube 5. There is no serum in Tube 6, which is set as the negative control. Add 0.5ml of *E. coli* suspension serum to each tube finally (Table 38-1).

Table 38-1 Adding sample table of tube agglutinatin test

Tube number	1	2	3	4	5	6
Normal salineml	0.9	0.5	0.5	0.5	0.5	0.5
E. coli immunized serum	0.1	0.5	0.5	0.5	0.5	—
E. coli suspension	0.5	0.5	0.5	0.5	0.5	0.5
Final serum dilution	1:20	1:40	1:80	1:160	1:320	control

3. Mix gently, incubate the tubes at 56℃ for 1~2h.

Results

1. If coagulation happened, the mixed flowing serum would become clarified from turbid, and red clot appeared. If the turbid condition maintained, there is no coagulation. (Table 38-2)

Table 38-2

Blood serum \ Blood type	A	B	AB	O
anti-A serum	+	−	+	−
anti-B serum	−	+	+	−

2. Results for the tube coagulationtest

(1) Negative Control (Tube 6), Turbid in supernatant, antigen aggregates in the tube bottom, homologous turbid solution could be achieved by gently shake the tube.

(2) Treatment group, Observe Tube 1~5, irregular clots appeared in the tube bottoms indicated the positive results.

(3) Classification of coagulation

++++: Clarified solution, clots sank to the tube bottom, large clots could be visualized by naked eyes.

+++: Light turbid solution, clots sank to the tube bottom, small.

++: Middle turbid solution, smaller clots could be visualized by naked eyes.

+: Turbid solution, tiny clots could be visualized by naked eyes.

−: Negative control.

(4) Measurement of coagulation titrates: Titrate usually represents the max dilution multiples for the obvious coagulation reaction.

Notes

1. Divide the slide into A and B sections. Keep the serum apart.

2. Do not stab the finger skin until the alcohol evaporated.

3. Observe the experiment result in time.

4. Discard the blood – containments to proper locations.

实验三十九 沉淀反应——琼脂双向扩散实验

【目的】

1. 掌握双向琼脂扩散实验原理。
2. 熟悉双向琼脂扩散实验方法。

【基本原理】

琼脂扩散是利用琼脂凝胶作为介质的一种沉淀反应。含有充分水分的半固体琼脂凝胶有如网状支架，允许可溶性抗原和相应抗体在其网间相遇，如有适量电解质存在，则在两者相遇且分子比例适合处形成白色沉淀线。以相应的抗原抗体分别放在半固体琼脂中的两个孔内，使其相互扩散，一定时间后，在两孔之间浓度比例合适的部位相遇时，即出现乳白色沉淀线，此为双向扩散法。

若在两孔内有两对或两对以上的抗原抗体，就能产生相应数量分离的沉淀线，一次利用此法可以进行抗原或抗体的纯度分析，并可在琼脂板上同时挖 4~6 个孔，进行不同来源的抗原或抗体比较。

此法操作简单、灵敏度高，是最为常用的免疫学检测方法。

【仪器与材料】

1. 试剂 正常人混合血清、羊抗人全血清 IgG、未知 Ab1 和 Ab2、生理盐水配制的 1% 琼脂。

2. 其他 载玻片、打孔器、湿盒、37℃恒温箱。

【方法与步骤】

1. 制备琼脂板 吸取 3ml 用生理盐水配制的 1.0% 液态的琼脂，浇于载玻片上，制成琼脂板。

2. 打孔 待琼脂凝固后，用打孔器打 5 个距离相近的孔，如图 39-1，并用针头挑去孔中琼脂（孔距为 5mm）。

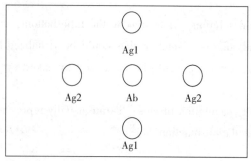

图 39-1 打孔加样示意图

Figure 39-1 Photo indicating of adding sample

3. 加样 在中心孔加入羊抗人全血清免疫球蛋白（标记为抗体），上下两孔中加入 Ag1，左右两孔中加入 Ag2。

4. 培养 将加好样品的琼脂板置于湿盒内，于37℃温箱内放置24h后观察结果。

5. 结果判断 沉淀线在中心孔和周边孔之间出现，即为阳性反应。表明抗体和相应抗原具有特异性。若没有沉淀线，为阴性反应。表明抗体和相应抗原没有特异性。

【实验内容】
学习双向琼脂扩散平板法，观察琼脂中形成的沉淀线。

【结果】
观察并绘图表示孔间沉淀线的数目与特征。

【注意事项】
1. 琼脂板应该一次铺成平整光滑的平板。
2. 倒置样品瓶，悬空加样。轻轻挤压，避免过多的液体溢出加样孔，影响形成沉淀线的形成。

Experiment 39 Precipitation Reaction Test

Objective

1. Be familiar with principles for the double agar diffusion test.
2. Be familiar with the protocols for the double agar diffusion test.

Principle

Agar diffusion is kind of precipitation test with the agar as media. The hydrated semi solid agar could be used as the scaffold, which may enable the soluble antigen and antibody to go through freely. If the antigen and antibody encountered somewhere, a white precipitation line could be formed and visualized. The antigen and antibody could be loaded in two separate wells in the semi solid agar and diffuse freely. After some time, a white precipitation line could be formed and visualized at the proper position. This method is called bidirectional diffusion.

If there are more than two pairs of antigen and antibodies, the proper precipitation lines would also be visualized. This method could be used to analyze the purity of the antigen and antibody, and to screen the best pair from different antigen and antibody by digging 4~6 wells in a plate at the same time.

This method is an easy, sensitive and commonly used immunological test method.

Apparatus and Materials

1. Reagents   Normal human serum, sheep anti human IgG, unknown Ab1 and Ab2, 1% agar in saline.
2. Others   Slide, hole puncher, humidify box, incubator.

Methods and Procedures

1. Agar plating   Take 3ml of 1.0% liquefied agar in saline on the slides to make agar

plate.

2. Well punching When agar solidified, punch 5 wells (Figure 39-1), and pick out the agar.

3. Loading samples Load sheep anti human IgG in the middle well, load Ag1 in the upper and lower wells, load Ag2 in the left and right wells.

4. Incubation Move the agar plate in the humid ifying box, incubate at 37℃ for 24h.

5. Results judgment If a precipitation line appears between the central well and peripheral wells, there should be the positive results, indicating the antibody binds the antigen specifically. If there is no precipitation line appears, there should be no specific combination for the antigen and antibody.

Experiment content

To be familiar with the double agar plate diffusion test. Observe the precipitation line in the agar plates.

Results

Observe and draw a result of precipitation line, indicating the numbers and characteristics of precipitation lines.

Notes

1. The agar plate should be made in a forming plate.
2. Load the samples drop by drop.

（周丽娜）

第三章 研究性与设计性实验

Chapter 3 Research and Designing Experiment

实验四十 土壤中抗生素产生菌的分离

【目的】

1. 掌握从土壤中分离与纯化放线菌的基本原理及常用方法。
2. 了解放线菌产生抗生素的抗菌谱测定方法。

【基本原理】

放线菌是重要的抗生素产生菌，许多临床应用的抗生素均由土壤中分离的放线菌产生的。放线菌一般在中性偏碱性、有机质丰富、通气性好的土壤中含量较多。由于土壤中的微生物是各种不同种类微生物的混合体，必须把各种放线菌从这些混杂的微生物群体中分离出来，从而获得某一放线菌的纯培养。根据放线菌对营养、酸碱度等条件要求，常选用合成培养基或有机氮培养基，也可采用选择性培养基或加入某种抑制剂，使细菌、霉菌出现的数量大大减少，从而分离土壤中的放线菌；再通过稀释法，使放线菌在固体培养基上形成单独菌落，并可得到纯菌株。

抗生素是放线菌的次级代谢产物，放线菌经液体培养后，其分泌的抗生素多数存在于离心所得的上清液中，可采用微生物方法进行检测，从而筛选到所需的抗生素产生菌。

【仪器与材料】

1. **菌种** 金黄色葡萄球菌（*Staphylococcus aureus*）和大肠埃希菌（*Escherichia coli*）牛肉膏蛋白胨液体培养物。
2. **培养基与试剂** 高氏1号合成培养基、肉汤琼脂培养基、灭菌的生理盐水。
3. **土壤** 校园土、空气中干燥、磨碎。
4. **仪器** 恒温培养箱、超净工作台、游标卡尺等。
5. **其他** 镊子、药敏试纸片、无菌平皿、无菌玻璃铲、无菌涂布棒、无菌移液管等。

【方法与步骤】

（一）土壤中放线菌的分离

1. 采集土样 土壤的种类和自然条件影响着放线菌的种类和分布数量。链霉菌主要存在于干燥、偏碱而营养丰富的土壤里；而小单孢菌多分布在潮湿土壤或湖底泥土中。因此在采土时，应注意地区、时间和植被情况。采土季节以春秋二季为宜，雨季不宜采土。一般认为，南方地区的土壤和北方地区土壤相比，南方地区的土壤链霉菌的种类较多，而某些特殊土壤或地区也可能存在一些特殊菌种。采集的土样最好随即分离，否则要放在阴凉通风处，防止变潮生霉。

在选定的采土地点，用小铲除去 5cm 厚表层土，用乙醇棉球擦拭过的铁锹深挖 15~20cm 处，将采得的土放入一个无菌纸袋或培养皿内，并按要求记录有关土样的内容，包括土壤编号、采集日期、采集地点及土壤基本特征。

2. 土壤悬液梯度稀释

（1）将土样放入用乙醇擦拭过的乳钵中，除去石块、草根，研磨压碎后，称取 5.0g 放入盛有 50ml 灭菌的生理盐水的三角瓶中。振荡 5min，即 10^{-1} 的土壤悬液，静置 30s。

（2）取 1ml 土壤悬液，加入到 9ml 灭菌生理盐水中 10 倍稀释。

（3）按 1∶10 梯度稀释至 10^{-4}、10^{-5} 和 10^{-6}（图 40-1）。

3. 制备培养基平板

（1）准备 9 个无菌玻璃平皿，分别标记 10^{-4}、10^{-5} 和 10^{-6}，每个稀释度平行 3 皿。

（2）配制高氏 1 号琼脂培养基并于 115℃ 灭菌 20 min，待其冷却至 50℃ 左右后，将其倒入无菌平皿中，每个平皿加 15~20ml，待冷凝制成无菌平板，见图 40-1。

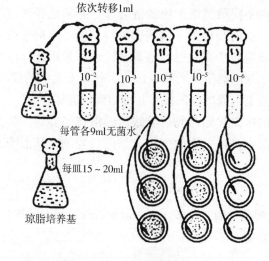

图 40-1 土壤放线菌的分离

Figure 40-1 Isolation of Actinomycetes from soil

4. 分离培养 从对应稀释度（10^{-4}、10^{-5} 和 10^{-6}）的土壤悬液中分别吸取 0.5ml 稀释液加到冷凝好的高氏 1 号平板中，用无菌涂布棒将加在平板培养基上的土壤稀释液在整个平板表面涂匀（每个稀释度换一支无菌涂布棒）。平板倒置于培养箱 28℃ 恒温培养 1 周。

5. 观察 分别挑取平板上的放线菌单菌落，接种于高氏 1 号琼脂培养基斜面或划线接种于高氏 1 号琼脂平板中，28℃ 恒温培养 1 周，获得放线菌的纯培养菌株，用于观察放线菌生长特征和产抗生素能力的测定。

（二）移块法粗筛抗生素产生菌

1. 在灭菌的空平皿中加入金黄色葡萄球菌肉汤培养液 4~5 滴，倒入溶化并冷却到 50℃ 左右的肉汤琼脂培养基约 20ml，制作混菌平板。

2. 在培养基表面贴上 1cm² 的放线菌琼脂块 5 块（中间为已知有抑菌作用的链霉菌块作为阳性对照，周围 4 个为未知的土壤中分离放线菌琼脂菌块），见图 40-2。

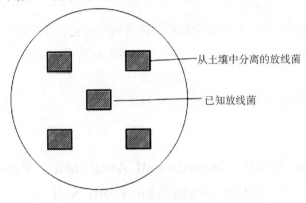

图 40-2　移块法

Figure 40-2　Moving-agar block method

3. 37℃，培养 48h，判定结果。如样品有抑菌作用，可在其周围出现不长菌的抑菌圈。

（三）抗菌谱测定

1. 发酵培养

（1）配制高氏 1 号液体培养基，分装 25ml 于 250ml 三角摇瓶中（每株放线菌对应 1 瓶），115℃ 灭菌 20min。

（2）将分离到的放线菌（可选择 4~6 株）分别接种于摇瓶中，28℃、240 r/min 恒温培养 1 周。

（3）发酵液过滤，滤液用于抗菌活性测定。

2. 抗菌活性测定

（1）将 10ml 灭菌的肉汤琼脂培养基加入到已灭菌的平皿中，冷却，制备底层平板。

（2）取培养 8h 的金黄色葡萄球菌和大肠埃希菌液体培养物各 2ml 分别加到 200ml 灭菌的肉汤琼脂培养基中（注意培养基应冷却到 50℃ 左右），震荡混匀，吸取 6ml 加入到底层平板上，分别制备含金黄色葡萄球菌和大肠埃希菌的双层平板（每种菌株平行做 2 皿）。

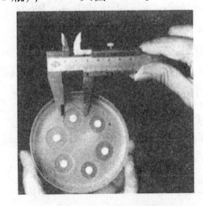

图 40-3　抗菌谱的测定

Figure 40-3　Determination of Antibiotic Spectra

（3）用无菌镊子将一系列吸附有放线菌发酵液的无菌滤纸片分别置于含金黄色葡萄球菌和大肠埃希菌的双层平板上，37℃ 培养 24h。

（4）观察并测量抑菌圈，判断放线菌是否产生抗生素，见图 40-3。

【实验内容】

1. 从校园土壤中分离培养放线菌。
2. 选择几种分离的放线菌采用移块法初筛抗生素产生菌。
3. 选择 4~6 种产抗生素的放线菌测定所产生的抗生素的抗菌谱。

【结果】

1. 记录各稀释度平板中分离到的放线菌菌落数并描述放线菌菌落的形态特征。
2. 记录 4~6 株分离的放线菌所产生的抗生素的抗菌谱。

【思考题】

在分离土壤中放线菌时,如何避免细菌和真菌的生长?

(苏　昕)

Experiment 40　Isolation of Antibiotic – Producing Microorganisms from Soil

Objectives

1. Learn the principles and methods how to isolate and purify Actinomycetes from soil.
2. Be familiar with the determination of anti – bacterial spectrum of antibiotics.

Principles

Actinomycetes are important antibiotic – producing bacteria. A lot of useful clinic antibiotics are derived from Actinomycetes in soil. Actinomycetes are usually found in the soil with neutral or alkaline pH, adequate organics and good ventilation. Since the microorganisms in soil are always mixture of different species, it is necessary for us to isolate different kinds of Actinomycetes from the colonies mixture to get certain pure culture of Actinomycetes. Synthetic medium or organic nitrogen culture is commonly used. We can use selective culture media or add certain inhibitor to reduce the number of bacteria and molds substantially, and finally we can isolate Actinomycetes from soil. After that, we can get individual colony of Actinomycetes on solid culture media by dilution, and then obtain pure strain.

Antibiotics are the secondary metabolites of Actinomycetes. After fermented in liquid media and centrifuged, antibiotics in the supernatant secreted by antibiotics – producing Actinomycetes can be determined with microbial method and screen the antibiotic – producing bacteria.

Apparatus and Materials

1. Specimens　Nutrient broth cultures of *Staphylococcus aureus* and *Escherichia coli*.
2. Media and Reagents　Gauss' No. 1 defined medium, nutrient agar media, sterilized physiological saline, etc.
3. Soil　Obtained from campus by air dry and powdering.
4. Apparatus　Incubator, clean bench, vernier caliper, etc.

5. Others Tweezers, paper slips, sterile dish, sterile glass spade, sterile glass spreader, sterile pipette, etc.

Methods and Procedures

Ⅰ. Isolation of Actinomycetes from Soil

1. Soil sample collection Actinomycetes species and distribution are varified by species and natural conditions of soil. Streptomycetes mainly exist in dry, alkaline and nutrient-rich soil, while Micromonospora mainly distribute in damp soil or lake-bottom clay. Therefore, region, time and condition of vegetation should be considered when sampling soil. It is appropriate to collect the samples in spring and autumn, but not in rainy days. The species of Actinomycetes in the South China are more abundant than those of the North. Certain special soil or region may contain some special species of Actinomycetes. It is better to isolate the strains from the sample soil right after sampling, or preserve the soil in cool places to prevent mildewing.

In the selected region, remove the surface soil of 5cm depth with a shovel and then dig into 15~20cm deep with alcohol cleaned spade. Put the collected soil in a sterile paper bag or plate, and record the number. with sample name, date, and place of sampling and the basic characteristic of the soil.

2. Prepare a serial dilution of the soil samples as follows

(1) Put the sample soil to mortar which is cleaned with alcohol, remove stones, grass roots, levigate the soil. Weigh 5.0g soil and put it into the flask containing 50ml sterile physiological saline. Mix thoroughly for 5min to make a uniform soil-water suspension.

(2) Transfer 1ml soil suspension to 9ml sterile physiological saline for 10 times dilution.

(3) Repeat the 1:10 dilution to obtain 10^{-4}, 10^{-5} and 10^{-6} diluted solution (Figure 40-1).

3. Prepare culture media plates

(1) Prepare 9 sterile glass plates, label three sets of sterile plate with the dilutions (10^{-4}, 10^{-5} and 10^{-6}).

(2) Make Gauss' No.1 agar media and sterilize at 115℃ for 20 min, cool to 50~55℃, add 15~20ml of it to each plate and curdle.

4. Isolate and incubate Transfer 0.5ml diluted solution of 10^{-4}, 10^{-5} and 10^{-6} to three sterile plates respectively and spread it. Incubate all plates in an inverted position at 28℃ for one week.

5. Aseptically isolate Actinomycetes colonies to starch agar plate respectively and incubate at 28℃ for one week for the observation growth feature of Actinomycetes and the determination of antibiotics.

Ⅱ. The moving-agar block method

This method is applicable to test disinfectant or inhibitory ability of ointment or agar strain, and it can be used to preliminarily screen Actinomycetes isolated from soil and its antibiotic-producing antibiotic ability.

1. Add 4~5 drops of broth culture of S. aureus to an empty plate, and then pour 15~20ml of agar media which has been melt and then cooled to 50℃. Mix them well.

2. Attach the surface of the culture media with five 1 cm^3 sizes of agar blocks of Actinomycetes isolated from soil. (Figure 40-2)

3. Incubate at 37℃ for 48h and then observe the inhibitory zone.

III. Determination of antibacterial Spectrum

1. Incubation

(1) For each strain of Actinomycetes, prepare starch media and pour 25ml to a 250ml shaking flask, sterilize at 115℃ for 20 min.

(2) Inoculating agar slant cultures of Actinomycetes to each shaking flask, incubate at 28℃ and 240r/min for a week.

(3) Filtrate the culture broth for the determination of antimicrobial activity.

2. Determination of Antimicrobial activity

(1) Pour 10ml sterile nutrient agar media to a sterile plate and cool to make the bottom plate.

(2) Transfer 2ml of 8h cultures of S. aureus and E. coli to 200ml sterilized nutrient agar media (50~55℃) respectively. For each bacterium, pour 6ml to the upper nutrient agar plate to prepare two layers agar plates.

(3) Add a series of paper slip absorbed the culture broth on the two layers agar plate for S. aureus and E. coli, incubate at 37℃ for 24h.

(4) Observe the inhibitory circle and determine whether Actinomycetes can produce antibiotics (Figure 40-3).

Experiment contents

1. Isolate Actinomycetes from soil obtained from campus.

2. Primary screens the antibiotic-producing Actinomycetes from several Actinomycetes from soil.

3. Select 4~6 kinds of the antibiotic-producing Actinomycetes to measure the antibacterial spectrum of antibiotics.

Results

1. Count the number of Actinomycete colonies in every plates and describe their characteristics.

2. Record the antibacterial spectrum of 4~6 kinds of antibiotics.

Questions

How to avoid the growth of bacteria and fungi when isolating antibiotic-producing Actinomycetes from the soil?

实验四十一 紫外线对枯草芽孢杆菌产淀粉酶的诱变效应研究

【目的】
1. 掌握紫外线的诱变原理。
2. 熟悉物理诱变育种的方法。
3. 了解紫外线对枯草芽孢杆菌产生淀粉酶的诱变效应。

【基本原理】

紫外线（ultraviolet ray，UV）作为物理诱变剂用于工业微生物菌种的诱变处理具有悠久的历史，尽管几十年来各种新的诱变剂不断出现和被应用于诱变育种，紫外线作为诱变因子还是有其特殊的意义。到目前为止，对于诱变处理后得到的高单位抗生素产生菌种中，有80%左右是通过紫外线诱变后经筛选而获得的，因此，对于微生物菌种选育工作者来说，紫外线作为首选诱变剂。

紫外线的波长在200~400nm之间，但对于诱变最有效的波长仅仅是在253~265nm。日常生活中，紫外灯是产生紫外线的主要方式，其波长为254nm。紫外线在空气中穿透能力强，在液体中穿透能力较弱，在固体介质中几乎不能穿透，所以诱变时，一定要将目标直接暴露在紫外线下，才能取得较好的诱变效果。

260nm的紫外线杀伤作用最强。其主要作用机制是：紫外线穿过微生物细胞时，DNA吸收了紫外线，引起DNA分子中相邻的胸腺嘧啶借助共价键形成胸腺嘧啶二聚体，影响了DNA复制和功能。

【仪器与材料】

1. 菌种 枯草芽孢杆菌 BF7658。

2. 培养基及试剂 牛肉膏蛋白胨固体培养基、淀粉培养基（见附录一，50）、碘液、含0.9% NaCl 的无菌生理盐水。

3. 仪器 培养箱、20W紫外灯、磁力搅拌器、离心机等。

4. 其他 无菌培养皿、无菌试管、无菌移液管、无菌三角瓶、无菌带玻璃珠的锥形瓶、量筒、烧杯、离心管等。

【方法与步骤】

（一）诱变

1. 菌悬液的制备

（1）取一支经活化已培养20h的枯草芽孢杆菌斜面。

（2）用5ml无菌0.85%生理盐水将菌苔轻轻洗下，此过程重复2次，并倒入盛有玻璃珠的锥形瓶中。

（3）强烈震荡10min，将此菌悬液用无菌移液管吸至10ml离心管中，3000r/min离心15min。

(4) 弃去上清液，将菌体用无菌生理盐水 10ml 洗涤 2 次（每次 3000r/min 离心 10min），最后用 10ml 无菌生理盐水制成菌悬液。

2. 菌悬液计数　用无菌移液管将菌悬液移入另一个锥形瓶中，调整细胞浓度为 10^8/ml。细胞菌悬液浓度可采用血球计数板直接计数，或采用平板活菌计数法测定，也可用光密度比浊法测定。

3. 平板制作　制备淀粉琼脂培养基灭菌备用，将其融化后，冷至50℃左右，倒平皿，凝固后待用。

4. 诱变处理

（1）正式照射前开启紫外灯，预热 20min。

（2）取制备好的菌悬液 4ml 移入直径为 6cm 的无菌培养皿中，在 30W 紫外灯下 30cm 处进行预照射 1min。

（3）开盖照射，诱变时间在 1~5min 之间（学生可自由选择 3 个时间梯度）。所有操作必须在红光灯下进行，以免发生光复活现象。

5. 稀释涂平板

（1）在红光灯下，分别将未诱变的菌悬液和已诱变的菌悬液进行系列稀释。

（2）分别取 10^{-5}、10^{-6} 和 10^{-7} 等 3 个稀释度的稀释液各 0.1ml，涂于淀粉培养基平板上；每个稀释度涂 3 个平板，用无菌三角爬涂匀。

（3）用黑布包好照射过的平板，37℃培养 48h。

注意：在每个平板背面要标明处理时间、稀释度、组别、姓名等。

（二）计算存活率及致死率

1. 存活率计算公式

$$存活率 = \frac{处理后\ 1ml\ 菌液中活菌数}{对照\ 1ml\ 菌液中活菌数} \times 100\%$$

2. 致死率计算公式

$$致死率 = \frac{对照\ 1ml\ 菌液中活菌数 - 处理后\ 1ml\ 菌液中活菌数}{对照\ 1ml\ 菌液中活菌数} \times 100\%$$

对照样品 1ml 中活菌数：将培养 48h 后对照平板取出，进行细胞计数。根据平板上菌落数，计算出对照样品 1ml 菌落中的活菌数。

处理后 1ml 样品中活菌数：同上，计算照射相应时间后样品 1ml 菌液中的活菌数。

（三）观察诱变效应

1. 观察菌落透明圈　对平板中菌落进行计数后，选择菌落数在 5 个左右的平板分别向平板内滴加碘液数滴，观察菌落周围出现的透明圈。

2. 计算比值（HC 值）　分别测量透明圈直径与菌落直径并计算比值（HC）。

3. 解读结果　与对照平板进行比较，根据结果说明紫外线对枯草芽孢杆菌产淀粉酶诱变的结果。

4. 接种纯培养　选取 HC 比值大的菌落，挑取单菌落并接种到新鲜牛肉膏斜面培养基上培养，备进一步复筛用。

【实验内容】

利用紫外线诱变枯草芽孢杆菌 BF7658，研究产生高淀粉酶活力的菌株的诱变效应。

【结果】

1. 记录各平板中的菌落数，并分别算出存活率、致死率。

2. 测量并记录经 UV 处理后的枯草芽孢杆菌菌落（6 个）周围的透明圈直径（mm）与菌落直径（mm），并计算比值（HC 值），与对照菌株进行比较。

$$HC\ 值 = \frac{透明圈直径（mm）}{菌落直径（mm）}$$

【注意事项】

1. 经紫外线照射后的样品需用黑纸或黑布包裹，一些必需的操作应在波长最长的红光下进行。

2. 照射处理后的细胞悬液不要放置太久，以免在黑暗中发生突变或被修复。

【思考题】

1. 紫外线诱变需注意的事项是什么？
2. 紫外线诱变的机制是什么？
3. 总结实验结果，哪一种照射时间的诱变效果最好？它的存活率、致死率和 HC 值各为多少？

（苏　昕）

Experiment 41　Screening and Ultraviolet Mutation Breeding of *Bacillus subtilis* Strains Producing Amylase

Objectives

1. Grasp the principle of Ultraviolet radiation.
2. Be familiar with and understand physical mutation breeding methods.
3. Understand the mutagenic effects of *Bacillus subtilis* using ultraviolet radiation.

Principles

Ultraviolet radiation, as a physical mutagen, has been used in industrial microorganism breeding for many years. Although all kinds of new mutagen appear constantly and are used in mutation breeding in recent decades, UV radiation has its special significance. So far, among the high-yield antibiotics producing strains induced, about 80% of them were obtained by UV radiation. So, UV radiation is the best choices for microbial breeders

Ultraviolet radiation (UV) ranges in wavelength from approximately 100nm to 400nm. It is most lethal from 253nm to 265nm. In everyday practice, the source of UV radiation is the germicidal lamp, which generates radiation at 254nm. Because UV radiation passes readily through air, slightly through liquids, and only poorly through solids, the objects to be induced must be directly exposed to it for better effects.

The most lethal UV radiation has a wavelength of 260nm. As UV radiation passes through some microbe, it is most effectively absorbed by DNA. The primary mechanism of UV damage is the formation of thymine dimers in DNA. Two adjacent thymines in a DNA strand are covalently joined to inhibit DNA replication and function.

Apparatus and Materials

1. Specimens *Bacillus subtilis* BF7658.

2. Cultures and Reagents Nutrient agar media, starch agar media (Appendix I, 50), iodine solution, 0.9% sterile normal saline.

3. Apparatus Incubator, 30W germicidal lamp, magnetic stirrer, centrifuge.

4. Others Sterile Petri plate, sterile tube, sterile pipette, sterile spreader rod, sterile flask with glass beads, cylinder, beaker, centrifuge tube.

Method and procedures

I. Mutation

1. Preparation of bacterial suspension

(1) Take a activated *Bacillus subtilis* BF7658 slant incubated for 20h.

(2) Scratched off lightly the lawn on the surface of the slant and wash it with 5ml 0.85% normal saline two times, and then transfer the suspension to a sterile triangular flask with glass beads.

(3) After it was shaken sufficiently for 10min, transfer the suspension to a 10ml centrifuge and centrifuge at 3500r/min for 10 min.

(4) Discard the supernatant. Wash and centrifuge two times with 10ml 0.85% normal saline, and the finial liquor volume reached 10ml.

2. Counting the bacterial suspension Using a sterile pipette, transfer some volume of the bacterial suspension above to another sterile flask, dilute it properly and then counted to 10^8 cells per milliliter.

3. Making the sterile starch agar plates Melt the sterile starch agar media and cool down to 50~55 ℃, add 15~20ml to each plate and curdle.

4. Mutation treatment

(1) The UV lamp (30 W) was opened 20 min before.

(2) 4ml of the suspension was set on the sterilized Petri plate (the diameter was 6cm). Set the Petri plates on the magnetic stirring apparatus which is 30cm (vertical distance) away from the UV lamp, irradiated for 1min at first.

(3) Uncap the Petri plate and treat the plate with ultraviolet for 1~5min (student may choose 3 of them at random) with shaking. All operation should be performed under red light.

5. Diluting and spreading plate

(1) Under the red light, dilute the mutated strains and the original bacteria suspension without irradiation into certain concentration gradient (10^{-4}, 10^{-5}, 10^{-6} and 10^{-7}).

(2) Transfer 0.5ml of diluted solution of 10^{-5}、10^{-6} and 10^{-7} to three sets of sterile plates respectively, and spread with sterile spreader.

(3) Wrap the plates with black paper, incubate away from light at 37℃ for more than 48h.

Notice: Label the bottom of the plate with your irradiation time, dilution, lab section, and your name.

II. Calculate the survival rate and the lethal rate

1. Formula for the survival rate

$$\text{Survival Rate} = \frac{\text{the viable counts of bacteria in check sample for one milliliter}}{\text{the viable counts of bacteria in control group for one milliliter}} \times 100\%$$

2. Formula for the lethal rate

$$\text{Lethal Rate} = \frac{\text{the viable counts of bacteria in check sample for one milliliter} - \text{the viable counts of bacteria in control group for one milliliter}}{\text{the viable counts of bacteria in control group for one milliliter}} \times 100\%$$

The viable counts of bacteria in check sample: After incubation for 48h, take the plates, count the number of the colony on each plate, and calculate the viable counts of bacteria in control group for one milliliter.

The viable counts of bacteria in tested group: After incubation for 48h, take the plates, count the number of the colony on each plate, and calculate the viable counts of bacteria in different tested groups for one milliliter.

III. Observe the mutagenic effects

1. Choose appropriate plates, on which, colony number is within five for each plate. Add afew drops of iodine solution onto the plate sand observe the zone transparent zone around the colony.

2. Message the transparent zone diameter and colony diameter. Calculate the ratio of flat transparent zone colony diameter (HC).

3. Interpretation of the results Compared with the control plates, explain the mutagenic effects of *Bacillus subtilis* using ultraviolet radiation.

4. Choose some colonies with a bigger HC value, pick the colonies and transfer them to fresh nutrient broth agar slants, incubate and preserve them for further secondary screening.

Experiment contents

The mutagenesis effects on the high activity of amylase will be investigated with *Bacillus subtilis* irradiated UV irradiation.

Results

1. Count the colony number on each plate and calculate the survival rate and the lethal rate.

2. Message and record the transparent zone diameter and colony diameter of 6 colonies. Calculate the ratio of flat transparent zone to colony diameter (HC), meanwhile compared with the control plates.

$$\text{HC value} = \frac{\text{the transparent zone diameter (mm)}}{\text{the colony diameter (mm)}}$$

Notes

1. Wrap the plates with black paper after mutation, some necessary operation should be performed under red light.

2. Do not set the mutated cell suspensions for too long time, in case the DNA would be repaired or generate other genetic mutation.

Question

1. What is the notice for UV irritation?

2. What is the mechanism for UV irritation?

3. Synthetically, which irradiation time has a better mutagenic effects? How about the survival rate, the lethal rate and the HC value?

（徐　威）

实验四十二　细菌氨基酸营养缺陷型菌株的筛选及鉴定

【目的】

1. 掌握营养缺陷型突变株筛选的原理和步骤。
2. 掌握紫外线的诱变方法。
3. 熟悉营养缺陷型突变株的筛选方法。
4. 了解营养缺陷型突变株的应用。

【基本原理】

突变通常是指 DNA 序列发生的稳定可遗传的改变。突变的发生有两条途径：自发突变和诱发突变。自发突变是在没有外来因素的作用下，在细胞中以极低频率发生的突变；诱发突变是微生物暴露在物理或化学诱变剂中产生的突变。诱变剂指的是能够引起 DNA 损伤，改变其化学性质或影响 DNA 损伤修复机制的任何因素，可以引发基因突变。

营养缺陷型是丧失合成某一物质（如：氨基酸、维生素、核苷酸）的能力，而必须从环境中获得该物质才能生长的菌株。它们在基本培养基上不能生长，而必须在基本培养基中补充某些物质或在完全培养基上才能生长。而能够在基本培养基上生长的菌株成为原养型。

筛选营养缺陷型菌株的主要步骤：诱变处理、营养缺陷型浓缩、营养缺陷型检出、生长谱鉴定。筛选营养缺陷型突变株的基本步骤如下：

浓缩营养缺陷型的方法一般有抗生素法、菌丝过滤法、饥饿法和差别杀菌法等。细菌的缺陷型筛选常用青霉素法；酵母菌和霉菌的缺陷型筛选常用制霉菌素或两性霉素B。丝状菌的缺陷型筛选可用菌丝过滤法，能形成芽孢的缺陷型菌株的筛选常用差别杀菌法。

营养缺陷型的检出方法一般有四种方法：随机逐个检出法、夹层培养法、影印平板法、限量补给法。其中随机逐个检出法简便、易行，但效率较低。

营养缺陷型的生长谱鉴定一般分两步，先确定为哪一类（氨基酸、维生素、核苷酸）营养缺陷型，再具体鉴定为哪种营养缺陷型。

本实验用紫外线来诱变大肠埃希菌，使其发生基因突变。并用青霉素法淘汰野生型，最后经生长谱法鉴定产生的营养缺陷型突变株。

【仪器与材料】

1. 菌种 大肠埃希菌（*Escherichia coli*）、大肠埃希菌突变株M3、$M_{mut}-812$和$M_{mut}-802$。

2. 培养基与试剂

（1）完全液体培养基（CM液体）：牛肉膏5.0g、蛋白胨10g、NaCl 5g、蒸馏水1000ml，pH 7.0~7.2。0.1MPa，20~30min高压蒸汽灭菌。

（2）完全固体培养基（CM固体）：在完全培养基（液体）中加1.4%~1.6%的琼脂。

（3）二倍完全培养基（2E培养基）：除蒸馏水加量不变，完全液体培养基各组分加倍即可。

（4）无氮培养基（g/L）

葡萄糖	5.0
KH_2PO_4	1.0
K_2HPO_4	3.0
$MgSO_4 \cdot 7H_2O$	0.1
$MnSO_4 \cdot 4H_2O$	0.01
$FeSO_4 \cdot 7H_2O$	0.01
蒸馏水	1000ml
pH	7.2

（5）二氮培养基（2N培养基）：在无氮培养基中加入$2 \times 0.15\%$（NH_4）$_2SO_4$即可。

（6）基本液体培养基（MM液体培养基）：在无氮培养基加入（NH_4）$_2SO_4$ 0.15%，生物素3μg/ml，硫氨素10μg/ml。

（7）基本固体培养基（MM固体培养基）：在基本培养基中加入1.4%~1.6%的琼脂粉。

上述（4）~（7）培养基，55.21 kPa高压蒸汽灭菌30min。

（8）0.85%生理盐水：见附录四，1。

（9）青霉素钠盐溶液：100mg/ml，过滤除菌。

（10）混合氨基酸：分为6组，分别称取等量的各种L型氨基酸于干净的研钵中

（用 DL 型氨基酸，量需加倍），混合，在 60~70℃烘箱中烘数小时，趁干燥立即磨细，然后装入小瓶中，避光、干燥保存。

Ⅰ 赖氨酸、精氨酸、甲硫氨酸、胱氨酸、亮氨酸、异亮氨酸

Ⅱ 缬氨酸、精氨酸、苯丙氨酸、络氨酸、色氨酸、组氨酸

Ⅲ 苏氨酸、甲硫氨酸、苯丙氨酸、谷氨酸、谷氨酰胺、脯氨酸

Ⅳ 羟脯氨酸、胱氨酸、络氨酸、谷氨酸、天门冬氨酸、天门冬酰胺

Ⅴ 丙氨酸、亮氨酸、色氨酸、谷氨酰胺、天门冬氨酸、甘氨酸

Ⅵ 丝氨酸、异亮氨酸、组氨酸、脯氨酸、天冬酰胺、甘氨酸

3. 仪器 电子天平、台式离心机、涡旋混合器、恒温摇床、恒温培养箱、超净工作台等。

4. 其他 无菌培养皿、无菌移液管、无菌涂布棒、无菌滴管、无菌离心管、无菌牙签、接种环。

【方法与步骤】

1. 菌悬液制备

（1）取一环过夜培养的大肠埃希菌斜面培养物，接种于装有 5ml CM 液体培养基的三角瓶中，37℃，震荡培养 16~18h。

（2）在上述培养物中，加入 5ml CM 液体培养基，37℃继续培养 6h。

（3）将上述培养菌液倒入一个无菌的离心管中，3500 r/min，离心 10min。倒去上清液，打匀沉淀，加入 5ml 0.85% 生理盐水，混匀，制备菌悬液。

2. 紫外线诱变处理 吸取上述菌悬液 3ml 于 75mm 无菌培养平皿中，放在提前半小时打开的紫外灯下（15W、距离 30cm）。先照射 1min，然后打开平皿盖，振荡照射 50s–2min。照射后盖上平皿盖。

3. 增殖培养 将 3ml 二倍完全培养基（2E 培养基）加入到上述处理后的培养皿中，将培养皿用黑纸包好，37℃避光培养 20h 以上。

4. 营养缺陷型浓缩（或淘汰野生型）–（青霉素法）

（1）饥饿培养 吸 5ml 上述培养的菌液于无菌离心管中，3500r/min 离心 10min，倒去上清液，打匀沉淀。加入 0.85% 生理盐水洗涤菌体 3 次后，最后一次加入 3ml 0.85% 生理盐水，制作菌悬液；吸取菌悬液 0.1ml 于 5ml 无氮培养基中，37℃培养 12h。

（2）加青霉素 在经过饥饿培养 12h 的菌液中，加入 5ml 2N 培养基；同时，加入最终浓度为 500μg/ml 的青霉素钠盐，再在 37℃继续培养 12~24h。

（3）涂布平板

①制备菌悬液并稀释 将培养 12h 或 24h 的菌液，于 3500r/min 离心 10min，弃上清液，菌体用 0.85% 生理盐水洗涤一次，再加 10ml 生理盐水制备菌悬液（10^0），并稀释至 10^{-1}、10^{-2}、10^{-3}。

②涂平板 各取 0.1ml 10^0、10^{-1}、10^{-2} 和 10^{-3} 菌悬液，分别涂布于 MM 平板和 CM 平板上，用三角爬涂匀，每个浓度、每种培养基各涂布 3 个平皿。

于 37℃倒置培养 48~72h。

5. 营养缺陷型的检出 – 逐个点种法

培养72h后，进行菌落计数（图42-1），计算A值。

$$A = \frac{CM \text{ 上的菌落数} - MM \text{ 上的菌落数}}{CM \text{ 上的菌落数}}$$

取A值大的那一组，用牙签挑取CM平板上长出的50~100个菌落，分别点于MM平板和CM平板上（先点MM，后点CM），见图42-1，37℃培养24h。

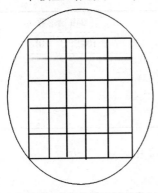

图42-1 牙签接种示意

Figure 42-1 Spot Inoculation

图42-2 划线复证

Figure 42-2 Streak to reexamination

6. 划线复证 培养24h以后，选在MM平板不长，CM平板上生长的菌落，再在MM平板和CM平板上划线复证。每皿划4个菌，见图42-2，37℃培养24h。

7. 扩大培养 在MM平板不生长，在CM平板生长的菌落可能是营养缺陷型，经划线复证后，将疑是菌株传CM斜面3~5支，37℃培养24h保存备用。

8. 营养缺陷型生长谱鉴定

（1）将疑是营养缺陷型的菌斜面培养物，接种一环于5ml CM液体培养基中，37℃震荡培养18h。

（2）将培养液转入离心管中，3500r/min离心10min，倒去上清液，打匀沉淀。然后用5ml 0.85%生理盐水洗涤、离心，此过程重复3次，最后加0.85%生理盐水3~5ml。

（3）吸取上述离心洗涤后的菌液1ml于无菌培养皿中，倒入融化好并冷却到50℃左右的MM固体培养基，摇匀，待凝，重复2皿。

（4）将培养皿的底部用记号笔分成若干等分，并编号（图42-3）；依次放入混合氨基酸（加入的量要少，否则会抑制菌的生长）。然后37℃培养24h左右，观察生长现象，并确定是哪种营养缺陷型。

（5）进一步确证 将被鉴定的缺陷型细菌制成MM双层固体培养基平板，用牙签挑取该种氨基酸粉末少许，点在平板中央，经培养后，观察在该氨基酸周围是否出现生长圈。

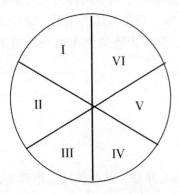

图42-3 平板分区示意图

Figure 42-3 Plate divided to 6 areas

【实验内容】

以大肠埃希菌为材料,采用紫外线诱变处理筛选氨基酸营养缺陷型突变株。

【结果】

1. 将青霉素淘汰法结果记录在表 42-1 中。

表 42-1

培养基	12h 菌落数				24h 菌落数			
	10^0	10^{-1}	10^{-2}	10^{-3}	10^0	10^{-1}	10^{-2}	10^{-3}
CM								
MM								
A 值								

2. 将生长谱法鉴定营养缺陷型结果记录于表 42-2 中。

表 42-2

缺陷型菌株编号	生长区	缺陷类型
1		
2		
3		
4		
5		

【注意事项】

紫外线对人体有害,尤以人的眼睛和皮肤最易受到紫外线损伤,所以,不要用用眼睛看紫外线,避免将手暴露在紫外线下。操作时需戴防护眼镜、手套等。

紫外线照射后造成的胸腺嘧啶二聚体损伤,在可见光照射下由于菌体内光复活酶的作用,可使其恢复突变,即具有光复活作用。因此,为了避免光复活作用,紫外线诱变处理时及处理后的操作都应在红光或黄光下进行,并且要将微生物放在黑暗下培养。

【思考题】

1. 诱变处理后,加入 2E 肉汤的目的是什么?
2. 加入无氮、二氮培养的目的各是什么?
3. 青霉素法淘野生型的作用原理是什么?
4. 用逐个点种法筛选营养缺陷型时,要先点种基本培养基,再点种完全培养基,为什么?

Experiment 42　Isolation and Identification of Amino Acid Auxotrophic Strain by UV

Objectives

1. Grasp the principle and procedure for selection of auxotroph strains.
2. Grasp UV mutagenesis method.
3. Be familiar with the selection method of auxotroph strains.
4. Understand the application of auxotroph.

Principle

Mutation is stable, heritable alterations in the DNA sequence and usually, but not always, produces phenotypic changes. Mutations may occur in two ways: spontaneous mutations or induced mutations. Spontaneous mutations arise occasionally in all cells and develop in the absence of any external agent. Induced mutations, on the other hand, are results of exposure of the organism to some physical or chemical agent called a mutagen. Mutagen is any agent that directly damages DNA, alters its chemistry, or interferes with repair mechanisms, it can induce mutation.

An auxotroph is a mutant that lacks the ability to synthesize some essential nutrients such as amino acids, vitamins and nucleotides and must obtain them from its surroundings. The mutant can grow on minimal medium (MM) with the necessary nutrients supplements or on complete medium (CM), but not on minimal medium. Microbial strains that can grow on MM are prototroph.

There are four main steps to isolate an auxotroph: mutagenesis, enrichment of auxotroph, detection of auxotroph and auxanogram identification of auxotroph. Here are the basic procedures for detection of auxotroph.

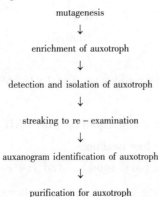

Four kinds of enrichment methods are usually used for isolating auxotroph including penicillin enrichment method, filtration enrichment method, starvation method and selective bactericidal method, and so on. Penicillin enrichment method can be used to enrich auxotroph of bacteria. Mycostatin or amphotericin B is used instead of penicillin in the case of mold and yeast. Filtration method is often used to enrich auxotroph of filamentous organisms. Selective bactericidal method is used to enrich auxotroph of spore – forming bacteria.

There are four kinds of detection and isolation method for auxotroph: random spot inoculation, layer plating method, replica plating and limited enrichment. The random spot inoculation is easy but with low efficiency.

The auxanogram identification of auxotroph includes two steps. Firstly, to determine which kind of nutrient (amino acids, vitamins and nucleotides) the auxotroph cannot synthesize. Secondly, to define which specific nutrient needs to be added.

In this experiment, ultraviolet is used as the mutagen to cause genetic mutations in *E. coli*. Penicillin is used to eliminate the wild type. Auxanogram is used to identify the new amino acid auxotrophic strains.

Apparatus and Materials

1. Specimens *Escherichia coli*; *E. coli* mutant strain M3, $M_{mut}-812$ and $M_{mut}-802$.

2. Cultures and Reagents

(1) Complete liquid medium (CM liquid): Beef extract 5.0g, peptone 10g, NaCl 5g, distilled water 1000ml, pH 7.0 ~ 7.2, 0.1MPa, autoclaving for 20 ~ 30min.

(2) Complete solid medium (CM solid): Add 1.4% ~ 1.6% agars in complete medium.

(3) 2 × complete medium (2E medium): Add twice as much as components in complete liquid medium except for distilled water.

(4) Nitrogen – free medium (g/L):

glucose	5.0
KH_2PO_4	1.0
K_2HPO_4	3.0
$MgSO_4 \cdot 7H_2O$	0.1
$MnSO_4 \cdot 4H_2O$	0.01
$FeSO_4 \cdot 7H_2O$	0.01
distilled water	1000ml
pH	7.2

(5) Dinitrogen medium (2N medium): Add 2 × 0.15% $(NH_4)_2SO_4$ in Nitrogen – free medium.

(6) Minimal liquid medium (MM liquid): Add $(NH_4)_2SO_4$ 0.15%, biotin 3μg/ml, thiamine 10μg/ml, in Nitrogen – free medium.

(7) Minimal solid medium (MM solid): Add 1.4% ~ 1.6% agar in minimal liquid medium.

(4) ~ (7) media mentioned above, 55.21 kPa autoclaving for 30min.

(8) 0.85% normal saline: Refer to Appendix Ⅳ, 1

(9) Penicillin salt solution: 100mg/ml, filtration sterilization.

(10) Amino acid mixture: Divided into six groups: weigh the same amount of each kind

of L-amino acids respectively and put them in a clean mortar (double the amount, if DL-amino acids used), mix well, dry at 60~70℃ in the oven for several hours, grind immediately, then file into vials and keep them in dark, dry place.

Ⅰ Lys, Arg, Met, Cys, Leu, Ile
Ⅱ Val, Arg, Phe, Tyr, Trp, His
Ⅲ Thr, Met, Phe, Glu, Gln, Pro
Ⅳ HyP, Cys, Tyr, Glu, Asp, Asn
Ⅴ Ala, Leu, Trp, Gln, Asp, Gly
Ⅵ Ser, Ile, His, Pro, Asn, Gly

3. Instruments   Electronic scales, desk centrifuge, turbine mixer, shaker, incubator, bechtop and so on.

4. Others   Sterile Petri plate, sterile pipette, sterile spreading rod, sterile dropper, sterile centrifuge tube, sterile toothpick, inoculating loop.

Method and procedures

1. The preparation of *E. coli* suspension

(1) Transfer a loopful of overnight slant culture of *E. coli* and inoculate it to flask with 5ml CM liquid medium. Incubate at 37℃ for 16~18h with shaking.

(2) Add 5ml CM liquid media into the above cultures, incubate at 37℃ for another 6h.

(3) Transfer the above *E. coli* suspension to a sterile centrifuge tube, centrifuge at 3500 r/min for 10min. Discard the supernatant and suspend the cells in 5ml 0.85% normal saline, mix well.

2. Ultraviolet Radiation   Transfer 3ml of *E. coli* suspension to a 75mm sterile Petri plate, put the Petri dish 30cm under the UV lamp, opened half an hour before. At first irradiated for 1min, then uncap the plate and irradiate for 50s~2min with shaking. Cover the plate in the end.

3. Multiplication culture   Add 3ml 2E medium into the above plate with induced solutions. Wrap the plate with black paper, incubate away from light at 37℃ for more than 20h.

4. Enrichment of auxotroph (or Eliminate the wild type) – (penicillin enrichment method)

(1) Starvation incubation   Add 5ml multiplication culture into a aseptic centrifugal tube, centrifuge at 3500r/min for 10min. Discard the supernatant, and then whisk the precipitation. Wash the centrifuge with normal saline for three times. At last, suspend the cells in 3ml 0.85% normal saline.

Transfer 0.1ml of bacterial suspension to 5ml nitrogen-free medium, incubate at 37℃ for 12h.

(2) Adding penicillin   Add 5ml 2N medium to the Starvation Microbial culture. Add penicillin sodium to the culture to make the final concentration about $500\mu g/ml$. Incubate at 37℃ for another 12~24h.

(3) Spread plate (Figure 42 – 1)

①Prepare bacterial suspensions and their dilutions   After the bacteria has been cultivated for 12h or 24h, centrifuge at 3500r/min for 10 minutes, discard the supernatant, then wash the centrifuge with 0.85% normal saline once. Add 10ml normal saline to make the bacterial suspension (10^0), and dilute it to 10^{-1}、10^{-2}、10^{-3} respectively.

②Spread plate   Add 0.1ml 10^0、10^{-1}、10^{-2} and 10^{-3} bacterial suspensions respectively, spread them on MM medium and CM medium separately with a sterile spreader, three plates for each kind of medium. Incubate upside down at 37℃ for 48 ~ 72h.

5. Detection of auxotroph strains – random spot inoculation   After incubation for 72h, count the number of the colony on each plate, calculate A value.

$$A = \frac{\text{the number of colonies in CM medium} - \text{the number of colonies in MM medium}}{\text{the number of colonies in CM medium}}$$

Select the group that A value is bigger. Pick out 50 ~ 100 colonies from the CM medium with sterile toothpick, then inoculate them onto another MM medium and CM medium with Spot Inoculation method (MM medium is first), as shown in Figure 42 – 1, incubate at 37℃ for 24h.

6. Streak to re – examination   After incubation for 24h, select the colonies grow on the CM plate but not on the MM plate. Then streak them on MM and CM plates respectively. Four strains each plate, see Figure 42 – 2, incubate at 37℃ for 24h.

7. Enlarge cultivation   After incubation, the colony grows on CM plates still not grow on MM plates may be auxotroph. Incubate the suspicious auxotroph to CM agar slant and incubate at 37℃ for 24h for auxanogram identification.

8. Determination of growth spectrum of the auxotroph strein

(1) Transfer a loopful the suspicious auxotroph CM agar slant culture to a tube containing 5ml CM liquid medium, incubate with shaking at 37℃ for 18h.

(2) Transfer the culture solution to sterile centrifuge tube and centrifuge at 3500r/min for 10 min. Discard the supernatant. Wash and centrifuge three times with 0.85% normal saline. Then add 3 ~ 5ml 0.85% normal saline at last.

(3) Transfer 1ml the suspension to the bottom of MM plate, and pour 10 ~ 15ml MM solid medium melted and cooled down to 50℃. Mix well rapidly and let it harden. The suspension can also be transferred to a small flask containing 10 ~ 15ml MM solid medium melted and cooled down to 50℃. Mix well rapidly and pour it down onto a sterile plate until it hardens.

(4) Divide the back bottom of the plate into 6 areas (Figure 42 – 3) using a maker pencil. Drop the mixed amino acids solution of the 6 groups with micropipette, on the corresponding area respectively. Then incubate at 37℃ for 24h.

Examine growth phenomenon and identify which amino acid the auxotroph cannot synthesize.

(5) Further identification   Make the MM double – layer plate mixed with the auxotroph

obtained above. Pick the amino acid powder onto the surface of the plate. Observe the growth circle surrounding the amino acid after incubation.

Experiment contents

E. coli was used as material to screen amino acid auxotrophic strains by ultraviolet radiation.

Results

1. Fill Table 42 – 1 with the results of Penicillin method.

Table 42 – 1

Medium	Colony number (12h)				Colony number (24h)			
	10^0	10^{-1}	10^{-2}	10^{-3}	10^0	10^{-1}	10^{-2}	10^{-3}
CM								
MM								
A value								

2. Fill Table 42 – 2 with the results of identification of auxanogram.

Table 42 – 2

Strain No.	Growth region	Amino acid reqiirment
1		
2		
3		
4		
5		

Notes

1. Ultraviolet is harmful to human, especially to eyes and skin. Do not look at the ultraviolet light, and do not leave your hand exposed to it. Wear safety glasses and gloves.

2. When thymine dimers are exposed to visible light, photoreacting enzymo are activated; these enzymes split the dimers, restoring the DNA to its undamaged state. This is called light repair or photoreactivation.

3. In order to avoid the photoreactivation, the ultraviolet irradiation should be carried out under red or yellow light, and incubate bacteria in dark.

Questions

1. What is the purpose of adding 2E broth after mutagenic treatment?

2. What is the purpose of nitrogen – free or dinitrogen medium?

3. What is the principle of penicillin enrichment method?

4. You should inoculate first on the MM plate and then on the CM plate when isolating auxotroph using random spot inoculation method, why?

（徐　威）

实验四十三 微生物培养条件的优化

【目的】

1. 了解影响微生物生长繁殖速度与发酵产量的培养条件。
2. 熟悉优化微生物培养条件的常用方法。

【基本原理】

利用微生物发酵生产有用代谢产物，其培养基成分相当复杂，影响微生物生长繁殖和积累某产物的因素除培养基成分外，还包括 pH 值、温度等环境条件，可变因子较多。因而，微生物培养基及培养条件的优化工作显得尤为重要。数学统计中的多种优化方法已开始广泛地应用于微生物发酵条件的优化工作。

单因素试验法是在假设因间不存在交互作用的前提下，通过一次只改变一个因素且保证其他因素维持在恒定水平的条件下，研究不同试验水平对结果的影响，然后逐个因素进行考察的优化方法，是试验研究中最常用的优化策略之一。然而，对于大多数培养基而言，其组分相当复杂，仅通过单因素试验往往无法达到预期的效果，特别是在试验因素很多的情况下，需要进行较多的试验次数和试验周期才能完成各因素的逐个优化筛选，因此，单因素试验经常被用在正交试验之前或与均匀设计、响应面分析等结合使用。

正交实验法是安排多因素、多水平的一种实验方法，即借助正交表的表格来计划安排实验，并正确地分析结果，找到实验的最佳条件，分清因素和水平的主次，就能通过比较少的实验次数达到好的实验效果。

均匀设计法是一种考虑试验点在试验范围内充分均匀散布的试验设计方法，其基本思路是尽量使试验点充分均匀分散，使每个试验点具有更好的代表性，但同时舍弃整齐可比的要求，以减少试验次数，然后通过多元统计方法来弥补这一缺陷，使试验结论同样可靠，均匀设计一般采用二次型回归模型，由于每个因素每一水平只作一次试验，因此，当试验条件不易控制时，不宜使用均匀设计法。对波动相对较大的微生物培养试验，每一试验组最好重复 2~3 次以确定试验条件是否易于控制，此外，适当地增加试验次数可提高回归方程的显著性。

响应面优化设计法是一种寻找多因素系统中最佳条件的数学统计方法，是数学方法和统计方法结合的产物，它可以用来对人们受多个变量影响的响应问题进行数学建模与统计分析，并可以将该响应进行优化。它能拟合因素与响应间的全局函数关系，有助于快速建模，缩短优化时间和提高应用可信度。一般可以通过 Plackett – Burman（PB）设计法或 Central composite design（CCD）等从众多因素中精确估计有主效应的因素，节省实验工作量。

【仪器与材料】

1. 菌种 啤酒酵母菌（*Saccharomyces cerevisiae*）。

2. 培养基及试剂 按正交试验表配制相应培养基；药品包括葡萄糖、磷酸盐、蒸馏水等。

3. 仪器　培养箱、离心机等。

4. 其他　三角瓶、量筒、烧杯、离心管等。

【方法与步骤】

本实验建议选用正交实验设计法进行培养基及培养条件的优化，通常包括以下几个步骤：

1. 所有影响因子的确认，包括培养基组成成分及培养条件如温度、pH 等。

2. 影响因子的筛选，可预先采用单因素试验法以确定各个因子的影响程度，排除作用不大的因素将其固定在同一水平作为试验的基础条件，抓住影响指标的限制因素进行优化试验，并根据实际可能和菌种的需求确定各因素的水平。

3. 根据影响因子和水平数的要求，选用适当的正交表［每因素都取二水平应选 $L_4(2^3)$；三水平的应选 $L_9(3^4)$ 等］。

4. 实施试验方案，按照规定试验内容制备培养基及培养条件，根据优化目的进行结果测定（一般一个培养基要重复 2~4 瓶）。

5. 对实验结果数学或统计分析，以确定其最佳条件；正交试验结果的分析方法有极差分析和方差分析（具体方法可参考文献）。

6. 最佳条件的复证。

【实验内容】

本实验运用正交实验法对啤酒酵母菌的培养条件进行优化，测定葡萄糖浓度、磷酸盐用量及培养温度这三个因素在不同水平（建议选 3 个水平）对发酵过程和啤酒酵母菌产量的影响，并应用生物统计学的计算方法分析处理实验数据，求得什么样的葡萄糖浓度、磷酸盐用量及培养温度对发酵效果最好，并确定影响实验的关键因素及最适条件。

【结果】

1. 对每一组三角瓶中的培养液进行离心称量菌体湿重及干重量并记录。

2. 查阅参考文献后进行试验结果极差分析，根据极差 R 值的大小排出因素作用的主次顺序。

3. 计算酵母产率（质量分数/葡萄糖）；比较各因素作用，选取最好水平组合即最优化条件。

（苏　昕）

Experiment 43　Optimization of Microbial Culture Conditions

Objectives

1. Understand the culture conditions which affect the microbial proliferation rates and yield of fermentation products.

2. Be familiar with the methods to optimize the microbial culture conditions.

Principles

A microbial culture is a method of multiplying microorganisms by letting them reproduce in predetermined culture media under controlled laboratory conditions. Many factors such as medium composition, pH value, temperature, etc. They are critical to the microbial growth and secondary metabolites accumulation. Thus, it is necessary to optimize medium composition and culture conditions by using statistical analysis.

Single factor design, a study designed with only one independent factor of treatment in which the factor is manipulated at multiple levels, is often used to determine the effect of a certain treatment. These designs are only suitable for experiments where the experimental treatment is homogeneous. If the experimental factors are highly variable or complicated (e.g. medium composition), the use of an orthogonal design, uniform design or response surface design should be considered otherwise the experiment will need to be large or it will lack power.

Orthogonal design is an experimental design used to test the comparative effectiveness of multiple intervention components – referred to here as "conditions"——each of which takes on two or more variants at multiple levels. By using an orthogonal design matrix, it allows the researcher to test the effectiveness of many variants simultaneously in a single experiment (and possibly identify some of their interactions) with far fewer experimental units than it would take to exhaust all possible variants combinations. This feature makes it particularly valuable for testing the best way to implement complex variants with many facets.

Uniform design seeks design points that are uniformly scattered on the domain. The uniform design minimizes the number of runs of experiments by using multivariate statistical analysis with quadratic regression model, yielding useful results of a simple and precise estimate of the integral statement. However, a uniform design does not guarantee even spacing of the factor levels, if the experimental units are highly variable. In case of complex microbial culture experiment, it is better to repeat 2~3 times to make sure whether it is easy to control. Moreover, appropriately increasing the number of runs could improve the significance of the regression equation.

Response surface design (RSD), a mixture of mathematic and statistic analysis, explores the relationships of between several explanatory variables and one or more response variables. The main idea of RSD is to use a sequence of designed experiments to obtain an optimal response. The critical advantage of RSD is that it helps optimizing the microbial culture condition with fast modeling, short time and increased credibility. Plackett – Burman (PB) and Central composite design (CCD) are suggested to do this, leading to the optimal response with conservation effort.

Apparatus and Materials

1. Specimens *Saccharomyces cerevisiae*.

2. Cultures and reagents Media according to the orthogonal design matrix, glucose, phosphate, distilled water, etc.

3. Apparatus Incubator, centrifuger.

4. Others　Flask, cylinder, beaker, centrifuge tube, etc.

Methods and Procedures

The orthogonal design is suggested in this experiment to optimize the medium composition and culture condition. The steps are listed as follows:

1. Identify all the variants, such as medium composition, temperature, pH value, etc.

2. Screen the possible variants. Determine the extent of impact of all the variants by using the single factor design. Factors with fewer impacts will be fixed at the same level as the basic condition, while limiting factors will be tested at multiple levels for optimization.

3. Choose the appropriate matrix based on the numbers of variants and levels tested (L4 -2^3 for double levels; L9 -3^4 for triple levels).

4. Prepare media and test under certain conditions as the matrix shows (normally one kind of medium with 2~4 flasks).

5. Define the optimum medium composition and culture conditions with methods of range analysis and variance analysis (refer to references in detail).

6. Verification of the optimum condition.

Experiment contents

In this experiment, the impact of glucose concentration, temperature and phosphate dosage, which significantly affect the process of fermentation and production of *Saccharomyces cerevisiae*, is investigated at three different levels. Data are collected and analyzed by biostatistical methods to obtain the optimum condition of yeast fermentation.

Results

1. Measure and record the wet and dry weight of the inoculum in every flask of each group using centrifugation.

2. Range analysis is suggested here (refer to references in detail), and the value of R represents the extent of impact of each factor.

3. Calculate the yield of yeast (mass fraction/glucose); compare the impact of each factor to determine the optimum condition of yeast fermentation.

实验四十四　影响微生物生长的物理因素

【目的】

1. 掌握温度、pH、紫外线等物理因素对微生物生长影响的原理。
2. 熟悉某些物理因素对微生物生长影响的检测方法。

【基本原理】

由前面实验已知道微生物的生长受到营养物质水平、种类的限制。此外，微生物的生长还受环境各种物理因素的影响。研究环境条件对微生物生长的影响，有助于控制有害微生物的生长，了解微生物的生态分布。

本实验将简要介绍环境中一些重要的物理因素是怎样影响微生物生长的。

1. 环境温度可明显影响微生物的生长繁殖。在一定温度范围内，基体的代谢活动与生长繁殖随着温度的上升而增加，当温度上升到一定程度，细胞生长速度减缓，如再继续升高，则细胞将急剧死亡。

2. 众所周知，pH 可以影响微生物的生长，每种微生物都有其合适的 pH 范围和生长最佳的 pH 值。pH 变化对微生物生命活动的影响是：破坏细胞膜结构变化，导致微生物死亡；抑制酶活性或膜转运蛋白活性进而杀死微生物；环境的 pH 变化有时还能引起物质电荷的变化，从而影响营养物质的运输。

3. 紫外线作用机制及作用特点已经在实验 39 有详细介绍。

紫外线除了能够直接改变 DNA 的结构之外，紫外辐射还能产生一种被称之为自由基的光化学毒物。这种具有高度活性的分子可以和细胞的 DNA、RNA 和蛋白质结合，进而杀伤微生物细胞。

紫外线的杀菌效果，因菌种及生理状态而异，照射时间、距离和剂量的大小也对其有影响。

由于紫外线的穿透能力差，因此一般适于某些不耐热的溶液，医院手术室空气、无菌室空气、食品加工区、物体表面等消毒。

【仪器与材料】

1. 菌种　建议选择大肠埃希菌、金黄色葡萄球菌、酿酒酵母菌、枯草芽孢杆菌。

2. 培养基与试剂　建议使用肉汤液体培养基、肉汤琼脂培养基、沙氏培养基或麦芽汁培养基。

3. 仪器　恒温培养箱、分光光度计等。

4. 其他　无菌培养皿、无菌滤纸、无菌试管、无菌三角涂棒、黑纸片等。

【方法与步骤】

设计参考（本实验选用单因素考察，多因素多水平的考察可参考实验 41 的设计思路）

（一）温度实验

1. 微生物生长的最适温度

（1）制备肉汤液体培养基，过滤后分装入 8 支试管（5ml/管），灭菌后分别标明 20℃、28℃、37℃和 45℃四种温度，每种温度 2 管。

（2）向每管接种经活化培养 18~20h 的大肠埃希菌液 0.1ml 混匀，将各管分别置于相应温度下，进行震荡培养 24h。观察结果，与未接菌的空白培养基对照，根据菌液的浑浊度，判断试验菌的最适生长温度。

2. 微生物对高温的抵抗能力

（1）同法制备 8 支肉汤试管，灭菌后按顺序 1~8 编号。

（2）向单号管中接入大肠埃希菌液 0.1ml 混匀，向双号管中接入枯草芽孢杆菌菌液 0.1ml 混匀。

（3）将 8 支试管同时放入 100℃水浴中，10min 后取出 1~4 号管，再过 10min 后取出余下的 5~8 号管，立即用冷水或冰浴冷却。

（4）将各管置于37℃恒温培养箱中培养24h。观察结果，与未接菌的空白培养基对照，根据菌液的混浊度，判断检测菌对高温的抵抗力。

（二）pH实验

1. 不同pH对细菌生长的影响

（1）配制肉汤液体培养基，分别调至pH3、5、7、9和11，每种pH分装3管，每管装培养液5ml，灭菌备用。

（2）取培养18~20h的大肠埃希菌斜面1支，加入无菌水4ml制成菌悬液。

（3）向每管中接种大肠埃希菌悬液1滴（或0.1ml），摇匀，置37℃恒温培养箱中培养24h后观察结果。根据菌液的浑浊程度判定细菌在不同pH的生长情况。

2. 不同pH对酵母菌生长的影响

（1）配制麦芽汁普通液体培养基，分别调至pH3、5、7、9和11，每种pH分装3管，每管装培养液5ml，灭菌备用。

（2）按上法制成酿酒酵母菌悬液，向上述每管培养基中接种1滴（或0.1ml）菌液，摇匀，置28℃温箱中培养48h后观察结果。根据菌液的浑浊程度判定酵母菌在不同pH的生长情况。

（三）紫外线杀菌实验

（1）将已经灭菌并冷却到50~55℃左右的肉汤琼脂培养基15~20ml倒入无菌培养皿中，水平放置待凝固。

（2）用无菌吸管吸取0.1ml培养18h的实验对象菌种（建议可选用金黄色葡萄球菌209P）加入上述平板中，涂布均匀。

（3）以无菌操作的方法将黑色纸片放入已涂菌的培养基平板表面，打开培养皿盖，在紫外灯下照射20~30min后盖上皿盖，然后在红光灯下无菌操作取出黑色纸片，平板用黑纸包好，在37℃恒温箱中培养24h后观察结果。

【实验内容】

1. 大肠埃希菌最适生长温度的测定；大肠埃希菌与枯草芽孢杆菌对高温抵抗力的比较。
2. 测定不同pH条件对细菌和酵母菌生长的影响。
3. 紫外线对金黄色葡萄球菌的杀菌作用测定。

【结果】

1. 记录在不同温度下微生物生长的状况并填写下面表格。

	对照	20℃	28℃	37℃	45℃
大肠埃希菌					

	100℃处理10min				100℃处理20min			
	1	2	3	4	5	6	7	8
大肠埃希菌								
枯草芽孢杆菌								

（以"-"表示不生长，"+"=轻微生长，"++"=中度生长，"+++"=重度生长，"++++"=生长量最大）

2. 记录不同 pH 对微生物生长状况，并填写表 44-2。

表 44-2

试验菌	pH3	pH5	pH7	pH9	pH11
大肠埃希菌					
酿酒酵母					

（以"-"表示不生长，"+"=轻微生长，"++"=中度生长、"+++"=重度生长，"++++="生长量最大）

3. 描述紫外杀菌平板中微生物的生长情况，判断紫外线的杀菌效果和特点。

【思考题】

1. 在本次实验中，为什么选用大肠埃希菌、金黄色葡萄球菌和枯草芽孢杆菌作为试验菌？说明原因。

2. 为什么经紫外线照射过的区域仍然有少量菌生长？

（苏　昕）

Experiment 44　Influence of Physical Factors on Microbial Growth

Objectives

1. Understand the principle of the influence of some physical factors such as temperature, pH and ultraviolet radiation on microbial growth.

2. Be familiar with the detection methods for some physical factors on microbial growth.

Priciples

In this experiment, we shall briefly review how some of the most important environmental factors affect microbial growth.

As we have known before, microorganisms must be able to respond to variations in nutrient levels, and particularly to nutrient limitation. The growth of microorganisms is also greatly affected by physical factors in their surroundings. An understanding of environmental influence aids the control of microbial growth and the study of the ecological distribution of microorganisms.

1. Environmental temperature profoundly affects microorganisms. At low temperature a temperature rise increases the growth rate to double for every 10℃ rise in temperature. Beyond a certain point, further increase actually slower the growth and sufficiently high temperatures are even lethal.

2. It is known that pH dramatically affects microbial growth. Each species has a definite pH growth range and optimum pH for growth. Drastic variation in cytoplasmic pH can harm microorganisms by disrupting the plasma membrane or inhibiting the activity if enzymes and membrane transport proteins. Changes in the external also might alter the ionization of nutrient molecules and thus reduce their availability to the organism.

3. The principle and characteristics of Ultraviolet radiation has been introduced in details in Experiment 39.

In addition to altering DNA directly, UV radiation also disrupts cells by generating toxic photochemical products called free radicals. These highly reactive molecules interfere with essential cell processes by binding to DNA, RNA, and protein.

The bactericidal effects of UV radiation differ with species, physiological states, irradiation time, distance from UV, irradiation doses of ultraviolet, and so on.

As a sterilizing agent, ultraviolet radiation is limited by its poor penetrating ability. It is used to sterilize some heat – labile solutions, the surface, to decontaminate hospital room, operating rooms and food – processing areas.

Apparatus and Materials

1. Specimens *Escherichia coli*, *Staphylococcus aureus*, *Saccharomyces cerevisiac*, and *Bacillus subtilis* are suggested.

2. Cultures Nutrient broth medium, nutrient broth agar medium, Salouraud Liquid Medium and malt extract medium are suggested.

3. Apparatus Incubator, spectrophotometer.

4. Others Sterile Petri plate, sterile filter paper, sterile tubes, sterile spreading rod, black paper.

Method and procedures

A reference to design this experiment: A single factor experiment is developed here, multi – factor tests will be performed in Experiment 41.

Ⅰ. **Temperature experiment**

1. The optimal temperature for microbial growth

(1) Prepare nutrient broth medium and filtrate it. Transfer 5ml of the medium to 8 tubes respectively. After sterilization, label the broth tubes with 20℃、28℃、37℃ and 45℃.

(2) Transfer 0.1ml of *E. coli* suspension incubated for 18 ~ 20h to each tube and mix well. Set the tubes at appropriate temperature, incubate for 24h with shaking. Observe your results. Determine the optimal growth ranges according to the bacteria suspension turbidity, comparing the effects with the blank broth medium.

2. Microbial resistance ability to high temperature

(1) Prepare sterile nutrient broth medium. Transfer 5ml of the medium to 8 tubes respectively. Label the broth tubes with 1, 2, 3, 4, 5, 6, 7 and 8.

(2) Transfer 0.1ml of *E. coli* suspension to each odd number tube and mix well. Add 0.1ml of *B. subtilis* suspension to each even number tube and mix well.

(3) Place the eight broth tubes into the water bath (100℃). After 10min, remove the tubes labeled with 1, 2, 3 and 4. After another 10min, remove the tubes labeled with 5, 6, 7 and 8. Each time you should make the tubes cool rapidly in ice bath when you remove them from the 100℃ water bath.

(4) Incubate all tubes at 37℃ for 24h. Observe your results. Evaluate the heat tolerance of the tested bacteria according to the bacteria suspension turbidity, comparing to the effects with the blank broth medium.

II. pH experiment

1. The influence of different pH values on the bacterial growth

(1) Prepare sterile nutrient broth medium and adjusting pH unit to 3, 5, 7, 9 and 11. Transfer 5ml of the medium to 15 tubes respectively, with three tubes for each pH unit.

(2) Prepare *E. coli* suspension by dding 4ml of sterile water to *E. coli* nutrient agar slant incubated for 18~20h. Mix well.

(3) Transfer 0.1ml of *E. coli* suspension to each broth tube and mix. Incubate at 37℃ for 24h. Observe your results. Determine the optimal pH according to the bacteria suspension turbidity, comparing the effects with the blank broth medium.

2. The influence of different pH value on the yeast growth

(1) Prepare sterile malt extract liquid medium and adjusting pH unit to 3, 5, 7, 9 and 11. Transfer 5ml of the medium to 15 tubes respectively, with three 3 tubes for each pH unit.

(2) Prepare *S. cerevisiac* suspension as the same way as *E. coli*. Transfer 0.1ml of *S. cerevisiac* suspension to each broth tube and mix. Incubate at 28 ℃ for 48h. Observe your results. Determine the optimal pH according the yeast suspension turbidity, comparing the effects with the blank broth medium.

III. Ultraviolet radiation sterilization experiment

(1) Melt the sterile nutrient broth agar medium and cool down to 50~55 ℃, transfer 15~20ml to each plate, waiting to cool down.

(2) Using a sterile pipette, transfer 0.1ml of tested bacterial suspension (suggested *Staphylococcus aureus*) to the plates above and spread.

(3) Aseptically put a sterile black paper on the surface of the inoculated Petri plates. Place each plate directly under the ultraviolet light at 30cm distance from the light for 20~30min with the lid off. Remove the covering materials, replace the lid, and incubate at 37℃ in a dark environment.

Experiment contents

1. Determination of the optimal growth temperature for *E. coli*. Compare the resistance abilities to high temperature between *E. coli* and *B. subtilis*.

2. Determine the influence of different pH unit on bacteria and yeast.

3. Determine the UV sterilization on *Staphylococcus aureus*.

Results

1. Record the appearance of growth at different temperature and fill Table 44-1.

Table 44 – 1

	Blank control	20℃	28℃	37℃	45℃
E. coli					

	Treated at 100℃ for 10min				Treated at 100℃ for 20min			
	1	2	3	4	5	6	7	8
E. coli								
B. subtilis								

(" – " = no growth, " + " = minimum growth, " + + " = moderate growth, " + + + " = heavy growth, " + + + + " = maximum growth)

2. Record the appearance of growth at different pH and fill Table 44 – 2.

Table 44 – 2

Tested strains	pH3	pH5	pH7	pH9	pH11
E. coli					
S. cerevisiae					

(" – " = no growth, " + " = minimum growth, " + + " = moderate growth, " + + + " = heavy growth, " + + + + " = maximum growth)

3. Describe the appearance of growth on the plates of UV sterilization experiment. Determine the UV sterilization effects and characteristics of UV disinfection.

Question

1. Why you choose *Escherichia coli*, *Staphylococcus aureus* and *Bacillus subtilis* as tested strains, give your explanation.

2. Why are there still some colonies growing in the areas exposed to ultraviolet light?

（徐 威）

实验四十五 药物的体外抗菌试验

【目的】

1. 掌握琼脂扩散渗透法测定药物体外抗菌活性的原理。
2. 熟悉常用的药物体外抗菌活性的测定方法。
3. 了解液体稀释法测定抗生素的最低抑菌浓度（MIC）。

【基本原理】

药物体外抗菌活性的测定方法很多，一般有两大类：琼脂扩散渗透法和系列浓度稀释法。广泛应用于新药研究和指导临床用药，如抗菌药物筛选、药物的抗菌谱测定、药敏试验、血药浓度测定等。

琼脂扩散渗透法是利用药物能够渗透到琼脂培养基的性质，将试验菌混入琼脂培养基后倾注倒平板，或将试验菌涂布于琼脂平板的表面，然后用不同的方法将药物置于已含试验菌的琼脂平板上。根据加药的操作方法不同分为滤纸片法、打洞法、挖沟法、管碟法和移块法等，经适宜温度培养后观察药物的抑菌能力。本实验主要介绍其中的滤纸片法和挖沟法。

系列浓度稀释法常用于测定药物的最低抑菌浓度（minimal inhibitory concentration，MIC）或最低杀菌浓度（minimum bactericidal concentration，MBC）。药物的最低抑菌浓度是指药物能够抑制微生物生长的最低浓度；可杀灭细菌的最低药物浓度为最低杀菌浓度。MIC（MBC）可以评价药物抑菌作用的程度，常以 μg/ml 或 U/ml 表示。其值愈小，则抑菌作用愈强。常见的测定方法有：试管稀释法、平板稀释法、斜面混入法和微孔板法等。本试验主要介绍试管稀释法和微孔板法。

【仪器与材料】

1. 菌种 表皮葡萄球菌、金黄色葡萄球菌和大肠埃希菌临床菌株的 8h 牛肉膏蛋白胨肉汤培养物，金黄色葡萄球菌（ATCC25925）、大肠埃希菌（ATCC25922）标准菌株的 8h 肉汤培养物。（注：若无临床株也可只用标准株）

2. 培养基及试剂 肉汤琼脂培养基等；待测药物：青霉素 G（β-内酰胺类）、链霉素（氨基糖苷类）、阿奇霉素（大环内酯类）和左旋氧氟沙星（喹诺酮类）、0.1% 新洁尔灭、0.1% 龙胆紫、2.5% 碘液、0.85% 生理盐水、板蓝根浸煮剂等。

3. 仪器 恒温培养箱、超净工作台、酶标仪、微量进样器、游标卡尺等。

4. 其他 无菌平皿、无菌吸管、无菌试管、无菌接种铲、无菌滤纸片、接种环、镊子等。

【方法与步骤】

（一）滤纸片法

滤纸片法是琼脂扩散法中最常用的方法，适用于新药的初筛试验（初步判断药物是否有抗菌作用）及临床的药敏试验（细菌对药物的敏感性试验，以便临床选择治疗用药参考）。可进行多种药物或一种药物的不同浓度对同一种试验菌的抗菌试验。

以一定直径（6~8mm）的无菌滤纸片，蘸取一定浓度的被检药液，将其紧贴在含菌平板上，如果纸片上含有药液，便会沿琼脂向四周扩散，且对该试验菌有抑制作用，经一定时间培养后，就可在滤纸片周围形成不长菌的透明圈。见图 45-1。

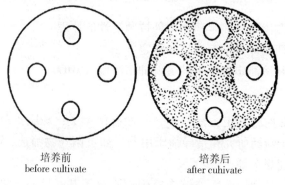

培养前　　　　　培养后
before cultivate　　after cultivate

图 45-1 滤纸片法
Figure 45-1 Paper-disk diffusion test

1. 制备含菌平板 用滴管分别取金黄色葡萄球菌和大肠埃希菌（临床菌株）肉汤培养物4~5滴，加到两个灭菌的空平皿中，每皿加入20ml已溶化并冷却至50℃左右的培养基，制成含菌平板，冷凝备用。

2. 加待检药液 用无菌镊子夹取滤纸片，分别浸入0.1%新洁尔灭、0.1%龙胆紫、2.5%碘液、0.85%生理盐水中，在盛药平皿内壁上除去多余药液后，分别贴在含菌平板表面，并做好标记，37℃培养20h。

3. 观察抑菌圈 观察滤纸片周围的抑菌圈。滤纸片边缘到抑菌圈边缘的距离在1mm以上者为阳性（+），即微生物对药物敏感；反之为阴性（-），即微生物对药物不敏感。

（二）挖沟法

本法适用于半流动药物或中药浸煮剂的抗菌试验。可在同一平板上试验一种药物对几种试验菌的抗菌作用。本实验建议选用板蓝根浸煮剂。

1. 在琼脂平板中央，用无菌接种铲挖一条长沟，将沟内琼脂全部挖出。
2. 将待测药物加入此沟内，以装满不流出为限。
3. 在沟两侧垂直划线接种各种试验菌，见图45-2。

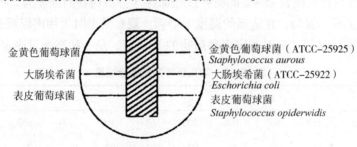

图45-2 挖沟法

Figure 45-2 Trenching test

4. 若为细菌则37℃培养24~48h；若为放线菌或真菌则28℃培养48~72h。
5. 观察沟两边所生长的试验菌离沟的抑菌距离，从而判断待测药物对这些菌的抗菌能力。

（三）试管稀释法

该方法是在一系列试管中，用液体培养基按照几何级数或数学级数连续稀释药物，然后在每一试管中加入一定量的试验菌，经培养后，肉眼观察能抑制试验菌生长的最低浓度即为该药物的MIC。本法所用药物可选做链霉素，试验菌可选用大肠埃希菌。

1. 取10支小试管，编号1~10。
2. 用5ml无菌移液管取肉汤培养基1.8ml加到第1管中，其余各管各加1ml。
3. 用1ml无菌移液管吸取待测药物溶液（1280μg/ml）0.2ml加入第一管内混匀，从第一管取出1ml加到第二管内，混匀，其他依次稀释，9号管混匀后取出1ml扔掉，10号管为空白对照，不加药物。
4. 用另一支1ml无菌移液管，分别吸取1∶10000的试验菌稀释液0.1ml加到含有不同浓度药液和对照小试管中,加入顺序为从对照管(10号)开始依次从药液浓度低向药液浓度高的试管进行。稀释过程见表45-1。

5. 37℃培养20h,对照管中菌应正常生长,液体变为混浊。观察其他管中药物对测试菌生长的抑制作用,以抑制细菌生长的最低药物浓度记录为MIC值。

表45-1 试管稀释法系列稀释过程

管号	1	2	3	4	5	6	7	8	9	10
肉汤培养基(ml)	1.8	1.0	1.0	1.0	1.0	1.0	1.0	1.0	1.0	1.0
药液(ml)	0.2→	1.0→	1.0→	1.0→	1.0→	1.0→	1.0→	1.0→	1.0→	弃掉
试验菌(ml)	0.1	0.1	0.1	0.1	0.1	0.1	0.1	0.1	0.1	0.1
药物终浓度(μg/ml)	128	64	32	16	8	4	2	1	0.5	0
每管总体积(ml)	1.1	1.1	1.1	1.1	1.1	1.1	1.1	1.1	1.1	1.1

(四)微孔板法

微孔板是一种聚氯乙烯塑料板,通常用96孔板,可用于同时测定多种药物的最低抑菌浓度。用微量进样器按表45-2的稀释过程加样(待检药物的浓度为2560μg/ml),然后再加100μl菌液,混匀,在适当的温度下培养。观察结果时可用肉眼观察,也可采用酶标仪检测。本法可用于同时测定多种药物的MIC值,是一种高通量测定法。

表45-2 微孔板法系列稀释过程

孔号	1	2	3	4	5	6	7	8	9	10
肉汤培养基(μl)	180	100	100	100	100	100	100	100	100	100
药液(μl)	20→	100→	100→	100→	100→	100→	100→	100→	100→	弃掉
试验菌(μl)	100	100	100	100	100	100	100	100	100	100
药物终浓度(μg/ml)	128	64	32	16	8	4	2	1	0.5	0
每管总体积(μl)	200	200	200	200	200	200	200	200	200	200

【实验内容】

1. 采用滤纸片法测定0.1%新洁尔灭、0.1%龙胆紫、2.5%碘液、0.85%生理盐水对金黄色葡萄球菌和大肠埃希菌的抑菌作用。

2. 采用挖沟法测定板蓝根浸煮剂的抗菌作用。

3. 采用试管稀释法链霉素对大肠埃希菌的MIC值。

4. 采用微孔板法测定青霉素G(β-内酰胺类)、链霉素(氨基糖苷类)、阿奇霉素(大环内酯类)和左旋氧氟沙星(喹诺酮类)对大肠埃希菌的MIC值。

【结果】

1. 各种化学药品对金黄色葡萄球菌的抑菌作用。测量抑菌圈填入表45-3。

表 45-3

化学药品	抑菌圈直径（mm）
0.1% 新洁尔灭	
0.1% 龙胆紫	
2.5% 碘液	
0.85% 生理盐水	

2. 观察挖沟法实验中沟两边所生长的试验菌离沟的抑菌距离，判断药物对这些试验菌的抑菌能力大小。

3. 观察试管稀释法实验中各试管菌液的生长状况，记录第几号管液体刚好澄清，即能完全抑制试验菌生长的最高稀释倍数管，其抗生素浓度即为最低抑菌浓度 MIC。

【思考题】

1. 在本实验中，化学药物对微生物所形成的抑菌圈未长菌部分是否说明微生物已经被杀死？如何通过实验加以确定，请自行设计实验。如何判断试验中导致细菌不生长的药物浓度是抑菌还是杀菌？

2. MIC 大，还是 MBC 大？

Experiment 45　Determination of Antimicrobial Activity

Objectives

1. Be familiar with the principles of determination of antimicrobial activity by disk diffusion method.

2. Be familiar with the common methods of determination of antimicrobial activity in vitro.

3. Understand the principles for the determination methods of the minimum inhibitory concentration (MIC) by double dilution on agar plate and series dilution in test tube.

Principles

There are many methods to determine antimicrobial activity in vitro, which are mainly classified as: the disk diffusion test and the serial dilution test. They are widely used in the research of new drug and clinical treatment, such as screening of antibiotics, determination of antimicrobial drug spectrum, susceptibility of antibiotics and measurement of drug concentrations in the blood.

The principal behind the disk diffusion test is as follows: when the antibiotic is placed on agar by some ways previously inoculated with the tested bacteria, the antibiotic diffuses radially outward through the agar, producing an antibiotic concentration gradient. A clear zone or ring is present around the antibiotic after incubation if the agent inhibits bacterial growth. The wider the zone surrounding the antibiotic, the more susceptible the bacterium is. It is divided into the paper-disk diffusion test, the trenching test, the cup-plate diffusion test and the moving-agar block test. We introduce the paper-disk diffusion test and the trenching test.

The serial dilution tests are usually used to determine drug minimal inhibitory concentration (MIC) and minimum bactericidal concentration (MBC). MIC is the minimum concentration of a drug that can inhibit the growth of microorganisms; MBC is the minimum concentration of a drug that can kill the microorganisms. We can assess the extent of drug inhibiting bacteria by MIC (MBC) and is usually represented by μg/ml or U/ml. The smaller MIC or MBC is, the higher antimicrobial activities is. The commonly used methods are disk dilution test, broth dilution test, the infiltrating – agar slant test, the microwell plate test and so on. We introduce the disk dilution test and the broth dilution test. The disk dilution test is to dilute the drug 2 fold to proportional extent in an agar media plate. The test bacteria are inoculated on the agar plate containing series antibiotic concentrations. The lowest concentration of antibiotic resulting in inhibition and no growth after 16 to 20 hours of incubation is the minimal inhibitory concentration (MIC) and minimal bactericidal concentration (MBC) respectively. The broth dilution test is to continuously dilute drugs with broth medium based on geometric or mathematic series in test tubes, and then add a certain amount of test bacteria culture to each tube. After incubation, the minimum concentration of drugs that can inhibit the growth of test bacteria with naked eyes is defined as MIC.

Apparatus and Materials

1. Strains   Nutrient broth cultures of clinical strains of *Staphylococcus aureus*, *Staphylococcus epidermidis*, *Escherichia coli*, and standard strains of *Staphylococcus aureus* (ATCC25925) and *Escherichia coli* (ATCC25922).

2. Cultures and Reagents   Nutrient agar media; Drugs to be determined: penicillin G (β – lactam), streptomycin (aminoglycosides), azithromycin (macrolides) and levofloxacin (quinolones), 0.1% benzalkonium bromide, 0.1% gentian violet, 2.5% iodine solution, 0.85% physiological saline, isatis root.

3. Apparatus   Incubator, clean bench, microplate reader, microsyringe, vernier caliper, etc.

4. Others   Sterile plates, sterile pipettes, sterile test tubes, sterile glass spade, sterile paper slips, inoculating loop, tweezers, etc.

Methods and Procedures

A. The paper – disk diffusion test

This is the most commonly used method in agar diffusion and is applicable to preliminarily screen for new drugs (to evaluate whether the drug has the antimicrobial effect) and to test clinical drug susceptibility. It also can be used to carry out antimicrobial experiment a test bacterium which is subjected to many drugs or one drug of different concentrations.

1. Transfer 4 ~ 5 drops of *S. aureus* and *E. coli* broth culture respectively to two sterile empty plates, pour 15 ~ 20ml nutrient agar medium (50℃) for each plate and mix by gentle rotation.

2. Add a series paper absorbed 0.1% Benzalkonium Bromide, 0.1% Gentian Violet,

2.5% Iodine solution, 0.85% physiological saline respectively on the plates, and then label them. Incubate at 37℃ for 20h (Figure 45 – 1).

3. Observe the inhibitory bacteria circle around the paper. If the distance between rim of the filter paper and the inhibitory bacteria circle is over 1mm, it is a positive result, which means the microorganisms are sensitive to the drug, otherwise it will be negative.

B. The trenching test

This method is applicable to the antimicrobial test of semi – fluid or the soak of Chinese medicinal materials. It can examine the effect of one drug on several different test bacteria on the same agar plate.

1. Dig out all agar of a ditch in the center of the agar plate with a sterile spade.

2. Add the testing drug to the ditch to its full capacity. Note: do not spill over the ditch.

3. Inoculate a variety of test bacteria to both side of the ditch. (Figure 45 – 2)

4. For bacteria, incubate for 24 ~ 48h at 37℃, for Actinomycetes or Fungi incubate for 48 ~ 72h at 28℃.

5. Observe the inhibitory bacterial distance between ditch and test bacteria on both sides of the ditch, and then assess antimicrobial ability of test drugs.

C. The broth dilution test

In a series of test tubes, the testing drugs are doubly diluted with broth culture medium. Add a certain amount of test bacteria to each test tube and incubate. The minimum concentration of the drug that can inhibit the growth of bacteria observed by bare eyes is MIC. We select one species – *Escherichia coli* and one drug – Streptomycin.

1. Take out 10 test tubes, label with 1 ~ 10.

2. Add 1.8ml broth culture medium to No.1 tube and add 1ml broth culture medium to other tubes.

3. Add 0.2ml test drug solution (1280μg/ml) to No.1 tube, mix evenly, draw 1ml from No.1 tube and add it to the second tube for double times dilution, mix evenly, repeat the double dilution from No.2 to No.9 tube and throw the last 1ml away. The No.10 is the control, no need to add the drug.

4. Add 0.1ml diluted solution of test bacteria whose concentration is 1:1000 to the ten tubes, start from lower drug concentration to higher concentration.

5. Incubate for 20h at 37℃. The bacteria in the control tube should grow well and the liquid becomes opaque. Observe the antimicrobial activity of test drug in other tubes and record MIC. (Table 45 – 1)

Table 45 – 1 Broth dilution test

No. of tube	1	2	3	4	5	6	7	8	9	10
broth culture (ml)	1.8	1.0	1.0	1.0	1.0	1.0	1.0	1.0	1.0	1.0
drug sample (ml) diluting	0.2→	1.0→	1.0→	1.0→	1.0→	1.0→	1.0→	1.0→	1.0→	lost

Continue

No. of tube	1	2	3	4	5	6	7	8	9	10
Staphylococcus aureus (ml)	0.1	0.1	0.1	0.1	0.1	0.1	0.1	0.1	0.1	0.1
concentration of drug (μg/ml)	128	64	32	16	8	4	2	1	0.5	0
total volume (ml)	1.1	1.1	1.1	1.1	1.1	1.1	1.1	1.1	1.1	1.1

D. The microwell plate test

This method is similar to the broth dilution test in tube. The difference is to add solution with the bacterial culture to the microwell instead of the tube (Table 45 – 2). Add 100μl drug diluted solution and 10μl bacterial solution to each hole, mix evenly and then incubate at appropriate temperature. The microwell plate is a plastic plate made of polyethylene which has 96 holes. It can be used to determine the minimum inhibitory concentration of many drugs.

Table 45 – 2 Microwell plate test

No. of hole	1	2	3	4	5	6	7	8	9	10
broth culture (μl)	180	100	100	100	100	100	100	100	100	100
drug sample (μl) diluting	20→	100→	100→	100→	100→	100→	100→	100→	100→	lost
E. coli (μl)	100	100	100	100	100	100	100	100	100	100
concentration of drug (μg/ml)	128	64	32	16	8	4	2	1	0.5	0
total volume (μl)	200	200	200	200	200	200	200	200	200	200

Experiment contents

1. Determination of antimicrobial activity for 0.1% Benzalkonium Bromide, 0.1 Gentian Violet, 2.5% Iodine solution and 0.85% physiological saline by the paper – disk diffusion test.

2. Determination of antimicrobial activity for isatis root by the trenching test.

3. Determination of the MIC of Streptomycin by the broth dilution test.

4. Determination of all these MIC of Penicillin G, Streptomycin, Azithromycin and Levofloxacin by the microwell plate test.

Results

1. The effect of chemical drugs on S. aureus, fill in Table 45 – 3.

Table 45 – 3

chemical drugs	Diameter of inhibition zone (mm)
0.1% Benzalkonium Bromide	
0.1% Gentian Violet	
2.5% Iodine solution	
0.85% physiological saline	

2. Observe the distance between ditch and test bacteria grown on both sides of the ditch, assess the inhibitory ability of the drugs.

3. Observe the growth of bacteria solution in each tube, record MIC.

Questions

1. In this experiment, there is no growth of bacteria in the inhibition areas, is it concluded that those bacteria are killed? How to prove it? And design another experiment to examine whether the drug exhibit inhibitory or bactericidal effect?

2. Which value is larger between MIC and MBC?

实验四十六　抗生素效价的测定

【目的】

1. 掌握用杯碟法测定抗生素的效价的原理。
2. 熟悉微生物法测定抗生素效价的方法。

【基本原理】

抗生素的效价常采用微生物学方法测定，它是利用抗生素对特定的微生物具有抗菌活性的原理来测定抗生素效价的方法，以抗生素的抑菌力和杀菌力作为衡量效价的标准，与临床使用具有平行相关性。微生物学法测定抗生素效价，一般可分为液体稀释法、比浊法和琼脂扩散法，以琼脂扩散法中的杯碟法最为常用。

杯碟法也称管碟法是根据抗生素在琼脂平板培养基中的扩散渗透作用，比较标准品和检品两者对试验菌的抑菌圈大小来测定检品的效价。基本原理是在含有高度敏感性试验菌的琼脂平板上放置小钢管（也称牛津杯）（内径 6.0mm ± 0.1mm，外径 8.0mm ± 0.1mm，高 10mm ± 0.1mm），管内放入标准品和检品的溶液，经 16~18h 恒温培养，抗生素扩散的有效范围内则产生透明的无菌生长的抑菌圈。抑菌圈的直径大小与抗生素的浓度相关，也与抗生素的扩散系数、扩散时间、培养基的厚度及抗生素的最低抑菌浓度等因素有关，比较抗生素标准品与检品的抑菌圈大小，可计算出抗生素的效价。管碟法的特点是灵敏度高，能直接显示抗生素的抗菌活性，因此作为国际通用的方法被列入各国药典法规中。

管碟法测定抗生素的效价又分为一剂量法、二剂量法和三剂量法。其中二剂量法最为常用，又称四点法。将抗生素标准品和检品各稀释成一定浓度比例（2∶1 或 4∶1）的两种溶液，在同一平板上比较其抗菌活性，再根据抗生素浓度对数和抑菌圈直径成直线关系的原理来计算检品效价。取含菌的双层平板培养基，每个平板表面放置 4 个小钢管，管内分别放入检品高、低剂量和标准品高、低剂量溶液。

先测量出四点的抑菌圈直径，按下列公式计算出检品的效价。

(1) 求出 W 和 V

W = (SH + UH) − (SL + UL)

V = (UH + UL) − (SH + SL)

式中，UH—检品高剂量之抑菌圈直径；
　　　UL—检品低剂量之抑菌圈直径；
　　　SH—标准品高剂量之抑菌圈直径；
　　　SL—标准品低剂量之抑菌圈直径。

（2）求出 θ

$$\theta = D \cdot antilog(IV/W)$$

式中，θ—检品和标准品的效价比；
　　　D—标准品高剂量与检品高剂量之比，一般为1；
　　　I—高低剂量之比的对数，即 log2 或 log4。

（3）求出 Pr

$$Pr = Ar \times \theta$$

式中，Pr—检品实际单位数；
　　　Ar—检品标示量或估计单位。

二剂量法也可利用放线图，查出抗生素的效价，这样可以节省计算时间并便于核对（具体原理可参考相关理论教材）。

药典中规定在测定抗生素效价时，对所用的标准品、试验菌、培养基、培养条件及药物的浓度范围都有相应的要求。本试验参照2010年版的药典选择测定四环素检品的效价。

【仪器与材料】

1. 菌种　藤黄微球菌［CMCC（B）28001］。

2. 培养基及药品　效价检定用培养基（见附录）；抗生素：四环素检品及标准品（高剂量、低剂量，高剂量与低剂量之比为2∶1，浓度范围：10.0~40.0U/mg），无菌磷酸盐缓冲液 pH∶6.0。

3. 仪器　恒温培养箱、超净工作台、游标卡尺等。

4. 其他　灭菌生理盐水、无菌平皿、牛津杯（小钢管）、无菌陶土盖、无菌吸管、镊子、滴管等。

【方法与步骤】

1. 制备试验菌悬液　取藤黄微球菌［CMCC（B）28001］接种于营养琼脂斜面，26~27℃培养24h，用0.9%灭菌氯化钠溶液将菌苔洗下备用。

2. 供试品溶液的制备　按药典所要求药物浓度范围精密称取适量四环素标准品和检品，用无菌磷酸盐缓冲液溶解后，稀释成高、低两种剂量，分别标记为"SH"、"SL"和"UH"、"UL"备用。

3. 制备双层平板　取直径90mm、高16~17mm的无菌平皿4个，分别加入20ml加热熔化的无菌检定用培养基，放置水平台上凝固后作为底层。另取检定用培养基适量，加热溶化后，冷却至50℃左右，加入上述制备好的试验菌悬液适量（以使标准品的高剂量所致抑菌圈直径在18~22mm为宜），摇匀后，在底层平板上加5ml并使之均匀摊平，放置水平台上凝固后作为含菌层。

4. 检定　在平板底部四边，分别对角注明"SH"、"SL"和"UH"、"UL"标记。镊子火焰灭菌3次，然后夹取4个牛津杯垂直放置在平板中标志附近（注：勿使牛津杯陷入培养基内，各小杯之间尽量等距）。以无菌滴管分别在每个牛津杯内加入相应的抗生素溶液至满但不溢出管外，且4杯内液面高度相同，换上陶土盖，静置30min。

5. 培养　将平皿置于35℃~37℃恒温培养箱培养16~18h。

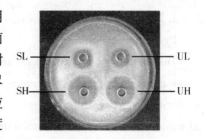

图46-1　管碟法

Figure 46-1　Cylinder plate method

6. 观察结果　用玻璃盖换下陶土盖，倒出牛津杯，用卡尺量取各抑菌圈直径（SH、SL、UH、UL），以mm为单位，4个平板中误差不超过0.1mm，代入公式计算检品效价。见图46-1。

【实验内容】

采用管碟法的二剂量法测定四环素的效价。

【结果】

1. 将抑菌圈直径记录于下表（已知四环素标准品1000U/mg）。

碟号	U_H (mm)	U_L (mm)	S_H (mm)	S_L (mm)
1				
2				
3				
4				
Σ				

2. 根据实验原理中的公式计算四环素的效价。

【注意事项】

管碟法是根据抑菌圈的直径大小来计算抗生素的效价，除与抗生素浓度有关之外，也与抗生素的扩散系数、扩散时间、培养基的厚度及抗生素的最低抑菌浓度等因素有关，要求操作熟练，所用实验材料和器具要规范，且要设最少3个平行对照。

【思考题】

1. 在实际操作中，哪些因素对生物效价测定有影响？
2. 抗生素的效价测定，除了微生物法外还有哪些方法？

（苏　昕）

Experiment 46　Biological Assay of the Potency of Antibiotics

Objectives

1. Grasp the basic principles of biological assay of antibiotics.

2. Determine the antibiotics concentration with the cylinder plate method.

Principles

The antibiotics concentrations are usually determined by microbial method which is based on the antimicrobial activity of antibiotics. We use inhibitory and bactericidal activity as assessment standard for antibiotic potency, which is parallel to clinical application. The microbiological methods to determine the antibiotics concentration include: the dilution method, the turbid metric method and agar diffusion method, of which the cylinder plate method is the most commonly used.

Cylinder plate method is based on the disk diffusion. Antibiotics concentration can be determined according to the inhibition zone diameter compared with standard sample. As a cylinder plate, four stainless tubes (inner diameter of 6.0mm ± 0.1mm, outer diameter of 8.0mm ± 0.1mm, height of 10.0mm ± 0.1mm) are placed on the agar plate previously inoculated with the sensitive bacteria. Then antibiotics sample and standard antibiotics solution are added into the tubes. The antibiotics in the tubes will diffuse radically and a clear zone or ring will present around an antibiotic disk incubated at 37℃ for 16 ~ 18 hours, which is named inhibition zone. The antibiotics concentration can be determined by comparing the diameter of inhibition zone between samples and standards because it is associated with the antibiotics concentration within the tube. The characteristics of cylinder plate method include its high sensitivity and exhibiting the antibiotic activity directly, so it is included in the pharmacopoeia of many nations as an internationally common method.

The cylinder plate method to determine antibiotics concentration can be classified as one-dosage method, two-dosage method and three-dosage method. The two-dosage method is the most commonly used one, and is also called four dots method. Standard antibiotics and samples are diluted to the concentration ratio of 2∶1 or 4∶1 and compare their antimicrobial activity with antibiotics sample on the same plate. The antibiotics concentration is calculated according to the linear relationship between logarithm amount of antibiotic concentration and the inhibition zone diameter. In this experiment, four stainless tubes are placed on double layer agar plate containing test bacteria in upper layer, and two tubes for high concentration and the other for low concentration.

After the determination of the inhibition zone diameter, we can calculate the antibiotics concentration as follows.

(1) Calculate W and V

W = (SH + UH) - (SL + UL)

V = (UH + UL) - (SH + SL)

Where UH and UL are the inhibitory diameters of high and low concentration of antibiotics sample, SH and SL are the inhibitory diameters of high and low concentration of standard antibiotics.

(2) Calculate θ

θ = D × antilog (IV/W)

Where θ is the concentration ratio of antibiotics sample and standard antibiotics, D is the high concentration ratio of antibiotics sample and standard antibiotics which is usually one, I is the logarithm of high and low concentration ratio which is usually log2 or log4.

(3) Calculate Pr

Pr = Ar × θ

Where Pr is the determined concentration of antibiotics sample and Ar is the assumed initial concentration.

The antibiotics standard, antibiotics sample, culture medium, culture condition and concentration of antibiotics are strictly required in the pharmacopoeia when determining antibiotics concentration. This experiment refers to determination of tetracyclines concentration in 2010 edition of pharmacopoeia.

Apparatus and Materials

1. Strains   8 hours broth cultures of *Micrococcus luteus* [CMCC (B) 28001].

2. Cultures and Reagents   Tetracycline sample and standard (the ratio of high dosage and low dosage is 2∶1, concentration range of Tetracycline is 10.0 ~ 40.0 U/mg), sterile phosphate buffer pH 6.0, sterilized physiological saline.

3. Apparatus   Incubator, clean bench, vernier caliper, etc.

4. Others   Sterile plates, Oxford cup (inner diameter of 6.0mm ± 0.1mm, outer diameter of 8.0mm ± 0.1mm, height of 10.0mm ± 0.1mm), sterile clay cover, sterile pipette, tweezers, dropper, ruler, etc.

Methods and Procedures

1. Prepare test bacteria suspension   Inoculate *Micrococcus luteus* [CMCC (B) 28001] to slant agar, then incubate at 26 ~ 27℃ for 24h. After incubation, rinse the lawn with 0.9% sterile NaCl solution and prepare the test bacteria suspension.

2. Prepare antibiotics solution   Precisely weigh a certain amount of the Tetracyclines standard and sample according to the pharmacopoeia, dissolve with sterile phosphate buffer solution, dilute the ratio of high dosage and low dosage is 2∶1, label with "SH" "SL" and "UH" "UL".

3. Prepare two layers plates   Take out four sterile plates whose diameter is 90mm and height is 16 ~ 17mm, add 20ml sterile nutrient agar medium respectively to make it as bottom layer after cool. Dilute the spore suspension of *Micrococcus luteus* with sterilized physiological saline to an appropriate concentration in order to get an inhibition zone diameter of 18 ~ 22mm for standard antibiotics solution. Absorb 1ml of spore suspension to 200ml upper layer nutrient agar incubated at 50℃ and mix. Absorb 5ml rapidly into the plate above and cool.

4. Determination   Label SH, SL, UH and UL on the back of the plate. Sterilize tweezers on fire 3 times and pick up 4 sterile Oxford cups on the plate. (Note that don't let the cups

into the media and keep the same distance between the cups.) Add drug solutions into 4 cups respectively, don't let them spill over to keep the same height of solutions within the cups, then cover the plate with the clay cover and stay it for 30 minutes.

5. Incubate   Incubate the plate at 37℃ for 16 ~ 18 hours.

6. Observe the results   Replace the clay cover with a glass one, take out the Oxford cups and then measure the diameter of each inhibition zone (mm) with a ruler (SH、SL、UH、UL). The error among the four plates should be no more than 0.1mm and then calculate the potency according to the formula (Figure 46 – 1).

Experiment contents

Determination of the potency of tetracycline sample by the method of two – dosage – cylinder – plate.

Results

1. Record diameter of inhibition zone and calculate antibiotic activity (standard tetracycline: 1000U/mg).

No.	UH (mm)	UL (mm)	SH (mm)	SL (mm)
1				
2				
3				
4				
Σ				

2. Calculate the potency of tetracycline.

Notes

The potency of antibiotics was determined and calculated by cylinder – plate method according to diameter of inhibition zone which relates to the concentration of antibiotics, diffusion coefficient, diffusion time, the thickness of agar plate and the MIC of antibiotics. It is necessary for skillful operation, normative materials and appliances as well as establishing three parallel controls.

Questions

1. What are the factors affecting biological activity in practical operation?

2. What kinds of method sexclude microbiological methods could be used to detect the potency of antibiotics?

（苏　昕）

第四章　综合性实验

Chapter 4　Comprehensive Experiments

实验四十七　大肠埃希菌噬菌体的分离及效价测定

【目的】
1. 掌握噬菌体效价的测定方法。
2. 熟悉大肠埃希菌噬菌体的分离原理。
3. 了解噬菌体的培养特征。

【基本原理】

噬菌体分离纯化的基本原理见第二章，实验二十噬菌体的分离与纯化。

噬菌体的检测和效价测定，可采用肉汤澄清方法和噬菌斑形成方法，测定待测样品中的噬菌体的存在和效价。

肉汤澄清方法：以能够引起宿主菌裂解的噬菌体溶液的最高稀释度表示。

噬菌斑形成单位测定法：一般采用琼脂叠层法或双层琼脂平板法（agar layer method）测定噬菌体效价。由于含有特异宿主细菌的平板上，噬菌体可繁殖并裂解细菌产生肉眼可见的噬菌斑，因此可以进行噬菌体计数。理论上一个噬菌斑是由一个噬菌体感染宿主形成，但也有些噬菌斑是由几个噬菌体颗粒感染相应的宿主形成，所以，为了准确表达噬菌体的浓度，一般不用噬菌体的绝对数量，而是用噬菌斑形成单位表示。

噬菌体的效价是指 1ml 培养液中所含活噬菌体的数量，以噬菌斑形成单位/ml（plaque forming units，PFU/ml）表示。

例如，稀释度为 10^{-3} 时，在 0.1ml 噬菌体试样中有 65 个噬菌斑，则该噬菌体原悬液的效价为：

$$\frac{65}{0.1 \times 10^{-3}} = 6.510^5 \text{PFU/ml}$$

【仪器与材料】

1. 菌种　大肠埃希菌 18~24h 斜面培养物、大肠埃希菌 18h 培养液、大肠埃希菌噬菌体稀释液（10^{-2}）。

2. 培养基

（1）上层半固体培养基（附录一，3,）。

（2）底层肉汤琼脂培养基（含琼脂 1.4%~1.6%）。

（3）三倍浓缩的肉汤培养基（附录一，4）。

（4）5 支小试管（内装 0.9ml 无菌肉汤液体培养基）、肉汤琼脂平板（10ml 培养基/

皿）5个、含3ml半固体琼脂培养基的试管5支。

3. 仪器 离心机、细菌过滤器、真空泵、抽滤装置、水浴锅。

4. 其他 无菌涂布棒、无菌吸管、无菌培养皿、三角瓶、阴沟污水等。

【方法与步骤】

（一）噬菌体的分离与纯化

见实验二十 噬菌体的分离与纯化。

（二）噬菌体效价的测定

1. 稀释噬菌体

（1）试管编号 取5只盛肉汤培养基试管（每管0.9ml），分别编号"1（10^{-3}）"，"2（10^{-4}）"，"3（10^{-5}）"，"4（10^{-6}）"和"5"（对照）。

（2）采用无菌操作技术，用1ml移液管吸0.1ml 10^{-2}大肠埃希菌噬菌体，注入1号试管（10^{-3}）中，旋摇试管，使混匀。

（3）用另一支无菌移液管从1号试管中吸0.1ml加入2号试管（10^{-4}）中，旋摇试管，使混匀，依次类推，稀释至4号试管（10^{-6}）。

2. 接种的半固体琼脂制备 用1ml移液管分别吸0.1ml对数期大肠埃希菌液，加入5支半固体琼脂试管中，放置在50℃水浴锅保存。

3. 倒底层琼脂平板 取无菌平皿5只，每皿倒入10ml肉汤琼脂培养基（底层培养基），并在皿底依次表明10^{-3}、10^{-4}、10^{-5}、10^{-6}和对照组。

4. 制备噬菌体和敏感菌的混合液 操作从标记5的肉汤试管开始。采用无菌操作，吸取0.1ml 5号试管的稀释液，加到含有敏感菌的半固体琼脂试管中，立即搓试管充分混匀，并倒在标注对应稀释度的底层琼脂平板表面，平置待凝。然后，用同一个移液管吸取0.1ml标记4（10^{-6}）的噬菌体稀释液，加到含有敏感菌的半固体琼脂试管中，立即搓试管充分混匀，并倒在标注对应稀释度（10^{-6}）的底层琼脂平板表面，平置待凝。其他平皿的操作方法依此类推。

5. 培养 将上述平皿于37℃倒置培养至噬菌斑出现（约6~8h）。

6. 观察平板中的噬菌斑 将每一稀释度的噬菌斑数目记录于实验报告表格内，选取噬菌斑数目在25~250个的平板，计算每毫升未稀释的原液的噬菌体数（效价）。

【结果】

1. 在噬菌斑法测定效价过程中，选择一个噬菌斑数在25~250的平板，绘图描述实验结果。

2. 将平板中各稀释度的噬菌斑记录于下表，计算噬菌体的效价。

噬菌体稀释度	10^{-3}	10^{-4}	10^{-5}	10^{-6}	对照
噬菌斑数					
效价					

【思考题】

1. 如何证实新分离到的噬菌体滤液确有噬菌体存在？

2. 测定噬菌体的效价时，哪些操作决定测定准确性？
3. 哪些因素影响噬菌斑的大小？

（周丽娜）

Experiment 47 Isolation and Titration of *Escherichia coli* Phage

Objectives

1. Grasp the assay method of determine the tilter of bacteriophage sample.
2. Be familiar with the principle of isolating *Escherichia coli* phage.
3. Understand the cultural characteristics of phage.

Principles

The basic principle for bacteriophage isolation, see Chapter 2, Experiment 20 the Isolation and Purification of Bacteriophages.

You will measure the viral activity in your sample by performing sequential dilution of the viral preparation and assaying for the presence of viruses. The assay method of titer includes broth clearing assay and plaque forming method.

In the broth clearing assay, the end point is the highest dilution (smallest amount of viruses) producing lysis of bacteria and clearing of the broth. The titer, or concentration, that results in a recognizable effect is the reciprocal of the endpoint.

In the plaque forming method, agar layer method is usually used. The tilter is determined by counting plaques. Each plaque theoretically corresponds to a single infective virus in the initial suspension. Some plaques may arise from more than one virus particles, and some virus particles may not be infectious. Therefore, the titer is determined by counting the number of plaque – forming units (PFU).

The titer, plaque – forming units per milliliter, is determined by counting the number of plaques and dividing by the amount plated times the dilution.

For example, 65 plaques with 0.1ml plated of a $1:10^3$ dilution is equal to

$$\frac{65}{0.1 \times 10^{-3}} = 6.510^5 \text{PFU/ml}$$

(Note: When counting the number of plaque – forming units, choose the plates with 10 to 100 plaques/plate.)

Apparatus and Materials

1. Specimens *Escherichia coli* slant culture for 18~24h, *Escherichia coli* broth culture for 18h, Diluent of bacteriophage culture (10^{-2}).

2. Cultures

(1) Semi – solid agar tube (or upper medium, see Appendix Ⅰ, 3).

(2) Nutrient broth agar medium (substratum or bottom medium, containing 1.4%~

1.6% agar).

(3) 3× nutrient broth medium (see Appendix I, 4).

(4) 4 sterile tubes (with 0.9ml of sterile broth medium), 5 plates with nutrient broth agar medium (10ml each plate), 4 soft agar tube (3ml / tube).

3. Apparatus   Centrifuges, bacterial filters, vacuum pumps, sterile membrane filter assemblies (0.45μm), water bath.

4. Others   Sterile glass spreader rod, sterile pipette, sterile Petri dishes, erlenmeyer flasks, sewage, etc.

Methods and Procedures

I. Isolation and purification of Bacteriophage

See Chapter 2 Experiment 20: Isolation and Purification of Bacteriophages.

II. Titer Detection of Bacteriophage

1. Serial dilutions of bacteriophage

(1) Label the broth tubes (0.9ml/tube) "1 (10^{-3})", "2 (10^{-4})", "3 (10^{-5})", "4 (10^{-6})" and "5" (blank control).

(2) Aseptically add 0.1ml of E. coli phage suspension (10^{-2}) to tube 1 (10^{-3}). Mix by carefully aspirating up and down three times with the pipette.

(3) Using a different pipette, transfer 1ml to the second tube, mix well. Continue until the fourth tube (10^{-6}).

2. Inoculated semi-solid agar medium   With a pipette, add 0.1ml of E. cioli to the semi-solid agar tubes and place them back in the water bath.

3. Pouring the substratum agar plate   Prepare 5 sterile Petri dishes, label them with the dilution $10^{-1} \sim 10^{-4}$ and blank control. Melt the nutrient agar media (as the substratum) and cool to 50~55 ℃, add 10ml of it to each plate and curdle.

4. With a pipette, start with broth tube 4 and aseptically transfer 0.1ml from tube 5 to a semi-solid agar tube, mix by swirling, and quickly pour the inoculated semi-solid agar evenly over the surface of Petri plate 4. Then, using the same pipette, transfer 0.1ml from tube 3 to a semi-solid agar tube, mix and pour over plate 3. Continue until you have completed tube 1.

5. Incubation   Incubate all plates in an inverted position at 37℃ until plaques develop (about 6~8h).

6. Observe the plaques on the plates   Record your results. Select a plate with between 25~250 plagues. Count the number of plaques, and calculate the number of plaque-forming units (PFU) per milliliter.

Experiment contents
Results

1. With Plaque-forming Assay, choose one plate with 25~250 plaques, draw what you observed.

2. Fill the table with the number of plaques /plate, and calculate the PFU/ml.

Dilution	10^{-3}	10^{-4}	10^{-5}	10^{-6}	Comtrol
the number of plaques					
PFU/ml					

Question

1. How to confirm the newly isolated phage filtrate indeed has the phages?
2. Which step determine the accuracy when determination the titer?
3. Which factor affects the plaque's size?

实验四十八 口服药细菌数的测定及大肠菌群的测定

【目的】

1. 掌握检验口服药中细菌数的测定方法。
2. 熟悉口服药大肠菌群的测定方法。

【基本原理】

细菌总数的测定是检查每克或每毫升被检药品内所含有的需氧菌的活菌数,以判断供试药物被细菌污染的程度。检测结果是该药物生产过程卫生学评价的一个重要依据。细菌总数的测定方法常采用平皿法或薄膜过滤法。

大肠菌群是指在35℃下能发酵乳糖,产酸产气(produces acid and gas)的一群需氧或厌氧的革兰阴性无芽孢杆菌。

大肠菌群,包括大肠埃希菌,是肠道菌科的成员。在自然状态下,药品中的大肠埃希菌易死亡或变异。如在药品中未检出大肠埃希菌,不能排除药物未被粪便污染的可能。采用大肠菌群为控制菌,则其检验范围更广,大肠菌群作为药品被粪便污染的指标菌,具有重要的卫生学和实际意义。

【仪器与材料】

1. 培养基与试剂 营养琼脂培养基、胆盐乳糖发酵培养基(附录一,22)、待测药品、无菌生理盐水。

2. 仪器 培养箱。

3. 其他 无菌移液管、无菌平皿、无菌乳钵、无菌试管、三角瓶、酒精灯等。

【方法与步骤】

(一)细菌总数的测定

A. 平皿法

1. 供试品取样及供试液制备 除另有规定外,一般药品的检验量为10g或10ml。根据供试品的理化特性与生物学特性,采取适宜的方法制备供试液。水溶性供试品,一般取供试品10ml,加pH7.0无菌氯化钠-蛋白胨缓冲液至100ml,混匀,作为1∶10的供试液。

2. 样品稀释　采用无菌操作技术用无菌移液管和装有 9ml 无菌氯化钠-蛋白胨缓冲液的试管进行 10 倍梯度稀释，得到 10^{-2}、10^{-3}、10^{-4}、10^{-5} 稀释的样品（稀释倍数依样品含菌量而定，且注意不同稀释度用不同移液管）。

3. 加样、培养　取供试液 1ml，置直径 90mm 的无菌平皿中，注入 15～20ml 温度不超过 45℃ 的熔化的营养琼脂培养基，混匀，凝固，倒置培养。每稀释级每种培养基至少制备 2 个平板。

4. 阴性对照　取试验用的稀释液 1ml，置无菌平皿中，注入培养基，凝固，倒置培养。

每种计数用的培养基各制备 2 个平板。

5. 观察与计数

（1）阴性对照平板中应不长菌。

（2）细菌培养 3d，逐日观察菌落生长情况，点计菌落数。必要时，可适当延长培养时间至 7d 进行菌落计数并报告。

菌数报告规则：细菌宜选取平均菌落数小于 300CFU 的稀释级，作为菌数报告（取两位有效数字）的依据，以最高的平均菌落数乘以稀释倍数的值报告 1g、1ml 供试品中所含的菌数。

B. 薄膜过滤法

1. 灭菌　采用薄膜过滤法，滤膜孔径应不大于 0.45μm，直径一般为 50mm。滤器及滤膜使用前应采用适宜的方法灭菌。使用时，应保证滤膜在过滤前后的完整性。

2. 制备供试液　取相当于每张滤膜含 1g、1ml 或 10cm² 供试品的供试液，加至适量的稀释剂中，混匀，过滤；若供试品每 1g、1ml 或 10cm² 所含的菌数较多时，可取适宜稀释级的供试液 1ml 进行试验。

3. 冲洗滤膜和培养　用 pH7.0 无菌氯化钠-蛋白胨缓冲液或其他适宜的冲洗液冲洗滤膜，冲洗后取出滤膜，菌面朝上贴于营养琼脂培养基平板上培养。

4. 阴性对照试验　取试验用的稀释液 1ml 照上述薄膜过滤法操作，作为阴性对照。阴性对照不得有菌生长。

5. 观察与计数　培养条件和计数方法同平皿法，每片滤膜上的菌落数应不超过 100 个。

（二）大肠菌群的测定

1. 取样　取装有 10ml 的乳糖胆盐发酵培养基试管 3 支。

2. 加入供试液　分别加入 1：10 的供试液 1ml（含供试品 0.1g 或 0.1ml）、1：100 的供试液 1ml（含供试品 0.01g 或 0.01ml）、1：1000 的供试液 1ml（含供试品 0.001g 或 0.001ml）。

3. 阴性对照　另取 1 支乳糖胆盐发酵培养基管加入稀释液 1ml 作为阴性对照管。

4. 培养　30～35℃ 培养 18～24h。

5. 观察乳糖胆盐发酵结果　乳糖胆盐发酵管若无菌生长，或有菌生长但不产酸产气，判该管未检出大肠菌群；若产酸产气，应将发酵管中的培养物分别划线接种于伊红美兰琼脂培养基或麦康凯琼脂培养基的平板上，培养 18～24h。

6. 观察伊红美兰琼脂培养或麦康凯琼脂培养结果 若平板上无菌落生长，或生长的菌落与表 48-1 所列的菌落形态特征不符或为非革兰阴性无芽孢杆菌，判该管未检出大肠菌群；若平板上生长的菌落与表 48-1 所列的菌落形态特征相符或疑似，且为革兰阴性无芽孢杆菌，应进行确证试验。

表 48-1 大肠菌群菌落形态特征

培养基	菌落形态
伊红美兰琼脂	呈紫黑色、紫红色、红色或粉红色，圆形，扁平或稍凸起，边缘整齐，表面光滑，湿润
麦康凯琼脂	鲜桃红色或粉红色，圆形，扁平或稍凸起，边缘整齐，表面光滑，湿润

7. 确证试验 从上述分离平板上挑选 4~5 个疑似菌落，分别接种于乳糖发酵管中，培养 24~48h。若产酸产气，判该乳糖胆盐发酵管检出大肠菌群，否则判未检出大肠菌群。

8. 判断数量 根据大肠菌群的检出管数，按表 48-2 报告 1g 或 1ml 供试品中的大肠菌群数。

表 48-2 可能的大肠菌群数表

各供试品量的检出结果			可能的大肠菌群数 N（个/g 或 ml）
0.1g 或 0.1ml	0.01g 或 0.01ml	0.001g 或 0.001ml	
+	+	+	$> 10^3$
+	+	−	$10^2 < N < 10^3$
+	−	−	$10 < N < 10^2$
−	−	−	< 10

注：+ 代表检出大肠菌群；− 代表未检出大肠菌群

【实验内容】
1. 学习药品中细菌总数测定的平皿法或薄膜过滤法。
2. 学习药品中大肠菌群数的检测方法。

【思考题】
1. 为什么要测定药品中的细菌总数？
2. 为什么要测定药品中的大肠菌群数？

（周丽娜）

Experiment 48　Detection of Bacteria Counts and Coliforms in Oral Medicines

Objectives

1. Grasp the detection of bacteria counts in oral medicines.
2. Be familiar with the determination of coliform groups in oral medicines.

Principles

The concentrations of total bacteria are usually expressed bacteria count per gram (g) or milliliters (ml) of the tested medicine, as a aim for extent of contamination. The detection results is an important reason for hygienic evaluation during drug production process. The bacteria count method includes plate count method and membrane filtration method.

Coliforms are defined as facultatively aerobic or anaerobic, gram-negative, nonsporing, rod-shaped bacteria that ferment lactose with gas and acid formation within 48h at 35℃.

Coliforms, including *Escherichia coli* are members of the family *Enterobacteriaceae*. Under natural conditions, Coliforms in drugs are more susceptible to death or easy to variation. So, if *E. coli* were not checkout in some drugs, the possibility of contaminated with feces could not be ruled out. Coliforms, as indicator for contamination with feces, have widespread use and important hygiene significance.

Apparatus and Materials

1. Cultures and Reagents   Nutrient broth agar, Bile salt lactose fermentation (Appendix Ⅰ, 22), tested drug, sterile normal saline solution.

2. Apparatus   Incubator.

3. Others   Sterile pipette, sterile Petri plate, sterile mortar, sterile tubes, flask, alcohol burner, etc.

Methods and Procedures

Ⅰ. **Detection of total bacteria**

A. Plate count method

1. Selecting the sample and preparing of the sample   Unless otherwise specified, use samples of 10g or 10ml of the oral medicine to be examined for testing. Samples are prepared appropriately depending on their physico-chemical properties and biological characteristics.

For liquid products, take 10ml of the product, dilute to 100ml with sterile sodium chloride-peptone buffer (pH7.0), and mix well.

2. Making serial dilution   Aseptically dilute the sample solution to be tested with pH 7.0 sterilesodium chloride-peptone buffer to make serial dilution of 1:10, 1:100, 1:1000, 1:100000, etc. (Dilution of sample is determined by the number of bacteria contained in sample, and pay attention to use different pipette for different dilution solution.)

3. Sample and incubation   Transfer 1ml of the sample dilution to a sterile Petri dish (90mm in diameter), add 15~20ml nutrient agar medium (melted at not exceeding 45℃), mix well, and allow the contents to solidify at room temperature. Incubate the Petri dish at 30~35℃, for 3 dayes. For each dilution, make at least 2 Petri dishes.

4. Negative control   Transfer 1ml of the dilution to a sterile Petri dish, 15~20ml nutrient agar medium, mix well, and allow the contents to solidify at room temperature. Incubate the Petri dishes at 35℃ for 3 dayes. Make at least 2 Petri dishes.

5. Observing and counting

(1) No evident growth of microorganisms occurs in either of the Priti dishes.

(2) Count the number of colonies every day, report the number on 72h; extend the incubation time to 7 days if necessary.

Microbial number of report rule: Select the dilution in which the average number of colonies of bacteria is less than 300 CFU. Report the result with the highest average number of colonies multiplying by the dilution folds to express the bacteria count per gram (g) or milliliters (ml) of the tested medicine.

Membrane filtration method

1. Sterilization filter   Use membranes having a nominal pore size not great than 0.45μm, and a diameter of appropriately 50nm. The filter unit and membrane are sterilized prior to use by appropriate means. Guarantee the performance characteristic of the filter during the test process.

2. Preparation of the sample   Take representing 1g, 1ml or 10cm^2 of the oral medicine, or 1ml suitable dilution solution of the sample if the product include large numbers of microorganisms, mix well and filter.

3. Rinse the membrane and incubation   Rinse the membrane with pH7.0 sterile sodium chloride – peptone buffer solution or other suitable rinsing solution. Transfer the membrane onto a nutrient agar medium, and incubate.

4. Negative controls   Take 1ml of dilution, and carry out the test as described above. No evident growth of microorganisms occurs in the negative control.

5. Observing and counting   Carry out the test as described above for plate count method. The number of microorganisms on each membrane is not more than 100CFU.

II. The determination of coliform

1. Take 3 tubes each containing an appropriate amount of Bile salt lactose fermentation culture medium (not less than 10ml).

2. Respectively add 1ml of 1:10 (containing 0.1g or 0.1ml of the oral medicine), 1:100 (containing 0.01g or 0.01ml of the product), 1:1000 (containing 0.0011g or 0.001ml of the oral medicine) dilution.

3. Negative controls   Add 1ml of diluting solution to another tube containing Bile salt lactose fermentation culture medium as negative control.

4. Incubate at 30~35℃ for 18~24h.

5. Observing Bile salt lactose fermentation results   The oral medicine passes the test if no microbial growth occurs or no gas bubbles or acid forms in Bile salt lactose fermentation tube. If the formation of acid and gas bubbles is observed, incubate the cultures on Eosin methylene blue agar medium plate or MacConkey agar medium plate, and incubate for 18~24h.

6. Observing the results on Eosin methylene blue agar medium or MacConkey Agar medium. If no growth of microorganisms occurs on the plate, or the appearance of the microbial

colonies does not match the descriptions in Table 48 – 1. Or the colonies are not gram – negative bacilli; the oral medicine passes the test. If the morphology of colonies matches the descriptions in Table 48 – 1, and they are gram – negative bacilli without spores, confirm the result by doing confirmatory tests.

Table 48 – 1   Morphologic characteristics of coliform of colonies

Medium	Morphology of colonies
EMB	Purple black, purple red, or pale red, circular, flat or slight convex, regular margin, smooth surface, moist
MacConkey Agar	Brilliant pink or pale red, circular, flat or slight convex, regular margin, smooth surface, moist

7. Confirmatory tests   Choose 4 ~ 5 suspect colonies from the plate, individually inoculate in tubes containing Bile salt lactose culture medium, and incubate for 24 ~ 48h. The formation of acid and gas bubbles indicates the presence of coliform. Otherwise, absence coliform in the oral medicine will be reported.

8. According to the number of coliform – positive tubes, and Table 48 – 2, record the probable number of coliform 1g or 1ml of the oral medicine.

Table 48 – 2   Probable number of coliform

Results of each quantity of the oral medicine			Probable number of coliform N
0.1g or 0.1ml	0.01g or 0.01ml	0.001g or 0.001ml	(per g or ml)
+	+	+	$> 10^3$
+	+	–	$10^2 < N < 10^3$
+	–	–	$10 < N < 10^2$
–	–	–	$N < 10$

Note: + represents coliform is detected; – not detected

Experiment contents

1. Study plate count method and membrane filtration method to determine the total number of bacteria in drugs.

2. Study detection of coliform bacteria in the drugs.

Question

1. Why determine the total number of bacteria in the drugs?

2. Why determine the number of coliform bacteria in the drugs?

(徐　威)

实验四十九　利用 Biolog 系统进行的分类鉴定

【目的】

1. 掌握利用 Biolog 自动微生物分析系统进行微生物鉴定的原理。

2. 熟悉利用 Biolog 自动微生物分析系统进行微生物鉴定的操作方法。

【基本原理】

Biolog 微生物自动分析系统是美国 Biolog 公司 1989 年研制开发的新型自动化快速微生物鉴定系统。以细菌对微平板上 95 种碳源的利用情况为基础从而进行微生物鉴定。

Biolog 微生物自动分析系统以微生物与 96 孔微平板上多种脱水碳源进行氧化实验和同化实验为基础进行鉴定，将菌体生长所产生的特征性代谢图谱与标准数据库做比对，从而得出鉴定结果。

该 96 孔板横排标记为：1、2、3、4、5、6、7、8、9、10、11、12；纵排为：A、B、C、D、E、F、G、H。A1 孔内为水，作为阴性对照，其他 95 孔是 95 种不同的碳源物质。96 孔中均含有四唑类氧化还原染色剂和胶质。

Biolog 系统利用微生物对不同碳源代谢率的差异，针对每一类微生物筛选 95 种不同碳源，配合四唑类显色物质，固定于 96 孔板上（A1 孔为阴性对照），接种菌悬液后培养一定时间（4h 或过夜培养），通过检测细菌细胞利用不同碳源进行新陈代谢过程中产生的氧化还原酶与显色物质发生反应而导致的颜色变化（吸光度）以及由于微生物生长造成的浊度差异（浊度），与标准菌株数据库进行比对，即可得出最终鉴定结果。细菌利用碳源进行呼吸时，会将四唑类氧化还原染色剂从无色还原成紫色，从而在微生物鉴定板上形成该微生物特征性的反应模式或"指纹"，通过读数仪来读取颜色变化，并将该反应模式或"指纹"与数据库相比就可在瞬间得到鉴定结果。而对于酵母和霉菌来说，还需结合读数仪读取同化的变化（也就是浊度的变化）进行微生物鉴定。

【仪器与材料】

1. 菌种 大肠埃希菌、枯草芽孢杆菌、啤酒酵母菌、产黄青霉。

2. 培养基与试剂 BUG 专用培养基（商品化的培养基）、BUG + M 培养基（BUG 琼脂培养基加 0.25% 麦芽糖）、BUG + B 培养基（BUG 琼脂培养基加 5% 脱纤维羊血）、Biolog 专用菌悬液稀释液、革兰阴性/阳性接种液。

3. 仪器 微孔板、浊度仪、浊度标准、Biolong 微生物分类鉴定系统及数据库、读数仪、培养箱、pH 计、光学显微镜、加样槽。

4. 其他 接种棉签、移液管、八道移液器、移液器头、矩阵多通道移液器头。

【方法和步骤】

1. 使用 Biolog 推荐的培养基，对待测微生物进行纯化培养 1～2 代

（1）使用 Biolog 推荐的 BUG + B 培养基，分离微生物纯培养。

（2）多数细菌的培养时间是 4～24h；可形成芽孢的革兰阳性细菌的培养时间应少于 16h，尽量减少芽孢的形成。

2. 根据待测微生物的革兰染色反应结果选择微孔板 革兰阳性菌采用 GP 微孔板，革兰阴性菌采用 GN 微孔板；真菌、酵母菌采用 YT 微孔板、霉菌采用 FF 微孔板。

3. 配制一定浓度的菌悬液 菌浓度决定待测微生物培养后的细胞浓度，在 Biolog 系统中，菌浓度是必须加以控制的关键参数。因此，接种物的准备必须严格按照 Biolog

系统的要求进行。如果是革兰阳性球菌和杆菌，则在菌悬液中加入 3 滴巯基乙酸钠和 1ml 100mmol/L 的水杨酸钠。使菌悬液浓度与标准悬液浓度具有同样的浊度。

4. 接种微孔鉴定板

（1）将菌悬液加入加样槽。

（2）用八道移液器，将菌悬液接种于微孔板的 96 孔中：一般细菌接种 150μl，芽孢菌接种 150μl，酵母菌接种 100μl，霉菌接种 100μl。给微孔板加盖。

5. 培养微孔鉴定板　将微孔鉴定板放入培养箱，30～35℃培养 3～36h。

6. 读取结果　将培养后的鉴定板放入读数仪中，利用 Biolog 自动微生物分析系统软件，人工自动读取微孔鉴定板结果。

7. 结果解释　软件将对 96 孔板显示出的实验结果按照与数据库的匹配程度列出 10 个鉴定结果，并在 ID 框中进行显示，如果第 1 个结果都不能很好匹配，则在 ID 框中就会显示"No ID"。

每个结果均显示三种重要参数：可能性（PROB）、相似性（SIM）、位距（DIS）% PROB 提供使用者可以与其他鉴定系统比较的参数；SIM 显示被鉴定的菌种（ID）与数据库中的种之间的匹配程度；DIS 显示 ID 与数据库中的种间的不匹配程度。

良好的鉴定结果 SIM 值在培养 4～6h 时应≥0.75，培养 16～24h 时应≥0.50。SIM 值越接近 1.00，鉴定结果的可靠性越高。DIS <5 表示匹配结果好。

如果你打算查看 10 个之外的结果，直接双击"Other 显示框"。在数据库中选中欲比较的种，就可以在电脑屏幕上显示出各种指标。

【实验内容】

利用 Biolog MicroLog 系统，评估微生物利用碳源的精确性，进而进行芽孢杆菌、霉菌和酵母等的精确分类鉴定。

【结果】

给出鉴定结果，并对系统读取结果进行详细说明。

【注意事项】

1. Biolog 自动微生物分析系统只适合微生物纯种的鉴定，不能进行混合样品的微生物鉴定。

2. 接种液和微孔板使用之前需要预热至室温。

【思考题】

评估鉴定结果的准确性，若鉴定结果不理想，分析其可能原因并解决问题。

（周丽娜）

Experoiment 49　Microbial Identification and Classification with Biolog Analysis System

Objective

1. Grasp the principle of Microbial Identification by Biolog Analysis System.

2. Be familiar with the operating method of using the Biolog Analysis System.

Principles

Biolog Microbial Identification System, which is developed by Biolog in 1989, is a new automated technology for rapid identification of microorganisms. The system is based around the assimilation of 95 carbon sources by bacteria on a microtiter tray.

The system is based around a 96 – well microtitre tray containing a range of dehydrated carbon sources for oxidation and assimilation tests by microorganism. The metabolic fingerprint provided by each isolate growth is compared to the standard profiles in the Biolog database and the identification result will be given.

With a 96 – well microtitre tray, the horizontal is labeled with 1, 2, 3, 4, 5, 6, 7, 8, 9, 10, 11, 12. The vertical is A, B, C, D, E, F, G, H. A1 is filled with water, as a negative control, the other holes are with 95 kinds of different carbon sources. Tetrazolium violetin is within 96 holes.

The Biolog MicroPlate is 96 – well dehydrated panel containing tetrazolium violet, a buffered nutrient medium, and a different carbon source for each well except the control, which does not contain a carbon source. The microwells are rehydrated with a cell suspension and read at either 4h or overnight (16 to 24h) for the ability of the bacteria to utilize the carbon source. Tetrazolium violet is a redox dye used to detect electrons donated by NADH to the electron transport system. Reduced tetrazolium violet is a purple formazan. When a carbon source is not used, the microwell remains colorless, as does the control well. The resulting pattern of purple wells yields a "metabolic fingerprint" of the bacterium tested. After incubation, the phenotypic fingerprint of purple wells compared to Biolog's extensive species library. If a match is found, a species level identification of the isolate is made. For yeasts and molds it is necessary to read the change of turbidity for a final classification and identification.

Apparatus and Materials

1. Specimens   *Escherichia coli*, *Bacillus subtilis*, *Saccharomyces cerevisiae*, *Penicillium Chrysogenum*.

2. Cultures and Reagents   BUG (Biolog Universal Growth Agar), BUG + M (BUG Agar with 0.25% maltose), BUG + B medium (BUG Agar with 5% sheep blood), BUG agar (Biolog Dehydrated Growth Agar), Inoculating Fluid, GN/GP – IF.

3. Apparatus   Microplate, turbidimeter, turbidity standards, Biolog Microbial Identification System and Data Base, microplate reader, incubator, pH instrument, optical microscope, reservoir.

4. Others   Sterile disposable inoculator swabs, transfer pipets, 8 channel electronic pipettor, sterile racked pipet tips, matrix multichannel pipette tips.

Methods and Procedures

1. Culture organism for 1 ~ 2 generations on Biolog Recommended agar media.

(1) Isolate a pure culture on Biolog recommended agar media (BUG + B) and incubate

at 33°C.

(2) The recommended incubation period for most organisms is 4~24h. Spore forming gram-positive bacteria should be grown for less than 16h to help minimize sporulation.

2. Select the type of MicroPlate according to Grams stain results of the tested microorganisms. The Biolog GN MicroPlate is for identification of gram-negative bacteria and the the Biolog GP MicroPlate is for identification of gram-positive bacteria. YT MicroPlate is for yeasts and FF MicroPlate is for mold.

3. Preparation of certain concentration of the bacterial suspension. Oxygen concentration determines the cell concentration after incubation. In Biolog system, oxygen concentration is the key parameter that must be controlled. So the preparation for the inoculum must be prepared under the request of Biolog system. If the checked bacteria are gram positive coccus or bacilli, three drops of sodium thioglycollate and 1ml 100mmol/L sodium salicylate should be added to the bacterial suspension. Only in this way, the concentration of bacterial suspension is the same turbidity as the turbidity standards. The inoculum should be used within 10 min.

4. Inoculate MicroPlate

(1) Pour the cell suspension into the multichannel pipet reservoir.

(2) Fill all wells precisely with 8-Channel Repeating Pipettor.

General bacteria are inoculated with 150μl, spore-forming bacilli are inoculated with 150μl, yeasts are inoculated with 100μl, and molds are inoculated with 100μl.

(3) Cover the MicroPlate with its lid.

5. Incubate MicroPlate   Place the MicroPlate into an incubator, for 3 to 36 hours. Incubate at 30~35℃.

6. Reading the Results   By using Biolog's Microbial Identification Systems software, put the incubated MicroPlate to Microplate Reader and read MicroPlates either manually or on a computer controlled microplate reader.

7. Interpretation of Results   The software compares the results 96-well plate displayed with the reference data in the database and provides 10 identification results based on matching degree. The results display in the ID box. If the first result is not a good match, then the ID box will display "No ID".

Each result shows three important parameters. PROB (probability), SIM (similarity), and DIS (distance of positions). %PROB provides parameters that the user can compare with other identification system; SIM displays matching degree between identified strain (ID) and the species in the database; DIS displays mismatching degree between ID and the species in the database.

With a better identification, SIM should be more than 0.75 after 4~6h incubation, ≥ 0.5 after 16~24h. The best identification is nearly to 1.00 for SIM value. DIS < 5, indicates a better result.

If you want to read the database beyond 10 results, just double click "Other display" box. Choose the species compared, then a variety of indexes will be displayed on the computer screen.

Experiment contents

This experiment is to estimate the accuracy of carbon source utilization tests with the Biolog system for the identification of selected species such as *Bacillus*, mold and yeast.

Results

Report the identification results for the selected experimental strains, and give a detailed interpretation for the result.

Notes

1. Pure cultures must be used to obtain identifications. The Biolog Analysis System is not designed to identify individual bacterial strains from within mixed cultures.

2. Prewarm the Inoculating Fluid and the MicroPlates to room temperature before use.

Questions

Assess the accuracy of the identification results, if the results are not satisfactory, analysis the possible reasons for trouble shooting.

（徐　威）

实验五十　药物无菌检查法

【目的】

1. 掌握药物制剂的无菌检查方法。
2. 熟悉药典要求的需无菌检查的规定范围。

【基本原理】

无菌检查法是用于检查药典要求无菌的药品、医疗器具、原料、辅料及其他品种是否无菌的一种方法，各种注射剂（如针剂、输液等）、手术眼科制剂等都必须保证无菌，符合药典相关规定。无菌检查应在环境洁净度10000级下的局部洁净度100级的单向流空气区域内或隔离系统中进行，其全过程应严格遵守无菌操作，防止微生物污染。无菌检查是用部分样品的测定结果推断整体的含菌情况，适当的试验操作技术能确保结果的科学、准确。

药物的无菌检查法包括直接接种法和膜过滤法。药典规定只要供试品性状允许，应优先采用膜过滤法。

药典中对无菌检查有如下几点要求：

1. 培养基的制备及培养条件的要求（参见附录）。
2. 稀释液、冲洗液配制后应采用验证合格的灭菌程序灭菌。
3. 培养基的适用性检查，无菌检查用的硫乙醇酸盐流体培养基及改良马丁培养基等应符合培养基的无菌性检查及灵敏度检查的要求。无菌性检查即每批培养基随机取不少于5支（瓶），培养14d，应无菌生长；培养基灵敏度检查所用的菌株传代次数不得超过5代,选用的阳性试验菌如下。

金黄色葡萄球菌(*Staphylococcus aureus*)〔CMCC(B)26003〕

铜绿假单胞菌(*Pseudomonas aeruginosa*)〔CMCC(B)10104〕

枯草芽孢杆菌(*Bacillus subtilis*)〔CMCC(B)63501〕

生孢梭菌(*Clostridium sporogenes*)〔CMCC(B)64941〕

白色念珠菌(*Candida albicans*)〔CMCC(F)98001〕

黑曲霉(*Aspergillus niger*)〔CMCC(F)98003〕

取每管装量为12ml的硫乙醇酸盐流体培养基9支,分别接种小于100CFU的金黄色葡萄球菌、铜绿假单胞菌、枯草芽孢杆菌、生孢梭菌各2支,另1支不接种作为空白对照,培养3d;取每管装量为9ml的改良马丁培养基5支,分别接种小于100CFU的白色念珠菌、黑曲霉各2支,另1支不接种作为空白对照,培养5d。逐日观察结果。空白对照管应无菌生长,若加菌的培养基管均生长良好,判该培养基的灵敏度检查符合规定。

4. 无菌检查方法的验证,当建立药品的无菌检查法时,应进行方法的验证,选用合适的阳性对照菌进行验证,以证明所采用的培养基和检验方法适合于该药品的无菌检查。方法验证试验中,应根据供试品特性选择阳性对照菌:无抑菌作用及抗革兰阳性菌为主的供试品,以金黄色葡萄球菌为对照菌;抗革兰阴性菌为主的供试品,以大肠埃希菌为对照菌;抗厌氧菌为主的供试品,以生孢梭菌为对照菌;抗真菌为主的供试品,以白色念珠菌为对照菌。这些验证可在供试品的无菌检查前或与供试品的无菌检查同时进行。做验证试验时,在加入被测样品之后,将一定量阳性对照菌(一般不多于100CFU)加在最后冲洗滤膜的稀释液中(膜过滤法)或检测培养基中(直接接种法),培养5d。与对照管比较,如含供试品各容器中的阳性试验菌均生长良好,则说明供试品的该检验量在该检验条件下无抑菌作用或其抑菌作用可以忽略不计,可照此检查方法和检查条件进行供试品的无菌检查。如含供试品的任一容器中的阳性试验菌生长微弱、缓慢或不生长,则说明供试品的该检验量在该检验条件下有抑菌作用,可采用增加冲洗量,或增加培养基的用量,或使用中和剂或灭活剂如 β - 内酰胺酶、对氨基苯甲酸,或更换滤膜品种等方法,消除供试品的抑菌作用,并重新进行方法验证试验。阴性对照的设置是取相应溶剂和稀释液、冲洗液同法操作,阴性对照不得有菌生长。

在药典中,对无菌检查所用供试品的检验数量和检验量也有严格规定;且针对不同性质的供试品要求做适当的处理。

【仪器与材料】

1. 阳性对照用试验菌

(1)金黄色葡萄球菌[*Staphylococcus Aureus*,CMCC(B)26003]菌液 取金黄色葡萄球菌的营养琼脂斜面新鲜培养物1白金耳(环),接种至需气菌、厌气菌培养基内,30~35℃培养16~18h后,用灭菌生理盐水稀释成10^{-6}。

(2)生孢梭菌(*Clostridium sporogenes*,CMCC(B)64941)菌液 取生孢梭菌的需气菌、厌气菌培养基新鲜培养物1白金耳(环),再接种至相同培养基内,30~35℃培养18~24h后,用灭菌生理盐水稀释成10^{-5}。

(3) 白色念珠菌 [*Candida albicans*, CMCC (F) 98001] 菌液 取白色念珠菌的霉菌琼脂培养基斜面新鲜培养物 1 白金耳（环），接种至霉菌培养基内，23～28℃培养 24h 后，用灭菌生理盐水稀释成 10^{-5}。

2. 培养基及试剂 需气菌、厌气菌培养基（硫乙醇酸盐流体培养基）、真菌培养基（改良马丁培养基），灭菌生理盐水（培养基配方见附录）。

3. 待检药物 0.9%氯化钠注射液（0.9% N.S）、注射用盐酸四环素。

4. 仪器 恒温培养箱、全封闭集菌培养器等。

5. 其他 试管、灭菌吸管等。

【方法与步骤】

（一）薄膜过滤法

本实验采用薄膜过滤法对注射用盐酸四环素做无菌检查。参照药典选用金黄色葡萄球菌为阳性对照菌。且无菌检查所用培养基及稀释剂均符合药典规定要求。

薄膜过滤法应优先采用封闭式薄膜过滤器（图 50-1），也可使用一般的薄膜过滤器。无菌检查用的滤膜孔径应不大于 0.45μm，直径约为 50mm。根据供试品及其溶剂的特性选择滤膜材质。滤器及滤膜使用前应采用适宜方法灭菌。

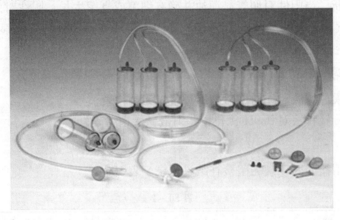

图 50-1 全封闭集菌培养器

Figure 50-1 Sealed sterility testing system

1. 准备培养基及供试液 配制 3 管硫乙醇酸盐流体培养基和 2 管真菌培养基（即改良马丁培养基）。参照药典规定样品用 0.9%灭菌氯化钠溶液 500ml 溶解制成供试液。

2. 薄膜过滤及接种 先用少量稀释剂润湿滤膜，摇匀样品，以无菌操作将供试液加入薄膜过滤器内，减压抽干；同法用 100ml 0.9%灭菌氯化钠溶液冲洗滤膜，重复 3 次，冲洗清除残留在滤筒、滤膜上的抗生素；直接将各种检验用培养基 50ml 加入封闭的滤筒内（封闭式薄膜过滤器）；或取出滤膜，分成 3 等片（一般的薄膜过滤器），取 2 片滤膜分别放在 2 管各装 50ml 硫乙醇酸盐培养基中，其中 1 管接种对照用菌液 1ml，供做阳性对照。另 1 片滤膜放在含 50ml 真菌培养基管中。另取 1 支硫乙醇酸盐培养基和 1 支真菌培养基做阴性对照。

3. 培养及观察 上述含培养基的容器按规定的温度培养 14d。（硫乙醇酸盐培养基培养温度 30~35℃，改良马丁基培养温度 23~28℃）培养期间应逐日观察并记录是否有菌生长。

4. 结果判定 当培养基中接种的阳性对照菌株培养 3~5d 生长良好，阴性对照管培养基 14d 后应澄清无菌生长；而供试品管均澄清，或虽显浑浊但经确证无菌生长，判供试品符合规定；若供试品管中任何一管显浑浊并确证有菌生长，判供试品不符合规定。

（二）**直接接种法**

本实验采用直接接种法对 0.9% 氯化钠注射液做无菌检查。参照药典选用金黄色葡萄球菌为阳性对照菌。

1. 以无菌操作吸取 1ml 阳性对照菌液或待测样品按表 50-2 所示加入含 15ml 的无菌试验培养基中，摇匀。

2. 需气菌、厌气菌培养基于 30~35℃培养 14d，真菌培养基于 23~28℃培养 14d，阳性菌对照管培养 1d，记录实验结果。

3. 培养期间应逐日检查是否有菌生长，阳性对照管 24h 内应有细菌或真菌生长。

4. 结果判断，当阳性对照管显浑浊并确有细菌或真菌生长，阴性对照管无菌生长，药物试验的需气菌、厌气菌及真菌培养基管均为澄清或显浑浊，但经显微镜检证明无菌生长，则判定被检测样品无菌试验合格。

【实验内容】

1. 选用薄膜过滤法对注射用盐酸四环素做无菌检查。
2. 采用直接接种法对 0.9% 氯化钠注射液做无菌检查。

【结果】

1. 将注射用盐酸四环素无菌检查结果记录于表 50-1。

表 50-1

管号	样品	接种	培养时间（d）	需气菌、厌气菌	真菌
1	需气菌、厌气菌培养基	金黄色葡萄球菌+滤膜 1 片	3~5		
2	需气菌、厌气菌培养基	阴性对照	14		
3	需气菌、厌气菌培养基	滤膜 1 片	14		
4	真菌培养基	阴性对照	14		
5	真菌培养基	滤膜 1 片	14		

*记录结果时，"+"为浑浊，"-"为澄清（培养基分装量 50ml，阳性对照菌接种量 1ml）

2. 将 0.9% 氯化钠注射液无菌检查结果记录于表 50-2。

表 50-2

管号	样品	接种	培养时间（d）	需气菌、厌气菌	真菌
1	需气菌、厌气菌培养基	金黄色葡萄球菌	1		
2	需气菌、厌气菌培养基	生孢梭菌	1		

续表

管号	样品	接种	培养时间（d）	需气菌、厌气菌	真菌
3	需气菌、厌气菌培养基	阴性对照	14		
4	需气菌、厌气菌培养基	0.9% N.S	14		
5	需气菌、厌气菌培养基	0.9% N.S	14		
6	需气菌、厌气菌培养基	0.9% N.S	14		
7	需气菌、厌气菌培养基	0.9% N.S	14		
8	需气菌、厌气菌培养基	0.9% N.S	14		
9	真菌培养基	白色念珠菌	1		
10	真菌培养基	阴性对照	14		
11	真菌培养基	0.9% N.S	14		
12	真菌培养基	0.9% N.S	14		
13	真菌培养基	0.9% N.S	14		
14	真菌培养基	0.9% N.S	14		
15	真菌培养基	0.9% N.S	14		

*记录结果时，"+"为浑浊，"-"为澄清（培养基分装量15ml，阳性菌接种量1ml）

注：因药典规定最少检验数量为4个，故本实验接种5个供试品管

【思考题】

1. 哪些药物需无菌检查？
2. 抗生素药物如何进行无菌检查？

（苏　昕）

Experiment 50　Sterility Test of Drugs

Objectives

1. Grasp the methods for sterility tests of pharmaceutical preparations.

2. Be familiar with the range of detecting products required in the Pharmacopoeia.

Principles

Sterility test is a method to detect whether pharmaceutical preparations, raw materials, medical devices or other articles, which are required to be sterile according to the Pharmacopoeia, are aseptic. All injections and ophthalmic and surgical preparations have to be sterile. Test for sterility should be carried out in a class 100 laminar – air flow cabinet located within a 10000 – class clean – room. The whole process should be performed under strictly aseptic conditions to avoid any microbial contamination. Sterility test is used to guestimate the contamination of a whole from the sample part. The testing results are ensured by suitable experimental techniques.

The test for sterility is carried out by the membrane filtration or direct inoculation methods. The membrane filtration one should be applied whenever the nature of the product permits.

In the Pharmacopoeia, followings could be found:

1. Culture media and incubation conditions (reference to the Appendix).

2. Diluents and rinsing fluids.

3. Suitability tests of media: The media used in the sterility tests (fluid thioglycollate medium and Modified Martin medium) should comply with the sterility and sensitivity tests. For sterility test, no less than 5 vessels of each batch of sterilized media should be incubated for 14 days. No microbial growth should occur. For media sensitivity test, strains no more than 5 generations required from the original one can be applied. Test strains are as follows.

Staphylococcus aureus [CMCC (B) 26003]

Pseudomonas aeruginosa [CMCC (B) 10104]

Bacillus subtilis [CMCC (B) 63501]

Clostridium sporogenes [CMCC (B) 64941]

Candida albicans [CMCC (F) 98001]

Aspergillus niger [CMCC (F) 98003]

Take 9 containers of fluid thioglycollate medium (12ml medium per container). Inoculate 2 containers of the medium with less than 100CFU test microorganisms of *Staphylococcus aureus*, *Pseudomonas aeruginosa*, *Bacillus subtilis* and *Clostridium sporogenes* respectively. The remaining uninoculated culture medium (in container) is used as blank control. Then incubate for 3 days. Take 5 containers of Modified Martin medium (9ml medium per container). Inoculate 2 containers of the medium with less than 100CFU test microorganisms of *Candida albicans* and *Aspergillus niger* respectively. The remaining uninoculated culture medium (in container) is used as blank control. Then incubate for 5 days. Observe the microbial growth every day during the incubation. No growth should be in the blank control. If a clearly visible growth of the microorganisms occurs in every inoculated medium, the medium meets the requirement of the sensitivity test of medium.

4. Validation test: In the course of establishing sterility test method for product to be examined, the method with appropriate positive strain must be verified to ensure that the adopted method is suitable for sterility test of the product. The microorganisms for positive control should be selected according to the nature of the product being examined. *Staphylococcus aureus* is used for the product possessing no antimicrobial activity or mainly anti – Gram positive bacteria activity. *Escherichia coli* is used for the product mainly possessing anti – Gram negative bacteria activity. *Clostridium sporogenes* is used for the product possessing anti – anaerobic bacteria activity. *Candida albicans* is used for the product possessing fungistatic activity. The validation may be performed before or simultaneously with the test for sterility. During validation test, after treating the testing specimen, add less than 100CFU of the positive test strain to the final portion of rinsing fluid (membrane filtration method) or testing media (direct inoculation

method) and incubate for 5 days. Compare with the control container, if clearly visible growth of each test microorganism is obtained in the test containers containing the product to be examined, either the product possesses no antimicrobial activity under the conditions of the test or such activity has been satisfactorily eliminated. The test for sterility of the product may then be carried out using the same method and conditions of the test. If the growth of the test microorganism is not obtained, or poor, or slow in any tested container, visually comparable to that in the control container without product, then the product with the specified quantity possesses antimicrobial activity under the conditions of the test. In this case, modify the conditions in order to eliminate the antimicrobial activity by using a large amount of rinsing fluid, or increasing the volume of media, or using suitable neutralizer or in-activator such as β-lactamase, p-aminobenzoic acid, or replacing the type of the membrane filter used etc. And repeat the validation test. A negative control should be performed with the same solvent and diluents as the test used. There must be no growth of microorganisms.

In Pharmacopoeia, number of products and quantity of product to be tested is strictly defined. Proper treatment of products with different natures needs to be performed.

Apparatus and Materials

1. Specimens test microorganisms as positive control

(1) Broth culture of *Staphylococcus aureus*, CMCC (B) 26003: Inoculate a loopful of *Staphylococcus aureus* from agar slant to fluid thioglycollate medium. Incubate at 30~35℃ for 16~18h, then dilute the inoculum to 10^{-6} with sterile normal saline.

(2) Broth culture of *Clostridium sporogenes*, CMCC (B) 64941: Inoculate a loopful of *Clostridium sporogenes* from fluid thioglycollate medium to a same medium. Incubate at 30~35℃ for 18~24h, then dilute the inoculum to 10^{-5} with sterile normal saline.

(3) Broth culture of *Candida albicans*, CMCC (F) 98001: Inoculate a loopful of *Candida albicans* from Modified Martin agar slant to Modified Martin medium. Incubate at 23~28℃ for 24h, then dilute the inoculum to 10^{-5} with sterile normal saline.

2. Cultures and reagents   Fluid thioglycollate medium (for culturing aerobic bacteria and anaerobic bacteria), Modified Martin medium (for fungi), 0.9% Sodium Chloride injection (0.9% N.S), tetracycline hydrochloride injection, sterile normal saline.

3. Apparatus   Constant temperature incubator, sealed sterility testing system.

4. Others   Test tube, sterile sucker, dropper, etc.

Methods and Procedures

1. Membrane filtration (tetracycline hydrochloride injection)   According to the Pharmacopoeia, *Staphylococcus Aureus* is used as positive control in this experiment. All the media and diluents for sterility tests comply with the stipulations in the Pharmacopoeia.

Sealed sterility testing system (Figure 50-1) is preferentially used in the method of membrane filtration. Other general membrane filter may also be adopted. The membrane used has a minimal pore size which is not greater than 0.45μm and a diameter about 50mm. Filtration membrane type should be selected according to the characters of product to be tested. The membrane and apparatus should be sterilized by appropriate ways before use.

(1) Preparation of media and product Prepare 3 containers of fluid thioglycollate medium and 2 containers of Modified Martin medium. The product (tetracycline hydrochloride injection) to be tested is dissolved in 500ml of 0.9% sterile N.S to make the sample.

(2) Filtration and inoculation Pre-wet the membrane with a small quantity of rinsing fluid. Mix the sample which is then added to membrane filter aseptically. Wash the membrane with 100ml sterile 0.9% NaCl solution for 3 times to remove the antibiotics residual on the filter. For sealed sterility test system, add 50ml of culture media to the filter system. For general membrane filter, aseptically remove the membrane and cut it into 3 equal parts. Add 2 parts into 2 containers containing 50ml of fluid thioglycollate medium respectively. And then inoculate one of the two containers with 1ml of test microorganism as positive control. Add the third part of the membrane into another container with 50ml of Modified Martin medium inside. Take another one container with fluid thioglycollate medium and one with Modified Martin medium only as negative controls.

(3) Incubation and observation Incubate the above containers for 14d at regular temperature (30~35℃ for fluid thioglycollate medium and 23~28℃ for Modified Martin medium). Observe and record each medium container for evidence of microbial growth every day during incubation period.

(4) Evaluation of results Except the positive control container, if no evidence of microbial growth is found in all of the test containers, the product to be examined complies with the test for sterility. If evidence of microbial growth is found in any one of the test containers, the product to be examined does not comply with the test for sterility.

2. Direct inoculation method This experiment will carry out the sterility tests of 0.9% N.S. *S. aureus* is used as positive test microorganism according to the Pharmacopoeia.

(1) Aseptically transfer 1ml of inoculum of positive test strain or testing sample into the containers containing 15ml of the culture media for sterility tests.

(2) Incubate bacteria at 30~35℃ for 14d and fungi at 23~28℃ for 14d. Incubate the container of positive control for 1d, and then record the results.

(3) Observe and record each medium container for evidence of microbial growth every day during incubation period. Microbial growth of the positive control should be obviously observed.

(4) Evaluation of results: Except the positive control container, if no evidence of microbial growth is found in all of the test containers, the product to be examined complies with the test for sterility. If evidence of microbial growth is found in any one of the test containers, the product to be examined does not comply with the test for sterility.

Experiment contents

1. Perform sterility tests on tetracycline hydrochloride injection with membrane filtration

method.

2. Perform sterility tests on 0.9% N.S with direct inoculation method.

Results

1. Record the results of sterility tests on tetracycline hydrochloride injection in Table 50 – 1.

Table 50 – 1

Container No.	Medium	Microorganisms	Incubation time (d)	Aerobic/anaerobic bacteria	Fungi
1	Media for aerobic and anaerobic bacteria	S. aureus + 1 part of filter membrane	3 ~ 5		
2	Media for aerobic and anaerobic bacteria	Negative control	14		
3	Media for aerobic and anaerobic bacteria	1 part of filter membrane	14		
4	Media for fungi	Negative control	14		
5	Media for fungi	1 part of filter membrane	14		

* " + " for turbid, " – " for clear (50ml of culture media, 1ml of inoculum for positive test microorganism)

2. Record the results of sterility tests on 0.9% N.S in Table 50 – 2.

Table 50 – 2

Container No.	Sample	Microorganisms	Incubation time (d)	Aerobic/anaerobic bacteria	Fungi
1	Media for aerobic and anaerobic bacteria	S. aureus	1		
2	Media for aerobic and anaerobic bacteria	C. sporogenes	1		
3	Media for aerobic and anaerobic bacteria	Negative control	14		
4	Media for aerobic and anaerobic bacteria	0.9% N.S	14		
5	Media for aerobic and anaerobic bacteria	0.9% N.S	14		
6	Media for aerobic and anaerobic bacteria	0.9% N.S	14		
7	Media for aerobic and anaerobic bacteria	0.9% N.S	14		
8	Media for aerobic and anaerobic bacteria	0.9% N.S	14		
9	Media for fungi	C. albicans	1		

Continue

Container No.	Sample	Microorganisms	Incubation time (d)	Aerobic/ anaerobic bacteria	Fungi
10	Media for fungi	Negative control	14		
11	Media for fungi	0.9% N.S	14		
12	Media for fungi	0.9% N.S	14		
13	Media for fungi	0.9% N.S	14		
14	Media for fungi	0.9% N.S	14		
15	Media for fungi	0.9% N.S	14		

* "+" for turbid, "-" for clear (15ml of culture media, 1ml of inoculum for positive test microorganism)

Because in the Pharmacopoeia, the minimum number of product to be tested in each batch is 4, this experiment takes 5 containers.

Questions

1. List the range of pharmaceuticals for sterility tests in Pharmacopoeia.
2. How to do sterility tests on antibiotics?

(马晓楠)

实验五十一　药物的微生物限度检查法

【目的】

1. 掌握控制菌之一大肠埃希菌的检验程序及其检验方法。
2. 熟悉平板菌落计数法测定口服药物中细菌、霉菌和酵母菌的总数，检查药物是否符合微生物限度标准。
3. 了解药物中控制菌的检查方法。

【基本原理】

微生物限度检查法系检查非规定灭菌制剂及其原料、辅料受微生物污染程度的方法。

口服药及外用药等均属于非规定灭菌制剂，在生产和临床使用过程中，不要求达到完全无菌，但是为了保证药品的质量，防止药品的污染，需要限制性控制微生物的数量和种类。我国2010年版药典规定的检查项目包括细菌数、霉菌数、酵母菌数及控制菌检查。控制菌的检验包括：大肠埃希菌、沙门菌、金黄色葡萄球菌、铜绿假单胞菌、破伤风梭菌等病原菌检查和活螨的检验等。细菌、酵母菌和霉菌总数的测定采用平皿法或膜过滤法；根据细菌的形态结构和生理生化特性来检查控制菌的存在；活螨可用显微镜法检验。《中国药典》中对微生物限度检查法有详细规定，包括检验量及样品的抽取、供试液的制备、培养基的适用性检查、检验方法的验证试验等，试验者可酌情参考。

1. 细菌数、霉菌数、酵母数的检查　细菌总数的测定是检查每克（g）或每毫升

（ml）被检药品内所含有的活菌数（实际是需氧菌的活菌数），以判断供试药物被细菌污染的程度；霉菌和酵母菌总数的测定是考察供试药物中每克或每毫升所含的活的霉菌和酵母菌的总数，以判明供试药物被真菌污染的程度；是对该药物生产过程卫生学总评价的一个重要依据。

计数方法包括平皿法和薄膜过滤法。检查时，按已验证的计数方法进行供试品的细菌、霉菌及酵母菌数的测定，且应按规定报告结果。平皿法菌数报告规则为：细菌、酵母菌宜选取平均菌落数小于300CFU、霉菌宜选取平均菌落数小于100CFU的稀释级，作为菌数报告（取两位有效数字）的依据。以最高的平均菌落数乘以稀释倍数的值报告1g、1ml或10cm²供试品中所含的菌数。如各稀释级的平板均无菌落生长，或仅最低稀释级的平板有菌落生长，但平均菌落数小于1时，以<1乘以最低稀释倍数的值报告菌数。而膜过滤法的菌数报告规则为每片滤膜上的菌落数应不超过100个，且以相当于1g、1ml或10cm²供试品的菌落数报告菌数；若滤膜上无菌落生长，以<1报告菌数（每张滤膜过滤1g、1ml或10cm²供试品），或<1乘以最低稀释倍数的值报告菌数。

2. 控制菌的检查 可根据细菌的形态结构和生理生化特性来检查，一般的检验程序如图51-1。

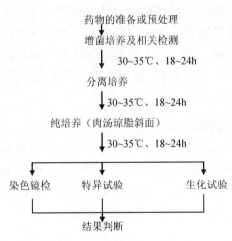

图51-1 一般检查程序

本实验只选作其中之一大肠埃希菌的检查。大肠埃希菌是口服药品的常规必检项目之一。大肠埃希菌是人和温血动物肠道内寄生的正常菌群，药品中的大肠埃希菌来源于人和温血动物的粪便。凡由供试品中检出大肠埃希菌时，表明该药品可能已被粪便污染，也就可能被存在于粪便的其他肠道致病菌所污染。患者服用这种药物后，就可能出现这些病原体感染的危险。因此，口服药品中不得检出大肠埃希菌。大肠埃希菌的检验程序如图51-2。

3. 药典规定 本检查法中细菌及控制菌培养温度为30~35℃；霉菌、酵母菌培养温度为23~28℃。检验结果以1g、1ml、10g、10ml、10cm²为单位报告，特殊品种可以最小包装单位报告。

4. 药物的微生物限度标准及控制菌限度标准 药物的微生物限度标准，见附录。

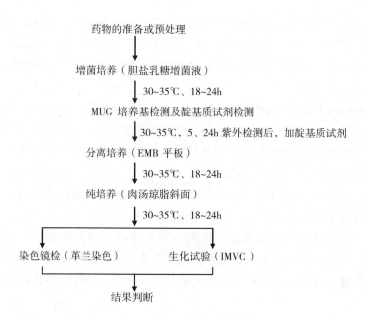

图51-2 大肠埃希菌的检查程序

控制菌的限度标准为口服药品中不得检出大肠埃希菌；凡外用药和眼科制剂不得检出金黄色葡萄球菌和铜绿假单胞菌；含动物组织来源的制剂不得检出沙门菌。抗细菌的口服抗生素制剂应检查霉菌，每克中不得超过100个；抗真菌的口服抗生素制剂应检查细菌，每克不得超过100个；霉变、长螨者均为不合格药品。

【仪器与材料】

1. 对照菌 大肠埃希菌液体培养物。

2. 培养基及试剂 营养琼脂培养基、玫瑰红钠琼脂培养基、酵母浸出粉胨葡萄糖琼脂培养基（YPD）、胆盐乳糖培养基（BL）、伊红美兰琼脂培养基（EMB）、IMVC生化试验各培养基等；革兰染色用相关试剂（见本书革兰染色实验）、IMVC生化试验用相关试剂（见本书生化反应实验部分）、灭菌生理盐水等。

3. 药物 咳嗽糖浆。

4. 仪器 恒温培养箱、菌落计数仪、紫外检测仪、超净工作台、显微镜等。

5. 其他 无菌试管、无菌吸管、无菌平皿等。

【方法与步骤】

（一）细菌、霉菌和酵母菌总数测定（平皿法）

1. 制备供试液 将待测咳嗽糖浆摇匀，用吸管吸取10ml加pH7.0无菌氯化钠-蛋白胨缓冲液至100ml，混匀，作为1：10的供试液。用稀释液稀释成1：10^2、1：10^3等稀释级的供试液。

2. 接种

（1）供试品分别吸取上述稀释的溶液各1ml于无菌平皿中，再加入15ml恒温于45℃的营养琼脂培养基（细菌计数），玫瑰红钠琼脂培养基（霉菌计数），酵母浸出粉胨葡萄糖琼脂培养基（酵母菌计数），混匀，待凝固后将平板于30~35℃（细菌）或

23~28℃（真菌）倒置培养，细菌培养3d，霉菌、酵母菌培养5d。每稀释级每种培养基至少制备2个平板。

（2）阴性对照：取试验用的稀释液1ml，置无菌平皿中，同法注入培养基作为阴性对照，凝固，倒置培养。每种计数用的培养基各制备2个平板，均不得有菌生长。

3. 计数菌落，写报告 计算各稀释级的平均菌落数，按规则报告细菌、霉菌和酵母菌含量。

（二）大肠埃希菌的检查

在对供试品进行控制菌检查时，应做阳性对照和阴性对照试验。阳性对照试验的加菌量为10~100CFU，方法同供试品的控制菌检查。阳性对照试验应检出相应的控制菌。取稀释液10ml照相应控制菌检查法检查，作为阴性对照。阴性对照应无菌生长。

1. 制备供试液 将待测咳嗽糖浆摇匀，用吸管吸取10ml加pH7.0无菌氯化钠－蛋白胨缓冲液至100ml作为供试液。

2. 增菌培养 准备3瓶内装100ml胆盐乳糖培养基的三角瓶。将待测咳嗽糖浆摇匀，分别取10ml加入到2份中，其中1份再加入对照菌液1ml做阳性对照，第3份100ml胆盐乳糖培养基中加入与供试液等量的稀释剂做阴性对照。于30~35℃恒温培养18~24h（必要可延至48h）。阴性对照瓶应无菌生长。其他2瓶培养液变浑浊表明有细菌生长。

3. MUG及靛基质检测试验 取上述培养物0.2ml接种至含5ml MUG培养基的试管内，培养，于5h、24h在366nm紫外线下观察，同时用未接菌的MUG培养基做本底对照。若管内培养基呈现荧光，为MUG阳性；不呈现荧光，为MUG阴性。观察后，沿培养基的管壁加入数滴靛基质试液，液面呈玫瑰红色，为靛基质阳性；呈试剂本色，为靛基质阴性。阴性对照应为MUG阴性和靛基质阴性。

如MUG阳性、靛基质阳性，判供试品检出大肠埃希菌；如MUG阴性、靛基质阴性，判供试品未检出大肠埃希菌；如MUG阳性、靛基质阴性，或MUG阴性、靛基质阳性，则应继续下述检查。

4. 平板分离培养与结果判定 将上述培养物对应划线接种于伊红美兰琼脂平板或麦康凯琼脂平板，于30~35℃恒温培养18~24h。

结果判定 当阳性对照的平板呈现阳性菌落时，供试品的平板无菌落生长，或有菌落但不同于表51-1所列的特征，可判为未检出大肠埃希菌。

表51-1 大肠埃希菌菌落形态特征

培养基	菌落形态
伊红美兰琼脂	呈紫黑色、浅紫色、蓝紫色或粉红色，菌落中心深紫色或无明显暗色中心，圆形，稍凸起，边缘整齐，表面光滑，湿润，常有金属光泽
麦康凯琼脂	鲜桃红色或微红色，菌落中心深桃红色，圆型，扁平，边缘整齐，表面光滑，湿润

如生长菌落与表51-1所列特征相符或疑似者，应挑选2~3个菌落分别接种于营养琼脂培养基斜面，培养18h，做以下检查。

5. 显微镜检 取上述斜面培养物，革兰染色后显微镜检，大肠埃希菌为革兰阴性无芽孢的短杆菌。

6. 生理生化反应 取上述斜面培养物，分别接种于乳糖发酵管，IMViC 各试验管，培养 24~48h 观察。

大肠埃希菌可发酵乳糖产酸产气。

大肠埃希菌的 IMViC 试验结果为"＋ ＋ － －"（IMViC 分别代表：吲哚试验、甲基红试验、乙酰甲基甲醇生成试验及柠檬酸盐利用试验）。

当空白对照试验呈阴性，供试品检查为革兰阴性无芽孢杆菌；乳糖发酵产酸产气或产酸不产气；IMViC 试验为阳性、阳性、阴性、阴性或阴性、阳性、阴性、阴性，判为检出大肠埃希菌。对可疑反应的菌株，应重新分离培养后，再做生化试验证实。

7. 观察结果，给出检查报告 咳嗽糖浆未检出大肠埃希菌时为控制菌检验合格。

【实验内容】

1. 采用平皿法检查口服药咳嗽糖浆的细菌、霉菌、酵母菌总数。
2. 口服药咳嗽糖浆的大肠埃希菌的检查。

【结果】

1. 计算各稀释级的平均菌落数，按规则报告细菌、霉菌和酵母菌含量；判断咳嗽糖浆的细菌、霉菌、酵母菌总数是否符合药物的微生物限度标准。
2. 记录大肠埃希菌各步检查结果，给出检验报告。

【思考题】

如何判断某一药物中控制菌符合微生物限度标准？举例说明。

（苏 昕）

Experiment 51　Microbial Limit Tests of Drugs

Objectives

1. Grasp how to examine the existence of *E. coli* in pharmaceuticals.
2. Be familiar with the plate method of oral drugs in microbial limit tests.
3. Understand the inspection methods of specified microorganisms in pharmaceuticals.

Principles

Microbial limit tests provide tests for the estimation of the number of viable microorganisms present in non-sterile pharmaceutical products of all kinds, including preparations, raw materials, excipients.

Oral drugs and drugs for external use are non-sterile pharmaceuticals which mean asepsis is not necessary during manufacture process and clinical use. But specified microorganisms are not permitted or number-limited to ensure the quality of medicines. In Pharmacopoeia of our country (edition 2010), bacteria, molds or yeasts count, as well as the specified bacteria, are tested. Specified microorganisms include *Escherichia coli*, *Salmonella species*, *Staphylococcus aureus*, *Pseudomonas aeruginosa*, *Bacillus tetani* and live mite. Plate method and membrane filtration method are used to count bacteria, yeasts and molds. Morphology and biochemical characteristics are applied to examine the existence of specified microbes. Mite can be identified by microscope. In Pharmacopoeia, quantity of products, preparation of the sample, media validation and method validation, etc. are all explicitly stipulated.

1. Bacteria, molds or yeasts count Total count of bacteria is the number of viable aerobic microorganisms in every gram (or ml) of tested product. This will help to determine the bacterial contamination of product. Total fungi count is the number of viable molds and yeasts in every gram (or ml) of tested product which indicates the fungal contamination degree.

Examination method includes plate count and membrane filtration method. Use validated methods to carry out the bacteria, molds or yeasts count of the product to be examined. Microbial number report rule of plate method is as follows: select the dilution in which the average number of colonies of bacteria and yeasts is no more than 300 CFU and of the molds is no more than 100 CFU. Calculate the number of CFU per g, per ml or per $10cm^2$ product to be tested by multiplying the highest average number of CFU by the dilution folds. When no microbial growth occurs in any of the dilutions, or it only occurs for the lowest dilution where the average microbial number of the CFU is less than 1, report the result with 1 multiplying by the lowest dilution folds. Microbial number report rule of membrane filtration is as follows: Multiply the average microbial number by dilution folds as the number of CFU per g or per ml ($10cm^2$) product to be tested. If no microbial growth occurs, report the result with less than 1 or 1 multiplying by the lowest dilution folds.

2. Tests for specified microorganisms According to bacterial morphological structure, testing methods is as Figure 51 – 1.

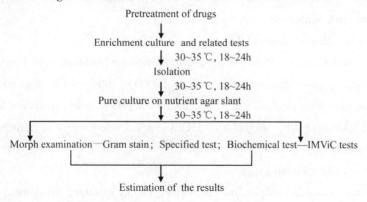

Figure 51 – 1

This experiment focuses on the examination of *E. coli* which is one of the specified microorganisms in microbial limit tests. *E. coli* is normal flora in the intestinal tract of human and warm – blooded animals and commonly found in feces. Once *E. coli* is detected in pharmaceuticals, the drugs may have been contaminated by feces and the possibility of contamination by other pathogens in the intestinal tract increases drastically. The risk of pathogenic infection may occur as well. Therefore, no *E. coli* is permitted to detect in oral drugs. The examination procedure is as Figure 51 – 2.

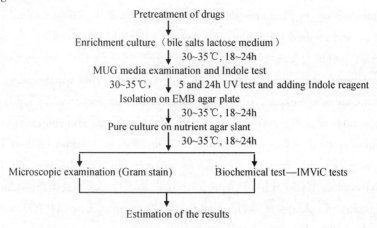

Figure 51 – 2

3. Incubate bacteria at 30 ~ 35℃, at 23 ~ 28℃ for molds and yeasts. The test result is reported in the unit of 1g, 1ml, 10g, 10ml or 10cm².

4. Microbial contamination limits of pharmaceutical preparations   For the preparations for oral administration, no *E. coli* should be present, as well as *Staphylococcus aureus* and *Pseudomonas aeruginosa* for the preparation for eye administration and external use, *Salmonella species* for preparations containing animal tissues. No more than 100 CFU per g of molds in anti – bacterial oral antibiotics is permitted. No more than 100 CFU per g of bacteria in anti – fungal oral antibiotics is permitted. Mite should be absent in pharmaceuticals.

Apparatus and Materials

1. Specimens   Broth culture of *E. coli*.

2. Cultures and reagents   Nutrient agar culture medium, sodium rose Bengal agar medium, yeast extracts peptone glucose agar medium (YPD), bile salts lactose medium (BL), eosin methylene blue agar medium (EMB), media for IMViC tests, reagents for Gram stain (reference to Experiment 7), reagents for IMViC tests (reference to Experiment 28 ~ 32), sterile normal saline

3. Testing sample   Cough syrup.

4. Apparatus   Constant temperature incubator, colony counter, UV lamp, clean bench, bright – field microscope.

5. Others   Sterile tubes, droppers, Petri plates.

Methods and Procedures

1. Bacteria, molds and yeasts count (plate method)

(1) Sample preparation: Mix the testing cough syrup and transfer 10ml of the product to 100ml of sterile sodium chloride – peptone buffer solution (pH7.0). Then mix well. Use this 1∶10 solution as the testing sample. Further dilution of 1∶10^2, 1∶10^3 is then performed.

(2) Inoculation

① Transfer 1ml of the sample (different dilutions) to a sterile Petri plate, then add 15ml melted nutrient agar medium (bacteria count), or sodium rose Bengal agar medium (molds count), or YPD (yeast count) (not exceeding 45℃). Mix well and incubate upside down after solidification at 30~35℃ (bacteria) or 23~28℃ (fungi) for 3d (bacteria) or 5d (fungi). For each dilution, use at least 2 Petri plates for each culture medium.

② Negative control: Transfer 1ml of the diluent to a sterile Petri plate. Add the culture medium and mix well as the above step. Solidify and incubate. Use at least 2 Petri plates for each culture medium. No evident growth of microorganisms occurs in either of the Petri plate.

(3) Colony count and report: After counting, calculate the number of colonies of each dilution of the product, report the result following the microbial number report rule in the principle part.

2. Examination of *E. coli*

Positive control: Add 10~100 CFU of the tested microorganisms of the positive control, and then carry out the test as the test for specified microorganism. The tested microorganism should be detected in the positive control.

Negative control: Transfer 10ml of diluting solution to a prescribed amount of culture media as negative control. Examine it as the test for specified microorganism. No microbial growth occurs in the negative control.

(1) Sample preparation: Mix the testing cough syrup and transfer 10ml of the product to 100ml of sterile sodium chloride – peptone buffer solution (pH7.0). Then mix well. Use this 1∶10 solution as the testing sample.

(2) Enrichment culture: Prepare 3 flasks (labeled as testing sample, positive control and negative control) of BL medium (100ml in every flask). Mix the testing sample and transfer 10ml to the testing sample and positive control flasks respectively. Add 1ml of *E. coli* inoculum to the positive control flask. Add 10ml of diluents to the negative control. Incubate the 3 flasks at 30~35℃ for 18~24h (48h if necessary). No microbial growth should occur in negative control. If the other 2 flasks become turbid, microbial growth occurs.

(3) MUG and Indole test: Inoculate 0.2ml of the above culture to a tube containing 5ml of MUG medium, and incubate. Observe under 365 nm UV light at the time of 5 hour and 24 hour respectively. Use blank MUG medium as the negative control. The result is MUG positive if the cultures give fluorescent light, or MUG negative if no fluorescent light is observed. Following observation, add several drops of Indole test solution. The result is Indole – positive if the liquid surface presents a rosy color, or Indole – negative of no color change takes place. The negative control is MUG – negative and indole – negative. A MUG – positive and Indole positive result indicates the presence of *E. coli* in the product. A MUG negative and Indole negative result indicates the absence of *E. coli* in the product. If the product is MUG – positive and Indole – negative, or MUG – negative and Indole – positive, continue the examination.

(4) Isolation: Inoculate the above culture on eosin methylene blue agar medium plate or MacConkey agar medium plate, and incubate at 30~35℃ for 18~24h.

When colonies of positive control show the same characteristics as *E coli* colonies and no growth of microorganism occurs on the plate of testing sample, or the appearance of the microbial colonies does not match the description in Table 51-1, *E. coli* is absent in the product.

Table 51-1 Morphologic characteristics of *E. coli* colonies

Culture medium	Colony characteristics
EMB agar medium	Purple black, light purple, bluish purple or pink, deep purple at the center of colony or no obvious dark center, circular, slight convex, regular margin, smooth surface, moist, metallic sheen under reflected light
MacConkey agar medium	Brilliant pink or pale red, deep pink at center of the colony, circular, flat, regular margin, smooth surface, moist

When the colonies on the sample plate share the similarities with *E. coli*, pick up 2~3 colonies to inoculate on agar slants and incubate for 18 hours for further tests.

(5) Microscopic examination: After incubation, examine cellular morphology with Gram stain. *E. coli* should be Gram negative brevis bacillus without endo-spore.

(6) Biochemical tests: Inoculate the bacteria from the slant culture to biochemical culture media (lactose fermentation and IMViC tests) and then incubate for 24~48h.

Positive control: *E. coli* can produce acid and gas when fermenting lactose. The IMViC tests (represent indole test, methyl red test, V-P test and citrate test) result should be "++--".

Negative control: No growth appears.

Testing sample: In Gram stain, if the sample is Gram negative bacterium without endo-spore, and produces acid with "++--" or "-+--" in IMViC tests, *E. coli* is detected. For the suspicious strain, isolate again and then carry out biochemical tests.

(7) Report the examination result: The testing cough syrup meets the microbial limits of *E. coli* or not.

Experiment contents

1. Examine bacteria, molds and yeasts count in oral drugs with plate method.
2. Examine the existence of *E. coli* in cough syrup.

Results

1. Calculate the average colony number of every dilution and report the bacteria, molds and yeasts count according to the report rules. Estimate if the examination results meet the microbial limits or not.
2. Record the results of *E. Coli* examination and report it.

Question

How to estimate a pharmaceutical preparation meets the microbial limits? Give an example.

（马晓楠）

附 录

一、微生物实验常用培养基配方

1. 牛肉膏蛋白胨培养基（又称营养肉汤培养基，主要用于培养细菌）（g/L）

牛肉膏（牛肉浸出粉）	5.0（3.0）
蛋白胨	10.0
氯化钠	5.0
蒸馏水	1000ml
pH（灭菌）	7.2±0.2

103.46kPa，15~30min 高压蒸汽灭菌。

2. 牛肉膏蛋白胨琼脂培养基（又称营养琼脂培养基，主要用于培养细菌）（g/L）

牛肉膏（牛肉浸出粉）	5.0（3.0）
蛋白胨	10.0
氯化钠	5.0
琼脂	14.0
蒸馏水	1000ml
pH（灭菌）	7.2±0.2

103.46kPa，15~30min 高压蒸汽灭菌。

3. 营养肉汤半固体培养基
制法：肉汤培养基中加入0.4%~0.6%琼脂即可。

4. 二倍/三倍浓缩营养肉汤培养基（2E/3E）
制法：肉汤培养基中各组分的量加倍即可。

5. 高氏合成一号合成培养基（主要用于培养放线菌）（g/L）

可溶性淀粉	20.0
磷酸氢二钾	0.5

七水硫酸亚铁	0.01
氯化钠	0.5
七水硫酸镁	0.5
硝酸钾	1.0
琼脂	14.0
蒸馏水	1000ml
pH（灭菌）	7.3±0.1

103.46kPa，15~30min 高压蒸汽灭菌。

制法：称量可溶性淀粉先加少量蒸馏水溶解，再将烧开的蒸馏水倒入其中，制成半透明液体，然后再与其他成分混合在一起。加热搅拌使之溶解后，补足蒸馏水至 1000ml。调 pH，分装，灭菌。

6. 马铃薯琼脂培养基（主要用于培养真菌）（g/L）

马铃薯	200.0
蔗糖（或葡萄糖）	15.0（20.0）
琼脂	14.0
蒸馏水	1000ml
pH	自然

68.95kPa，15~30min 高压蒸汽灭菌。

制法：将马铃薯洗净去皮切碎，称取 200g 放入烧杯内，加水 1000ml。煮沸半小时后，用双层纱布滤去马铃薯渣，滤液补加水量至 1000ml。在容器上将体积做出标记，加蔗糖和琼脂，加热搅拌使全溶后，补加水分至原体积，分装，灭菌。

7. 沙氏琼脂培养基（主要用于培养真菌）（g/L）

蛋白胨	10.0
葡萄糖	40.0
氯化钠	5.0
琼脂	14.0
蒸馏水	1000ml
pH	自然

68.95kPa，15~30min 高压蒸汽灭菌。

注：pH 低于 5 时，应调节至 5 左右，分装，灭菌。

8. 沙氏液体培养基（g/L）

蛋白胨	10.0
葡萄糖	40.0
氯化钠	5.0
蒸馏水	1000ml
pH	自然

68.95kPa，15~30min 高压蒸汽灭菌。

9. 葡萄糖-天门冬素琼脂培养基（主要用于培养小单孢菌 m-220）（g/L）

葡萄糖	10.0
天门冬素	0.5
磷酸氢二钾	0.5
琼脂	14.0
蒸馏水	1000ml
pH（灭菌）	7.3±0.1

68.95kPa，15~30min 高压蒸汽灭菌。

10. 单糖发酵培养基（g/L）

蛋白胨	10.0
氯化钠	5.0
糖（葡萄糖、乳糖、麦芽糖等）	5.0
0.4%溴麝香草酚蓝（BTB）指示液	6ml
蒸馏水	1000ml
pH（灭菌）	7.0±0.1

68.95kPa，15~30min 高压蒸汽灭菌。

注：（1）做细菌产酸产气实验时，将培养基分装到带一倒立小管的试管——德汉氏小套管中，每管装量约为4~5ml（以使倒立小管完全被培养基浸没为标准）。

（2）0.4%溴麝香草酚蓝（B.T.B）溶液：溴麝香草酚蓝0.1g，加0.05mol/L氢氧化钠溶液3.2ml使溶解，再加蒸馏水稀释至200ml。变色范围 pH6.0~7.6（黄→蓝）。

11. 甘露醇发酵培养基（g/L）

蛋白胨	10.0
氯化钠	5.0
甘露醇	5.0
0.5%酸性复红指示剂	10.0ml
（或0.4%BTB）	6.0ml
蒸馏水	1000ml
pH（灭菌）	6.9±0.1

68.95kPa，15~30min 高压蒸汽灭菌。

12. 蛋白胨水培养基（吲哚试验用）（g/L）

蛋白胨	10.0
氯化钠	5.0
蒸馏水	1000ml
pH（灭菌）	7.3±0.1

103.46kPa，15~30min 高压蒸汽灭菌。

13. 葡萄糖蛋白胨水培养基（主要用于甲基红和V.P试验）（g/L）

蛋白胨	5.0
葡萄糖	5.0
磷酸氢二钾	5.0
蒸馏水	1000ml
pH（灭菌）	7.3±0.1

68.95kPa，15~30min 高压蒸汽灭菌。

注：分装时，做V-P试验用的每管定量为2ml。

14. 柠檬酸（或枸橼酸）盐培养基（柠檬酸盐利用实验）（g/L）

氯化钠	5.0
硫酸镁	0.2
磷酸氢二钾	1.0

磷酸二氢铵	1.0
二水枸橼酸钠	2.0
0.4%溴麝香草酚蓝（B.T.B）溶液	20.0ml
琼脂（连续充分水洗3天）	14.0
蒸馏水	1000ml
pH（灭菌）	6.9±0.1

103.46kPa，15~30min 高压蒸汽灭菌。

制法：除 BTB 和琼脂外，混合上述成分，加热溶解，调 pH。加入琼脂加热溶化，再加入 BTB 溶液混匀，分装，灭菌。

注：所用琼脂应不含游离糖，用前可用水浸泡或冲洗除糖。

15. 醋酸铅培养基（产硫化氢试验用）（g/L）

营养肉汤半固体培养基	100ml
硫代硫酸钠	0.25
5%醋酸铅水溶液	1.0ml

68.94kPa，15~30min 高压蒸汽灭菌，直立静止待凝。

制法：将琼脂培养基加热溶化后，冷至60℃，加入硫代硫酸钠，再加5%醋酸铅水溶液，边加边搅拌至乳白色。分装小试管，每管2~3ml。

注：硫代硫酸钠是还原剂，能保持还原环境，使形成的硫化氢不再氧化。

16. 明胶液化试验培养基（g/L）

明胶	200.0
牛肉膏蛋白胨液体培养基	1000ml
pH（灭菌）	7.3±0.1

103.46kPa，15~30min 高压蒸汽灭菌，灭菌后立即浸入冷水中冷却。

17. 石蕊牛奶培养基（主要用于牛奶凝固与发酵试验）（g/L）

脱脂奶粉	10.0
10%石蕊溶液	0.65ml
蒸馏水	1000ml

68.94kPa，10min 高压蒸汽灭菌。

18. 淀粉培养基（主要用于链霉菌淀粉水解试验）（g/L）

可溶性淀粉	10.0
硝酸钾	1.0
磷酸氢二钾	0.3
碳酸镁	1.0
氯化钠	0.5
琼脂	14.0
水	1000ml
pH	7.4±0.2

103.46kPa，15~30min 高压蒸汽灭菌。

19. 纤维素培养基（主要用于纤维素上生长试验）（g/L）

硝酸钾	1.0
磷酸氢二钾	0.5
氯化钠	0.5
硫酸镁	0.5
蒸馏水	1000ml
pH	7.2±0.2

103.46kPa，15~30min 高压蒸汽灭菌。

20. Tresner's 培养基（主要用于硫化氢试验）（g/L）

蛋白胨	10.0
柠檬酸铁	0.5
琼脂	14.0
蒸馏水	1000ml
pH	7.2±0.2

68.95kPa，15~30min 高压蒸汽灭菌。

21. 碳源利用试验培养基（g/L）

硫酸铵	2.64
磷酸二氢钾	2.38
磷酸氢二钾	5.65
七水硫酸镁	1.0
五水硫酸铜	0.0064
七水硫酸亚铁	0.0011
四水氯化锰	0.0079
七水硫酸锌	0.0015
琼脂	14.0
蒸馏水	1000ml
pH	7.4±0.2

103.46kPa，15~30min 高压蒸汽灭菌。

22. 胆盐乳糖（B.L）增菌液（g/L）

蛋白胨	20.0
氯化钠	5.0
磷酸氢二钾（或三水磷酸氢二钾）	4.0（5.2）
磷酸二氢钾	1.3
牛胆盐	2.0
（或去氧胆酸钠）	（0.5）
乳糖	5.0
蒸馏水	1000ml
pH	7.4±0.2

68.95kPa，15~30min 高压蒸汽灭菌。

制法：除乳糖、胆盐（或去氧胆酸钠）外，将上述其他成分混合微温溶解，调 pH 为 7.4±0.2，煮沸，过滤，加乳糖，胆盐（或去氧胆酸钠）待溶后，摇匀，灭菌。

注：胆盐是抑菌剂，它对革兰阳性菌有良好的抑菌作用，但用量过大，则对大肠埃希菌亦有轻度抑制作用。每1000ml 培养基中胆盐用量如下：牛胆盐、去氧胆酸钠按上述配方量。猪胆 5g，羊胆盐 5g，三号胆盐（Bile SaltNO$_3$）1.5g。

23. 伊红美兰（EMB）琼脂（g/L）

灭菌营养琼脂培养基	100ml
20%乳糖溶液	5.0ml
2%伊红（Eosin）水溶液	2.0ml
0.5%美兰（Methylene Blue）水溶液	1.3~1.6ml

制法：将灭菌营养琼脂培养基加热溶化，冷却至60℃左右，按照无菌操作加入灭菌的其他3种溶液。

注：大肠埃希菌在伊红美兰琼脂平板上的菌落有时不典型，应注意使用的平板应表面干燥，接菌培养以不超过24h观察为宜。

24. 半乳糖–EMB培养基（g/L）

伊红	0.4
美兰	0.06
半乳糖	10.0
蛋白胨	10.0
磷酸氢二钾	2.0
蒸馏水	1000ml
pH	7.2±0.2

68.95kPa，15~30min 高压蒸汽灭菌。

制法：除伊红、美兰预先充分溶解外，将上述其他成分混合微温溶解，调pH后，再加入伊红、美兰溶解液，灭菌备用。

注：必须在加入伊红、美兰前调节pH。

25. 麦康凯（MacC）琼脂（g/L）

蛋白胨	20.0
乳糖	10.0
氯化钠	5.0
牛胆盐	5.0
琼脂	14.0
1%中性红指示液	3ml
蒸馏水	1000ml

68.95kPa，15~30min 高压蒸汽灭菌。

制法：除琼脂、乳糖、胆盐、中性红指示液外，将其他成分混合微温溶解。调pH为7.2±0.2，加入琼脂，煮沸至琼脂溶化后，再加入其他各成分，摇匀，分装，灭菌。

注：此培养基避光保存。

26. 亚碲酸钠肉汤培养基（g/L）

普通肉汤培养基	1000ml
10%亚碲酸钠溶液	2ml

103.46kPa，15~30min 高压蒸汽灭菌。

注：10%亚碲酸钠液，精确称取亚碲酸钠 0.1g，加在 10ml 蒸馏水中，微温溶解。亚碲酸钠液必须现用现配，不可久存。

27. 甘露醇高盐（g/L）

蛋白胨	10.0
牛肉膏	1.0
D-甘露醇	10.0
氯化钠	75.0
琼脂	14.0
酚红	25.0mg
蒸馏水	1000ml
pH	7.4±0.2

68.95kPa，15~30min 高压蒸汽灭菌。

制法：除酚红、琼脂外，其他成分加入蒸馏水中，微温溶解，调节 pH，加入 1%酚红水溶液 2.5ml 混匀后分装锥形烧瓶，按比例加入琼脂。

28. 十六烷三甲基溴化胺琼脂培养基（g/L）

牛肉膏	3.0
蛋白胨	10.0
氯化钠	5.0
十六烷三甲基溴化胺	0.3
琼脂	14.0
蒸馏水	1000ml
pH	7.4~7.6

103.46kPa，15~30min 高压蒸汽灭菌。

制法：除琼脂外，将上述成分混合加热溶解，调节 pH 后加入琼脂，灭菌备用。

29. 绿脓菌素测定用培养基（PDP 琼脂）（g/L）

蛋白胨	20.0
氯化镁（无水）	1.4
硫酸钾（无水）	10.0
琼脂	14.0
甘油	10.0ml
蒸馏水	1000ml
pH	7.4±0.2

103.46kPa，15~30min 高压蒸汽灭菌。

制法：称取蛋白胨、氯化镁和硫酸钾，加在蒸馏水中微温溶解。调节 pH，加入甘油及琼脂，加热溶解，分装，灭菌。

30. 硝酸盐胨水培养基（g/L）

蛋白胨	10.0
酵母膏	3.0
硝酸钾	2.0
亚硝酸钠	0.5
蒸馏水	1000ml
pH	7.4±0.2

68.95kPa，15~30min 高压蒸汽灭菌。

制法：称取蛋白胨和酵母膏，加入蒸馏水中微温溶解。调节 pH 后，加入硝酸钾和亚硝酸钠溶解混匀，分装在加有倒管的试管内，灭菌备用。

31. SS（Salmonella Shigella，SS）琼脂培养基（g/L）

蛋白胨	5.0
牛肉浸出粉	5.0
乳糖	10.0
牛胆盐	8.5
柠檬酸钠	8.5
柠檬酸铁铵	1.0

硫代硫酸钠	8.5
1%中性红溶液	2.5ml
0.1%亮绿溶液	0.33ml
琼脂	14.0
蒸馏水	1000ml
pH	7.4±0.2

103.46kPa，15~30min 高压蒸汽灭菌。

制法：先将蛋白胨、牛肉浸出粉、琼脂、蒸馏水制成普通琼脂培养基，调节 pH，灭菌，趁热加入其他成分，充分摇匀使溶解，制得平板后备用。

32. 三糖铁琼脂培养基（triple sugar iron，TSI）（g/L）

蛋白胨	20.0
牛肉浸出粉	5.0
乳糖	10.0
蔗糖	10.0
氯化钠	5.0
硫酸亚铁	0.2
硫代硫酸钠	0.2
0.2%酚红指示液	12.5ml
葡萄糖	1.0
琼脂	14.0
蒸馏水	1000ml
pH（灭菌）	7.3±0.1

68.95kPa，15~30min 高压蒸汽灭菌。

制法：除三种糖、0.2%酚红指示液、琼脂外，取上述成分，混合，微温溶解，调节 pH 为 7.3±0.1，加入琼脂，加热使溶化后，再加入其他成分，摇匀，分装，灭菌，制成高底层（2~3cm）短斜面。

33. 亚碲酸盐甘露醇酚红琼脂平板 Tellurite – Mannitol – Phenolred，TMP 琼脂平板（g/L）

蛋白胨	10.0
酵母膏	5.0
甘露醇	5.0

氯化钠	40.0
琼脂	14.0
1%酚红溶液	2.5ml
磷酸氢二钾	5.0
1%亚碲酸钠溶液	2.0ml
蒸馏水	1000ml
pH	7.4±0.2

68.95kPa，15~30min 高压蒸汽灭菌。

制法：除甘露醇、酚红、琼脂、亚碲酸钠外，称取其他成分加入蒸馏水中，微热使溶，调节 pH，加入琼脂，加热使溶化，过滤，加入甘露醇、酚红混匀后，灭菌。临用前于每 100ml 培养基中加入 1%亚碲酸钠溶液 0.2ml，混匀即得。冷至 60℃，制备平板备用。

34. 血浆培养基

抗凝剂：称取 0.5g 枸橼酸钠加入 10.0ml 生理盐水中溶解后装入试管，103.46kPa，15~30min 高压蒸汽灭菌，备用。

血浆培养基制法：无菌操作采取家兔（或羊、人）血 9ml 与 1ml 抗凝剂混合装入无菌试管，混匀数分钟，待血液不凝固时，离心分离血浆，吸取血浆后加入无菌试管中，冰箱保存备用。

注：临用前必须用已知血浆凝固酶试验阳性的金黄色葡萄球菌测试，证明血浆合格后，方可用于实验。

35. 0.001%TTC（氯化三苯基四氮唑）营养琼脂培养基

制法：营养琼脂培养基中加入 0.001%TTC（氯化三苯基四氮唑）。

36. 需氧菌培养基（营养琼脂培养基）（g/L）

蛋白胨	10.0
氯化钠	5.0
琼脂	14.0
牛肉浸液	1000ml
pH	7.0~7.5

68.95kPa，15~30min 高压蒸汽灭菌。

37. 硝酸盐还原培养基（g/L）

蛋白胨	10.0
酵母浸出粉	3.0
硝酸钾	2.0
亚硝酸钠	0.5
蒸馏水	1000ml
pH	7.4±0.2

68.95kPa，15~30min 高压蒸汽灭菌。

制法：称取蛋白胨和酵母膏，加在蒸馏水中微温使溶。调节 pH，加入硝酸钾和亚硝酸钠溶解混匀，分装在加有倒管的试管内，灭菌，备用。

38. 醋酸铅试纸

制法：将普通滤纸剪成 0.5cm 宽的纸条，长度根据试管与培养基高度而定。用 5% 的醋酸铅将纸条浸透，然后用烘箱烘干，放于培养皿中灭菌备用。

39. 硫化氢试验培养基 I（g/L）

蛋白胨	10.0
氯化钠	5.0
牛肉膏	10.0
半胱氨酸	0.5
甘油（化学纯）	10ml
蒸馏水	1000ml
pH	7.2±0.2

68.95kPa，15~30min 高压蒸汽灭菌。

注：分装试管，每管高度为 4~5cm。

40. 硫化氢试验培养基 II（g/L）

蛋白胨	10.0
氯化钠	5.0
牛肉膏	7.5
明胶	5.0

10%氯化亚铁	5.0ml
蒸馏水	1000ml
pH	7.2±0.2

制法：培养基灭菌后，在明胶尚未凝固时，加入新制备的过滤除菌的 $FeCl_2$，用无菌试管分装培养基，高度为 4~5cm，立即置于冷水中冷却凝固，供穿刺接种用。

41. 硫乙醇酸盐流体培养基（g/L）

酪胨（胰酶水解物）	15.0
酵母浸出粉	5.0
葡萄糖	5.0
氯化钠	2.5
L-胱氨酸	0.5
新配制的0.1%刃天青溶液	1.0ml
硫乙醇酸钠（硫乙醇酸）	0.5（0.3ml）
琼脂	0.75
蒸馏水	1000ml
pH（灭菌）	7.1±0.2

制法：除葡萄糖和刃天青溶液外，取上述成分混合，微温溶解，调节pH为弱碱性，煮沸，滤清，加入葡萄糖和刃天青溶液，摇匀，调节pH值为7.1±0.2。分装至适宜的容器中，其装量与容器高度的比例应符合培养结束后培养基氧化层（粉红色）不超过培养基深度的1/2，灭菌。在供试品接种前，培养基氧化层的高度不得超过培养基深度的1/5，否则，须经100℃水浴加热至粉红色消失（不超过20min），迅速冷却，只限加热一次，并防止被污染。硫乙醇酸盐流体培养基置30~35℃培养。

42. 改良马丁培养基（g/L）

蛋白胨	5.0
磷酸氢二钾	1.0
酵母浸出粉	2.0
硫酸镁	0.5
葡萄糖	20.0
蒸馏水	1000ml

制法：除葡萄糖外，取上述成分混合，微温溶解，调节 pH 值约为 6.8，煮沸，加入葡萄糖溶解后，摇匀，滤清，调节 pH 为 6.4±0.2，分装，灭菌。改良马丁培养基置 23～28℃培养。

43. 选择性培养基（g/L）

按上述硫乙醇酸盐流体培养基或改良马丁培养基的处方及制法，在培养基灭菌或使用前加入适宜的中和剂、灭活剂或表面活性剂，其用量同验证试验。

44. 改良马丁琼脂培养基（g/L）

按改良马丁培养基的处方及制法，加入 14.0g 琼脂，调节 pH 为 6.4±0.2，分装，灭菌。

45. 玫瑰红钠琼脂培养基（g/L）

蛋白胨	5.0
玫瑰红钠	0.0133
葡萄糖	10.0
琼脂	14.0
磷酸二氢钾	1.0
硫酸镁	0.5
蒸馏水	1000ml

制法：除葡萄糖、玫瑰红钠外，取上述成分，混合，微温溶解，滤过。再加入葡萄糖、玫瑰红钠，分装，灭菌。

46. 酵母浸出粉胨葡萄糖琼脂培养基（YPD）（g/L）

蛋白胨	10.0
琼脂	14.0
酵母浸出粉	5.0
葡萄糖	20.0
蒸馏水	1000ml

制法：除葡萄糖外，取上述成分混合，微温溶解，滤过，加入葡萄糖，分装，灭菌。

47. 4-甲基伞形酮葡糖苷酸（MUG）培养基（g/L）

蛋白胨	10.0
磷酸二氢钾（无水）	0.9
硫酸锰	0.5mg
磷酸氢二钠（无水）	6.2
硫酸锌	0.5mg
亚硫酸钠	40mg
硫酸镁	0.1
去氧胆酸钠	1.0
氯化钠	5.0
MUG	75mg
氯化钙	50mg
蒸馏水	1000ml

除 MUG 外，取上述成分，混合，微温溶解，调节 pH 为 7.3 ± 0.1，加入 MUG，溶解，每管分装 5ml，68.95 kPa，15~30min 高压蒸汽灭菌。

48. 微生物检定效价用培养基 I 配方为（g/L）

蛋白胨	5
牛肉浸出粉	3
磷酸氢二钾	3
琼脂	14.0
蒸馏水	1000ml

制法：除琼脂外，混合上述成分，调节 pH 使比最终 pH 略高 0.2~0.4，加入琼脂，加热溶化后过滤，调节 pH 为 7.8~8.0 或 6.5~6.6，分装，灭菌。

49. 微生物检定效价用培养基 II 配方（g/L）

蛋白胨	6
牛肉浸出粉	1.5
酵母浸出粉	6
葡萄糖	1

| 琼脂 | 14.0 |
| 蒸馏水 | 1000ml |

制法：除琼脂和葡萄糖外，混合上述成分，调节pH使比最终pH略高0.2~0.4，加入琼脂，加热溶化后过滤，加葡萄糖溶解后，摇匀，调节pH为6.5~6.6，68.95 kPa灭菌15~30min。

50. 淀粉琼脂培养基（紫外线对枯草芽孢杆菌产淀粉酶的诱变效应研究）（g/L）

琼脂	12.0
可溶性淀粉	10.0
牛肉膏	13.0
蒸馏水	1000ml
pH	7.5

103.46kPa，高压蒸汽灭菌15~30min。

51. 玉米粉培养基（g/L）

玉米粉	5
蛋白胨	0.1
葡萄糖	1
自来水	1000ml

103.46kPa，15~30min高压蒸汽灭菌。

52. 葡萄糖–醋酸盐培养基（g/L）

葡萄糖	1
酵母膏	2.5
醋酸钠	8.2
琼脂	14
蒸馏水	1000ml
pH	4.8

68.95kPa，15~30min高压蒸汽灭菌。

二、微生物实验常用菌种

1. 细菌

中文名	学名
炭疽芽孢杆菌	*Bacillus anthracis*
枯草芽孢杆菌	*Bacillus subtilis*
短小芽孢杆菌	*Bacillus pumilus*
生孢梭状芽孢杆菌	*Clostridium sporogenes*
巴氏梭菌	*Clostridium pasteurianum*
丙酮丁醇梭菌	*Clostridiumacetobutylicun*
破伤风梭菌	*Clostridium tetani*
北京棒杆菌	*Corynebacteriumpekinense*
谷氨酸棒杆菌	*Corynebacterium glutamicum*
肺炎双球菌	*Diplococcus pneumoniae*
大肠埃希菌 K_{12} 供体菌	*E. coli* K_{12} (λ) gal^+

（带有整合在半乳糖基因旁的原噬菌体的溶源性细菌）

大肠埃希菌 K_{12} 受体菌	*E. coli* K_{12} (S) gal^-

（染色体上半乳糖基因缺陷）

产气肠杆菌	*Enterobacter aerogenes*
大肠埃希菌	*Escherichia coli*
普通变形菌	*Proteus vulgaris*
铜绿假单胞菌	*Pseudomonas aeruginosa*
藤黄八叠球菌	*Sarcina lutea*
结核分枝杆菌	*Mycobacterium tuberculosis*
金黄色葡萄球菌	*Staphyloccus aureus*
金黄色葡萄球菌 209P	*Staphylococcus aureus* 209P
肺炎球菌	*Streptococcus pneumoniae*
链球菌属	*Streptococcus*
伤寒沙门菌	*Salmonellat*
鼠伤寒沙门菌	*Salmonella typhimurium*

链霉菌属	*Streptococcus*
小单孢菌属	*Micromonospora*
白孢链霉菌	*Streptomyces Albosporus*
黄色链霉菌	*Streptomyces Flavus*
球孢链霉菌	*Streptomyces Globisporus*
粉红孢链霉菌	*Streptomyces Roseosporus*
淡紫灰链霉菌	*Streptomyces Lavendulae*
青色链霉菌	*Streptomyces Glaucus*
烬灰链霉菌	*Streptomyces Cinerogriseus*
绿色链霉菌	*Streptomyces Viridis*
蓝色链霉菌	*Streptomyces Cyaneus*
灰红紫链霉菌	*Streptomyces Griseorubro violaceus*
灰褐链霉菌	*Streptomyces Griseofuscus*
金色链霉菌	*Streptomyces Aureus*
吸水链霉菌	*Streptomyces Hygroscopicus*
轮生链霉菌	*Streptomyces Verticallatus*
链霉菌 1787 – 3	*Streptomyces 1787 – 3*
链霉菌 12 – 21	*Streptomyces 12 – 21*
链霉菌 4.794	*Streptomyces 4.794*
诺卡氏 71 – N	*Nocardia 71 – N*
小单孢菌 220	*Micromonosporam 220*

2. 酵母菌

啤酒酵母	*Saccharomyces cerevisiae*
白假丝酵母菌	*Candida albicans*

3. 霉菌

黑根霉	*Rhizopus nigricans*
黑曲霉	*Aspergillusniger*
青霉	*Penicillium sp.*

三、常用染色液的配制

配制染色液时，一般先将染料溶解于乙醇或水中（乙醇原液较稳定，便于保存），配成饱和原液，应用时再以蒸馏水或适当溶液稀释使用。

饱和溶液配制方法：将一定的染料于乳体中研碎，再渐渐加入定量乙醇（95%），边加边研磨，充分研细使其溶解（底部必然有微量的色素沉淀），然后将上清液转入棕色瓶中密闭保存备用。常用染色液的溶解度和配制方法如表附-1。

表附-1 常用染料的溶解度及饱和染液配制

染料名称	溶解度（g/100ml）（于95%乙醇中）	饱和染液配制（g/100ml）（于95%乙醇中）	染料原液名称
碱性复红（Basic Fuchsin）	3.20	3.5	碱性复红原液
结晶紫（Crystal Violet）	13.87	14	结晶紫原液
龙胆紫（Gentian Violet）	15.21	15.5	龙胆紫原液
碱性美蓝（Basic Methylene blue）	1.48	1.5	碱性美蓝原液
沙黄（番红）（Safranine）	3.41	3.5	番红原液
孔雀绿（Malachite green, oxalate）	7.52	8.0	孔雀绿原液

1. 吕氏（Loeffler）碱性美蓝染液

A液：美蓝（亚甲基蓝）0.3g，溶于30ml 95%乙醇溶液中。

B液：氢氧化钾0.01g，蒸馏水定容至100ml。

分别配制A液和B液，混合备用。

（通常可将此混合液稀释4倍使用，即稀释美蓝染色液）

2. 齐氏（Ziehl）石炭酸复红染色液

A液：碱性复红（basic fuchsin）0.3g，在研钵中研细后，逐渐加入95%乙醇10ml，使其溶解。

B液：石炭酸5g，以95ml蒸馏水溶解。

混合A液及B液即成。

（通常可将此混合液稀释5~10倍使用，稀释液易变质失效，一次不宜多配）

3. 草酸铵结晶紫染液

A液：结晶紫2.5g，在研钵中研细后，逐渐加入95%乙醇25ml，使其溶解。

B液：草酸铵1.0g，以100ml蒸馏水溶解。

混合A液和B液，静置48h后使用。

4. 卢戈（Lugol）碘液

碘 1.0g，在研钵中研细。先将 2g 碘化钾溶于约 10~20ml 蒸馏水，再将碘溶于碘化钾溶液，待碘完全溶解后定容至 300ml。

5. 番红花红染色液

番红花 2.5g，以 95% 乙醇 10ml 溶解，取上述配好的乙醇溶液 10ml 与 90ml 蒸馏水混匀即可。

6. 孔雀绿染色液（芽孢染色液）

孔雀绿 7.6g，以 100ml 蒸馏水溶解。

7. 黑素溶液（荚膜染色液）

水溶性黑色素 5g，溶于 100ml 蒸馏水中，煮沸 10min。冷后，补加水到 100ml，加 0.5ml 40% 甲醛做防腐剂，混匀备用。

8. 硝酸银鞭毛染色液（鞭毛染色液）

A 液：单宁酸 5g，$FeCl_3$ 1.5g，溶解于 100ml 蒸馏水中。再加入 15% 福尔马林 2ml 和 1% NaOH 1ml，溶解，混匀。滤纸过滤后使用（最好当日使用）。

B 液：$AgNO_3$ 2g，用 100ml 蒸馏水溶解后，取出 10ml 备用，向其余的 90ml $AgNO_3$ 中滴入浓 NH_4OH，使之成为很浓的悬浮液，再继续滴加 NH_4OH，直到新形成的沉淀又重新刚刚溶解为止。再将备用的 10ml $AgNO_3$ 慢慢滴入，则出现薄雾，但轻轻摇动后，薄雾状沉淀又消失，再滴入 $AgNO_3$，直到摇动后仍呈现轻微而稳定的薄雾状沉淀为止。冰箱内保存通常 10d 内仍可使用。如雾重，则银盐沉淀析出，不宜使用。

9. 李夫森（Leifson）鞭毛染色液

A 液：碱性复红 1.2g，95% 乙醇 100ml。

B 液：单宁酸 3g，蒸馏水 100ml。

C 液：NaCl 1.5g，蒸馏水 100ml。

临用前，将 A、B、C 液等量混合均匀后使用。三种溶液分别保存于冰箱可保存数月。

10. 改良李夫森鞭毛染色液

A 液：20% 单宁酸 2ml。

B 液：20% 钾明矾 2ml。

C 液：4% 石炭酸溶液 2ml。

D 液：4% 碱性复红 95% 乙醇溶液 1.5ml。

将以上各液于染色前 1~3d，按下列顺序混合：B 加入 A 中，C 加入 A、B 混合液中，D 加入 A、B、C 混合液中，混合均匀，马上过滤 15~20 次，2~3d 内使用效果较好。

11. 乳酸石炭酸棉蓝染色液（观察真菌）

取石炭酸 10g，于 10ml 蒸馏水中加热溶化。再加入乳酸（比重 1.21）10ml、甘油 20ml，最后加入棉蓝 0.02g，将棉蓝溶于蒸馏水中，再加入其他成分，微加热使其溶解，冷却后使用。

12. 美蓝（Levowitz–Weber）染液

于含52ml 95%乙醇和44ml四氯乙烷的三角烧瓶中，慢慢加入0.6g氯化美蓝，旋摇三角烧瓶，使其溶解。5~10℃放置12~24h后加入冰醋酸4ml，混匀，滤纸过滤后使用。

13. 姬姆萨（Giemsa）染液（用于核型多角体和质型多角体的区别染色）

取姬姆萨染料0.5g于研钵中研细，边加入33ml甘油边继续研磨，再加入33ml甲醇混匀。56℃放置12~24h后，即为姬姆萨贮存液。临用前取1ml姬姆萨贮存液加入到pH7.2磷酸缓冲液20ml中，混匀，即为使用液。

14. 3%酸性乙醇

95%乙醇溶液97ml，加入浓盐酸3ml，混匀备用。

四、常用试剂和溶液的配制

1. 0.85%生理盐水溶液　取氯化钠8.5g，加水溶解使成1000ml，过滤，分装。

2. 0.1mol/L磷酸缓冲液

溶液A：称取27.8g NaH_2PO_4 溶于1000ml蒸馏水中，溶解，即为0.2mol/L NaH_2PO_4 溶液。

溶液B：称取53.65g $Na_2HPO_4 \cdot 7H_2O$ 或71.7g $Na_2HPO_4 \cdot 12H_2O$ 溶于1000ml蒸馏水中，溶解，即为0.2mol/L NaH_2PO_4 溶液。

制法：溶液A与溶液B按表附-2中比例混合后，用蒸馏水稀释至终体积为200ml。

表附-2　溶液配比表

溶液A（ml）	溶液B（ml）	pH	溶液A（ml）	溶液B（ml）	pH
93.5	6.5	5.7	45.0	55.0	6.9
92.0	8.0	5.7	39.0	61.0	7.0
90.0	10.0	5.9	33.0	67.0	7.1
87.7	12.3	6.0	28.0	72.0	7.2
85.0	15.0	6.1	23.0	77.0	7.3
81.5	18.5	6.2	19.0	81.0	7.4
77.5	22.5	6.3	16.0	84.0	7.5
73.5	26.5	6.4	13.0	87.0	7.6
68.5	31.5	6.5	10.5	89.5	7.7
62.5	37.5	6.6	8.5	91.5	7.8
56.5	43.5	6.7	7.0	93.0	7.9
51.0	49.0	6.8	5.3	94.7	8.0

3. 溴麝香草酚蓝（Bromothymol blue，BTB）溶液

制法：称取溴麝香草酚蓝0.1g，加0.05mol/L氢氧化钠溶液3.2ml使溶解，再加蒸馏水稀释至200ml，即得。变色范围pH6.0~7.6（黄→蓝）。

4. 柯氏（Kovacs）试剂（g/100ml）

成分：对二甲基氨基苯甲醛 3.0g，戊醇（或丁醇）75.0ml，浓盐酸 25.0ml。

制法：将对二甲基氨基苯甲醛加入纯戊醇内，使其溶解。将浓盐酸一滴滴慢慢加入，边加边摇，不能加得太快，以致温度升高溶液颜色变深。

5. 甲基红（Methyl red）指示剂（g/100ml）

成分：甲基红 0.02g，95% 乙醇 60ml，蒸馏水 40ml。

制法：先将甲基红溶于乙醇中，之后再加水定容。

6. 格利斯（Griess）亚硝酸试剂 I、II

试剂 I：称取磺胺酸 0.5g，溶于 150ml 醋酸溶液（30%）中，保存于棕色瓶中备用。

试剂 II：称取 α-萘胺 0.5g，加入 50ml 蒸馏水中，煮沸后，缓缓加入 30% 的醋酸溶液 150ml，保存于棕色瓶中备用。

7. 6% 萘酚

称取萘酚 6g，溶于 100ml 的无水乙醇溶液中，溶解备用。

8. 石蕊溶液（2.5%）

称取石蕊 2.5g，溶解于 100ml 蒸馏水中，过滤备用。

9. 稀释液、冲洗液及其制备方法

（根据供试品的特性，可选用其他经验证过的适宜的溶液作为稀释液、冲洗液。如需要，可在上述稀释液或冲洗液的灭菌前或灭菌后加入表面活性剂或中和剂等。稀释液、冲洗液配制后应需采用验证合格的灭菌程序灭菌）

（1）0.1% 蛋白胨水溶液：取蛋白胨 1.0g，加水 1000ml，微温溶解，滤清，调节 pH 值至 7.1±0.2，分装，灭菌。

（2）pH7.0 氯化钠-蛋白胨缓冲液：取磷酸二氢钾 3.56g，磷酸氢二钠 7.23g，氯化钠 4.30g，蛋白胨 1.0g，加水 1000ml，微温溶解，滤清，分装，灭菌。

（3）0.9% 无菌氯化钠溶液：取氯化钠 9.0g，加水溶解使成 1000ml，过滤，分装，灭菌。（仅用于上述两种溶液不适用时使用）

10. pH6.8 无菌磷酸盐缓冲液

取 0.2mol/L 磷酸二氢钾 250ml，加 0.2mol/L 氢氧化钠溶液 118ml，用水稀释至 1000ml，摇匀，过滤，68.95kPa 灭菌 30min。

11. pH7.6 无菌磷酸盐缓冲液

取磷酸二氢钾 27.22g，加水使之溶解成 1000ml。取 50ml，加 0.2mol/L 氢氧化钠溶液 42.4ml，加水稀释至 200ml，过滤，68.95kPa 灭菌 30min。

12. 洗液的配制

低浓度洗液：重铬酸钾 100g，水 750ml，硫酸 250ml。

中浓度洗液：重铬酸钾 60g，水 300ml，硫酸 460ml。

高浓度洗液：重铬酸钾 100g，水 200ml，硫酸 800ml。

制法：先将重铬酸钾倒入自来水中，然后加入浓硫酸，边加硫酸边用玻璃棒搅拌。由于加入浓硫酸后产生高热，故加酸时要慢，容器应用耐酸耐高温塑料或陶器制品。

注：配制好的清洁液，应存于有盖的玻璃容器内。需要浸泡的玻璃器皿一定要干燥，如果清洁液经过长期使用已呈黑色，表明已经失效，不宜再用。由于洗液有强腐蚀性，故操作时要十分注意。

五、微生物限度标准

非无菌药品的微生物限度标准是基于药品的给药途径及对患者健康潜在的危害而制订的。药品的生产、贮存、销售过程中的检验，原料及辅料的检验，新药标准制订，进口药品标准复核，考察药品质量及仲裁等，除另有规定外，其微生物限度均以本标准为依据。

1. 无菌制剂

制剂通则、品种项下要求无菌的制剂及标示无菌的制剂，应符合无菌检查法规定。

2. 口服给药制剂

细菌数：每 1g 不得过 1000CFU，每 1ml 不得过 100CFU。

霉菌和酵母菌数：每 1g 或 1ml 不得过 100CFU。

大肠埃希菌：每 1g 或 1ml 不得检出。

3. 局部给药制剂

（1）用于手术、烧伤及严重创伤的局部给药制剂，应符合无菌检查法规定。

（2）耳、鼻及呼吸道吸入给药制剂。

细菌数：每 1g、1ml 或 10cm^2，不得过 100CFU。

霉菌和酵母菌数：每 1g、1ml 或 10cm^2，不得过 10CFU。

金黄色葡萄球菌、铜绿假单胞菌：每 1g、1ml 或 10cm^2 不得检出。

大肠埃希菌：鼻及呼吸道给药的制剂，每 1g、1ml 或 10cm^2，不得检出。

（3）阴道、尿道给药制剂

细菌数：每 1g、1ml 或 10cm^2，不得过 100CFU。

霉菌数和酵母菌数：每 1g、1ml 或 10cm^2 应小于 10CFU。

金黄色葡萄球菌、铜绿假单胞菌、白色念珠菌：每 1g、1ml 或 10cm^2，不得检出。

（4）直肠给药制剂

细菌数：每 1g 不得过 1000CFU，每 1ml 不得过 100CFU。

霉菌和酵母菌数：每 1g 或 1ml 不得过 100CFU。

金黄色葡萄球菌、铜绿假单胞菌：每 1g 或 1ml 不得检出。

（5）其他局部给药制剂

细菌数：每 1g、1ml 或 10cm^2 不得过 100CFU。

霉菌和酵母菌数：每 1g、1ml 或 10cm^2 不得过 100CFU。

金黄色葡萄球菌、铜绿假单胞菌：每 1g、1ml 或 10cm^2 不得检出。

4. 含动物组织（包括提取物）的口服给药制剂，每 10g 或 10ml 还不得检出沙门菌。

5. 有兼用途径的制剂，应符合各给药途径的标准。

6. 霉变、长螨者，以不合格论。

7. 原料及辅料，参照相应制剂的微生物限度标准执行。

六、微生物实验室废品处理

微生物与免疫学实验不可避免产生一些废品，尽管实验所用的材料一般为非致病菌或条件致病菌，实验者都要在实验结束后对实验过程中产生的可能含有毒、有害物质进行妥善处理，将废品的危险及其对环境的危害降为零。微生物学实验室的废品主要有生物废品和化学废品。生物废品包括带菌培养基、发酵液及某些生物试剂如EB等及辅助材料（Eppendorf管，移液器吸头，一次性手套、帽子等）。化学废品包括有机试剂、酸碱液等。

根据不同的废弃物采用不同的处理方法，尽快消毒灭菌，严防污染扩散，消毒和处理废弃物品时，要注意穿工作服、戴口罩和戴橡胶手套，以防止消毒剂或废弃物对操作人员产生危害。下面简单介绍微生物学实验废弃物的处理。

（一）生物废品处理

1. 带菌培养基及培养菌过程中所用的各类相关实验废料（接菌的吸头等），必须经过高温灭活处理才能作为普通废物处理，即使是无害的工程菌，其携带的具有抗性基因的质粒也会造成耐药性的传播。高温灭菌的主要目的就是杀死包括细菌的芽孢在内的所有微生物，条件为：103.46kPa，20~30min 灭菌。

2. 所有不再需要使用的样本、培养物和其他生物性材料应弃置于专门的标记容器内，存放在指定安全地方，集中通过高压消毒和化学消毒处理。

3. 微生物接种培养过的琼脂平板或者试管斜面，经过高温灭菌后，趁热将琼脂培养基稀释处理，玻璃容器洗净，沥干，可以重复使用。

4. 染菌的可重复利用的相关实验材料如玻片、吸管，可煮沸 15min 或者用 1000mg/L 有效氯漂白粉澄清液浸泡 2~6h，消毒后用洗涤剂及流水刷洗、沥干后，可以重复使用。

5. 含有EB的凝胶等有毒废弃物处理时，先将凝胶放置在指定位置氧化，装入指定的袋子中，做好标记，交由相关部门处理（学生实验建议使用 Golden view）。

6. 接触过有毒有害物质的一次性用品如手套、帽子、口罩等使用后放入污物袋内集中烧毁。

（二）化学废品处理

1. 酸、碱等化学试剂必须经过中和反应，再足够的稀释后，消除其腐蚀性，方可废弃。

2. 对于有毒有害的有机溶剂废液，实验室一般没有处理资质，需要找有资质的回收商回收处理，实验室做的工作就是按照废弃物分类表和分类要求，将废弃物分类存放，并标示，做好收集记录。

3. 含有放射性物质的固体废弃物（一般实验室的放射性废弃物为低水平放射性废弃物），必须严格按照有关的规定，严防泄漏，谨慎地将放射性废物收集在专门的污物桶内，桶的外部标明醒目的标志，根据放射性同位素的半衰期长短，分别采用贮存一定时间使其自然衰变后掩埋处理。

七、实验室常用的化学消毒剂分类及其应用

化学消毒剂是利用化学药物渗透微生物的体内，使菌体蛋白凝固变性，干扰微生物酶的活性，抑制微生物代谢和生长或损害细胞膜的结构，改变其渗透性，破坏其生理功能等，从而起到消毒灭菌作用。按照作用水平可分为灭菌剂（可杀灭一切微生物使其达到灭菌要求的制剂）、高效消毒剂、中效消毒剂、低效消毒剂。

实验室常用消毒剂使用方法

消毒剂名称	消毒水平	适用范围及使用方法	注意事项
乙醇	中效	70%~80%的乙醇溶液作为消毒剂，主要用于皮肤和医疗器械消毒	易挥发，不宜用于黏膜及创面的消毒
碘酊	高效	2%溶液用于皮肤消毒，擦后20s再用75%乙醇脱碘	对皮肤有较强的刺激，不能用于黏膜消毒。不能与红药水同时涂于一处
新洁尔灭	低效	0.01%~0.05%溶液用于黏膜消毒。0.05%~0.1%溶液用于皮肤消毒。0.1%水溶液用于外科器械消毒。0.2%~0.5%用于公共场所等环境消毒	忌与肥皂或其他阴离子表面活性剂配伍。有吸附作用
洗必太（氯苯双胍己烷）	低效	0.02%溶液用于手的消毒浸泡。0.05%溶液用于创面消毒。0.1%溶液用于物体表面的消毒	洗必泰是阳离子活性物质，忌与肥皂或其他阴离子表面活性剂配伍
福尔马林（37%~40%的甲醛溶液）	高效	2%~4%的甲醛水溶液，用于器械消毒。15ml/m³加入等量水加热蒸发成气雾，用于室内消毒	对皮肤、黏膜有刺激和固化作用，并可使人致敏
环氧乙烷	灭菌剂	0.1%~0.8%用于皮毛、书籍、文字档案材料等消毒。密闭熏蒸6~24h	超过3%时易燃易爆。有一定毒性，注意密闭消毒。对皮肤及黏膜刺激性强
过氧乙酸	灭菌剂	0.2%~0.4%，皮肤消毒；0.5%，物品消毒；2%，15ml/m³室内喷雾空气消毒	刺激性及腐蚀性。高温容易引起爆炸。易氧化分解，故须现用现配
漂白粉	低效	0.02%蔬菜、水果浸泡30min，净水冲洗。0.1%手的消毒，洗刷1min后用洁净水冲洗。1%用于一般用具，浸泡20min	消毒作用受使用浓度、作用时间的影响
来苏尔	低效	0.3%~1%，室内空气喷雾消毒	为有机酸，禁止与碱性药物及其他消毒药物混用
双氧水（过氧化氢水溶液）	高效	3%清洗伤口（创伤、溃疡等）。1%口腔含漱	双氧水用于伤口清创会引发剧烈的疼痛
高锰酸钾	中效	0.1%溶液水果等物消毒。0.2%~0.5%溶液创伤洗涤、皮肤浸泡消毒	现用现配。不可与碘化物、有机物接触或并用
龙胆紫（紫药水）	中效	1%~2%的水溶液或乙醇溶液，用于皮肤、黏膜创伤、感染及溃疡，杀菌力强且无刺激性	近年来有研究发现，紫药水有极强的致癌性

（刘晓辉）